Microgrid Cyberphysical Systems

Microgrid Cyberphysical Systems

Renewable Energy and Plug-in Vehicle Integration

Edited by

Bidyadhar Subudhi
School of Electrical Sciences, Indian Institute of Technology Goa, Ponda, Goa, India

Pravat Kumar Ray
National Institute of Technology Rourkela (NIT Rourkela), Rourkela, Odisha, India

Elsevier
Radarweg 29, PO Box 211, 1000 AE Amsterdam, Netherlands
The Boulevard, Langford Lane, Kidlington, Oxford OX5 1GB, United Kingdom
50 Hampshire Street, 5th Floor, Cambridge, MA 02139, United States

ISBN: 978-0-323-99910-6

For information on all Elsevier publications
visit our website at https://www.elsevier.com/books-and-journals

Publisher: Charlotte Cockle
Acquisitions Editor: Graham Nisbet
Editorial Project Manager: Jose Paolo Valeroso
Production Project Manager: Prasanna Kalyanaraman
Cover Designer: Miles Hitchen

Typeset by STRAIVE, India

Contents

Contributors xi
Preface xiii

1. Denial-of-service attack resilient control for cyber physical microgrid system

Vivek Kumar and Soumya R. Mohanty

1. Introduction 1
2. Microgrid control design for DOS attack 5
2.1 Control objective under DOS attack 6
2.2 Secondary control 7
2.3 Preliminaries of graph theory 7
2.4 Consensus-based voltage error 8
2.5 Consensus-based frequency error 8
2.6 Sliding mode-based control strategy 9
2.7 Event-based SMC scheme in the absence of DOS attack 10
3. Event-based control in presence of DOS attack 11
3.1 DOS attack preliminaries time limitations 12
3.2 Parameter design for the convergence rate in absence of DOS 14
3.3 Parameter design for the divergence rate in presence of DOS 16
4. Verification and result analysis of the proposed scheme 20
4.1 Simulation results obtained from the proposed scheme and discussion 21
5. Conclusion 25
References 25

2. Distributed generating system integration: Operation and control

Priyatosh Mahish, Manas Ranjan Mishra, and Sukumar Mishra

1. Introduction 29
1.1 Research on distributed generating system integration: Focus area 29

1.2 IEEE standards for distributed generating system integration 31
2. **Grid supporting mode of VSI in distributed generating system** 32
2.1 Standards for distributed generating system integrations 32
2.2 Frequency Support 36
2.3 Voltage support 38
2.4 Inertial support 43
3. **Grid following mode of VSI in distributed generating system** 47
3.1 Structure of grid-imposed frequency VSC system 47
3.2 Real and reactive power controller 48
3.3 VSC current control-SRF (DQ-control) 49
3.4 Phase locked loop (PLL) 50
3.5 Experimental results 51
4. **Grid forming mode of VSI in distributed generating system** 52
4.1 Conventional droop control approach 52
4.2 Virtual synchronous machine (VSM) based control approach 55
5. **Converter control in EV-based distributed generating system** 57
5.1 Modified droop control for frequency regulation of EV-based system 57
5.2 Distributed control for voltage regulation of EV-based system 60
6. **Conclusions** 62
References 63

3. Short-term solar irradiance forecasting using ground-based sky images
Pravat Kumar Ray and Hoay Beng Gooi

1. **Introduction** 67
2. **Methodology adopted for forecasting of solar irradiance using sky images** 68
2.1 Preprocess sky images 69
2.2 Cloud pixel determination 69
2.3 Cloud movement determination 75
2.4 Forecast solar irradiance 77
2.5 Solar irradiance forecast 82
3. **Updating clearness index Kt using RLS algorithm** 83
4. **Updating clearness index Kt using variable leaky LMS (VLLMS) algorithm** 85
5. **Conclusions** 87
References 87

4. Design and experimental validation of robust PID control for a power converter in a DC microgrid application

R. Jeyasenthil, Tarakanath Kobaku, Bidyadhar Subudhi, Subham Sahoo, and Tomislav Dragicevic

1. **Introduction** 89
2. **QFT design of external disturbance minimization problem** 92
   2.1 QFT design for NMP system 93
3. **Application of the proposed QFT design to NMP boost-type DC-DC power electronic converter** 94
   3.1 Desired specifications 96
   3.2 Design frequency selection 97
4. **Simulation studies** 104
   4.1 Linear simulation studies 105
   4.2 Nonlinear simulation studies 106
5. **Experimental validation** 108
   5.1 Scenarios (a-b) 108
   5.2 Scenario (c) 110
   5.3 Scenarios (d-e) 110
6. **Conclusions** 112
   References 112
   Further reading 114

5. Control of PV and EV connected smart grid

Zunaib Ali, Komal Saleem, Ghanim Putrus, Mousa Marzband, and Sandra Dudley

1. **Introduction** 115
   1.1 PV market penetration 116
   1.2 EV market growth 118
2. **Impact of PV and EV on the electricity grid** 120
   2.1 Impact on power system stability 120
   2.2 Impact on power quality 122
   2.3 Impact on electricity market 123
3. **Smart microgrids, classifications, and interconnection of sources** 123
   3.1 Classification of microgrids 124
   3.2 Microgrid configurations and infrastructure 125
4. **Control architecture for efficient operation of microgrids** 126
   4.1 Grid connected system and control of the DC side 128
   4.2 Grid connected system and control of AC side 128
   4.3 Controllers for the grid-side inverter 135
5. **EV modeling and charging algorithms** 145
   5.1 EV modeling techniques 145
   5.2 Charging algorithms for EVs 145
6. **Digital real-time simulators** 147
   6.1 dSPACE 147
   6.2 OPAL-RT 151

6.3 Speedgoat 152
6.4 RTDS 152
7. Conclusions 153
Acknowledgment 153
References 153

6. Adaptive control of islanded AC microgrid
Sweta Panda, Gyan Ranjan Biswal, and Bidyadhar Subudhi

1. Introduction 159
2. Detailed modeling of islanded AC MG 162
3. Proposed adaptive control scheme 165
3.1 Adaptive voltage and reactive power control 166
3.2 Adaptive frequency and active power control 169
4. Simulation results and analysis 172
5. Conclusion 176
References 177

7. Proactive defense system for enhanced resiliency of power grids with microgrids
Najda Vadakkeyveettil Mohamed Anwar, Gopakumar Pathirikkat, and Sunitha Rajan

1. Introduction 179
2. Dependencies and energy management in deregulated grid 180
2.1 Market model for procurement of reactive power 182
2.2 Voltage control area formation 183
2.3 Automatic voltage control (AVC) in emerging grid 184
3. Renewable uncertainty instigated disturbance propagation 185
4. Enhancing the power grid resiliency against long-term voltage instability 187
5. Proactive defense methodology for enhancing power system resiliency 187
6. Implementation of presented methodology 189
7. Case studies and discussions 193
7.1 Case 1: Investigation of propagation characteristics of disturbance 194
7.2 Case 2: Investigation about the influence of concentrated reactive power support on disturbance propagation 194
7.3 Case 3: Investigation about the effectiveness of distributed reactive power support on disturbance propagation 197
7.4 Case 4: Influence of nonlinear long line reverberation characteristics on all the above three cases 197
8. Conclusions 200
References 200

8. Adaptive controller-based shunt active power filter for power quality enhancement in grid-integrated PV systems

Jitendra Kumar Sao, Pravat Kumar Ray, Ram Dayal Patidar, and Sushree Diptimayee Swain

1. **Introduction** 203
2. **Filter configuration** 205
   - 2.1 MPPT algorithm 205
   - 2.2 Reference current estimation 206
3. **Modeling of three-phase three-wire SAPF** 206
4. **PI controller for DC voltage stabilization** 209
   - 4.1 Transfer function for DC-link capacitor 210
   - 4.2 Transfer function of VSI 210
   - 4.3 Transfer function of PI controller 211
5. **MRAC for DC voltage stabilization** 213
6. **Simulation results** 216
   - 6.1 Case I: Balanced supply with steady-state load condition 216
   - 6.2 Case II: Balanced supply with varying load condition 220
   - 6.3 Case III: Balanced supply with parameter variation of SAPF 220
7. **Experimental results** 220
   - 7.1 Case I: Balanced supply with steady-state load condition 223
   - 7.2 Case II: Balanced supply with varying load condition 223
   - 7.3 Case III: Balanced supply with parameter variation of SAPF 227
8. **Conclusions** 231

References 231

9. Issues and challenges in microgrid protection

Nikhil Kumar Sharma and Subhransu Ranjan Samantaray

1. **Introduction** 233
2. **Microgrid protection challenges** 234
3. **Review of microgrid protection schemes** 240
   - 3.1 Overcurrent protection schemes 240
   - 3.2 Differential protection schemes 241
   - 3.3 Distance protection schemes 243
   - 3.4 New microgrid protection techniques 243
   - 3.5 Rate of angle difference-based protection scheme 243
   - 3.6 Integrated impedance angle-based protection scheme 246
   - 3.7 Impedance Difference-Based microgrid protection scheme 249
4. **Conclusions** 252

References 252

10. Protection schemes in microgrid

Kartika Dubey, Sanat, and Premalata Jena

1. **Introduction** 255
2. **Classification of microgrids** 257
   2.1 Classification of microgrids based on location 257
   2.2 Classification based on capacity 258
   2.3 Classification of microgrids based on power supply 259
3. **Hierarchical control architecture of microgrids** 262
4. **Protection issues in microgrids** 263
5. **Protection schemes for AC microgrids** 265
   5.1 Nonadaptive protection for microgrids 265
   5.2 Adaptive protection for microgrids 265
6. **Protection schemes for DC microgrids** 269
   6.1 Current-based protection scheme 270
   6.2 Voltage-based protection scheme 271
   6.3 Traveling wave-based protection scheme 271
   6.4 Interruption in DC current-based protection scheme 271
7. **Protection scheme for hybrid AC/DC microgrid** 271
   7.1 Fuse-based short-circuit protection of converter 271
   7.2 Protection scheme based on islanding detection 272
   7.3 Protection scheme for DC lines in hybrid AC/DC microgrid with multilevel converter 272
8. **Conclusion** 272
   References 273

Index 277

Contributors

Numbers in parentheses indicate the pages on which the authors' contributions begin.

Zunaib Ali (115), School of Engineering, London South Bank University, London, United Kingdom

Najda Vadakkeyveettil Mohamed Anwar (179), Department of Electrical Engineering, National Institute of Technology Calicut, Kerala, India

Gyan Ranjan Biswal (159), Department of Electrical & Electronics Engineering, VSS University of Technology, Burla, Odisha, India

Tomislav Dragicevic (89), Department of Electrical Engineering, Technical University of Denmark, Kongens Lyngby, Denmark

Kartika Dubey (255), Department of Electrical Engineering, Indian Institute of Technology Roorkee, Roorkee, India

Sandra Dudley (115), School of Engineering, London South Bank University, London, United Kingdom

Hoay Beng Gooi (67), School of Electrical and Electronic Engineering, Nanyang Technological University, Singapore, Singapore

Premalata Jena (255), Department of Electrical Engineering, Indian Institute of Technology Roorkee, Roorkee, India

R. Jeyasenthil (89), Department of Electrical Engineering, NIT Warangal, Warangal, India

Tarakanath Kobaku (89), School of Electrical Sciences, Indian Institute of Technology Goa, Farmagudi, Ponda, Goa, India

Vivek Kumar (1), Department of Electrical Engineering, Indian Institute of Technology (BHU), Varanasi, India

Priyatosh Mahish (29), Indian Institute of Technology Delhi, New Delhi, India

Mousa Marzband (115), Department of Physics, Mathematics and Electrical Engineering, Northumbria University, Newcastle, United Kingdom

Manas Ranjan Mishra (29), Indian Institute of Technology Delhi, New Delhi, India

Sukumar Mishra (29), Indian Institute of Technology Delhi, New Delhi, India

Soumya R. Mohanty (1), Department of Electrical Engineering, Indian Institute of Technology (BHU), Varanasi, India

Sweta Panda (159), Department of Electrical & Electronics Engineering, VSS University of Technology, Burla, Odisha, India

Gopakumar Pathirikkat (179), Department of Electrical Engineering, National Institute of Technology Calicut, Kerala, India

Ram Dayal Patidar (203), OPJU Raigarh (C.G.), Raigarh, India

Ghanim Putrus (115), Department of Physics, Mathematics and Electrical Engineering, Northumbria University, Newcastle, United Kingdom

Sunitha Rajan (179), Department of Electrical Engineering, National Institute of Technology Calicut, Kerala, India

Pravat Kumar Ray (67, 203), Department of Electrical Engineering, National Institute of Technology, Rourkela, India

Subham Sahoo (89), Department of Energy Technology, Aalborg University, Copenhagen, Denmark

Komal Saleem (115), School of Engineering, London South Bank University, London, United Kingdom

Subhransu Ranjan Samantaray (233), School of Electrical Sciences, Indian Institute of Technology Bhubaneswar, Bhubaneswar, India

Sanat (255), Department of Electrical Engineering, Indian Institute of Technology Roorkee, Roorkee, India

Jitendra Kumar Sao (203), OPJU Raigarh (C.G.), Raigarh, India

Nikhil Kumar Sharma (233), School of Electrical Sciences, Indian Institute of Technology Bhubaneswar, Bhubaneswar, India

Bidyadhar Subudhi (89, 159), School of Electrical Sciences, Indian Institute of Technology Goa, Farmagudi, Ponda, Goa, India

Sushree Diptimayee Swain (203), OPJU Raigarh (C.G.), Raigarh, India

Preface

The contents and structure of this book are organized considering the recent trends in the adoption of electric vehicles (EVs) and integration of renewable energy sources to the electric grid or a microgrid—*a small power network*. A number of uncertainties in generation and loads are encountered in the operation and control of a microgrid. Hence, suitable power electronic control and protection schemes are required for successful operation of microgrids. Further, supply and demand side management play a key role in the stability of a microgrid. Emphasis needs to be given for improving reliability and economic performance of microgrids, focusing on power quality, storage, and voltage and frequency control.

The book comprises 10 chapters. Each chapter provides an overview of the topics covered in the chapter and simulation/experimental results that allow readers to enhance their knowledge of real-time implementation of smart grids. Case studies on different issues are included to provide the readers with clear suggestions. Major emphasis is also given to experimental or test bed validations of the proposed algorithms or methods for the interest of the readers.

Chapter 1 discusses the prerequisite of cyber resiliency in a microgrid for smooth operation under cyber threats. Denial of service (DOS) in a microgrid-based network system has a high probability of cyber threat, where system stability is challenged by the interruption in the communication on control signal to the actuator or on the measurement of sensor data to the controller. Input to state stability (ISS) of the closed loop microgrid system is preserved by characterizing the DOS duration and frequency. ISS of the microgrid system is achieved under DOS through the determination of adequate arrangement in transmission triggering. The event trigger analysis with suitable inequality is presented both in the absence and in the presence of DOS attack to make the controller design practically possible for the secondary voltage-frequency regulation in an autonomous AC microgrid.

Microgrids facilitate the integration of various distributed energy resources (DERs) such as photovoltaics (PV), battery, wind, and fuel cells. For maintaining the harmony of generation between the grid and the DER, the control of voltage source inverters (VSIs) is important. The DER-led VSIs are operated in three modes, namely (i) grid following (GFL) mode, in which the VSI follows the voltage and frequency of the grid at the local point of common coupling (PCC); (ii) grid supporting mode, in which the VSI supports the grid; and

(iii) grid forming (GFM) mode, in which VSI operates in the islanded mode. All the modes of VSI control have been discussed in Chapter 2.

Forecasting solar irradiance based on microscale information, i.e., sky images captured by Total Sky Imager (TSI), is described in Chapter 3. The sky images are processed to determine the cloud cover and compared with the 1-minute-ago sky image to analyze cloud movement using an optical flow algorithm. The cloud condition in the next minute is predicted and, consequently, the solar irradiance values are calculated based on parts of clouds covering the sun. For more accuracy in forecasting, the clearness index is updated using Variable Leaky Least Mean Square (VLLMS) algorithms. Depending on different locations and applications, the same algorithm can be applied accordingly by calibrating the threshold value for the cloud determination algorithm and the clearness index value for solar irradiance calculation.

Chapter 4 describes the need for maintaining constant DC output voltage across the load terminals of a DC microgrid; in particular, the problem of controlling the load voltage of nonminimum phase DC-DC boost converter with the measured load voltage under voltage mode control is addressed. Quantitative feedback theory is used to synthesize a robust PID controller systematically with external disturbances and uncertainties.

Chapter 5 describes the implementation of controllers for EV charging and the use of the Vehicle to Grid concept, where EVs are used to supply power to the grid at times of peak demand. Local microgrids are used for incorporating distributed generating sources to actively support the main power grid by utilizing locally generated renewable energy. Smart control of EV battery charging can significantly help in reducing power exchange between the microgrid and the main grid, resulting in overall cost reduction for the grid as well as the microgrid (or a house) by effectively managing the local energy use and the power exchange with the utility grid.

Chapter 6 provides an overview of the design of a resilient adaptive secondary controller for a stand-alone AC microgrid system. Lyapunov functions are used to construct the proposed controller. Regardless of unforeseen disturbances, the proposed technique has the additional advantages of accurate active and reactive power sharing, along with the restoration of voltage and frequency reference values before the predetermined timeframe. The controller design involves a simple tuning procedure and has a straightforward mathematical formulation that makes it easy to apply. The proposed controller exhibits improved performance in terms of voltage and frequency tracking in the face of uncertainties and load variations.

Chapter 7 investigates the challenges posed by wide area propagation of disturbance instigated by renewable, dominant, uncertain microgrids in evolving deregulated market-oriented grids. Besides, a proactive defense system that utilizes the characteristics of this disturbance propagation to mitigate the spread of the disturbance is presented and discussed. This defense strategy caters to the ancillary service capabilities of the microgrid itself to mitigate the disturbance spread caused by other microgrids. A case study conducted on NY-NE 16 machine 68 bus system substantiates the wide area propagation characteristics of disturbance induced by uncertain, renewable, dominant microgrids and efficacy of the presented proactive defense strategy in alleviating the disturbance spread.

Chapter 8 highlights the implementation of a shunt active power filter (SAPF) for power quality improvement through harmonics compensation. In this chapter, an improved indirect current controller with model reference adaptive controller (MRAC) is proposed to enhance the performance of PV-integrated SAPF. The advantages of this proposed controller are evident by studying the harmonics compensation ratio for various load and filter parameter configurations. Performance enhancement of the proposed MRAC controller is evident by comparing it with conventional indirect current control with proportional integral controller and is also verified with a laboratory developed experimental setup.

New protection schemes for the fast detection of faults and isolation of faulty sections to minimize disruption in power supply to the consumers are presented in Chapter 9. The authors provide a systematic description of protection schemes for microgrids. The chapter mentions the challenges faced by the existing relaying schemes in microgrids and also discusses the development of new protection schemes for mitigation of the challenges.

Chapter 10 presents an insight on the importance and need of the microgrid in the distribution network. The various advantages offered by the microgrid are also discussed in detail and the classification of the microgrid based on size and capacity, electrical power supply, and geographical locations is discussed. Further, the challenges and issues faced by the microgrid are discussed in detail for AC, DC, and hybrid AC/DC microgrids.

It is expected the book will immensely benefit undergraduates, postgraduates, research scholars, faculty members, engineers, and scientists in electrical and computer engineering from different industries and organizations. In particular, the target audience is a broad domain of engineers in the power industry and computer scientists focusing on cyber security design and machine learning.

The book will help readers in enriching their technical knowledge on several aspects including power quality, techniques for smooth operation of microgrids, coordination control of distributed active power filters and STATCOMs in smart power networks, EV integration issues, and stability in situations of high penetration of renewables.

Bidyadhar Subudhi
Pravat Kumar Ray

Chapter 1

Denial-of-service attack resilient control for cyber physical microgrid system

Vivek Kumar and Soumya R. Mohanty
Department of Electrical Engineering, Indian Institute of Technology (BHU), Varanasi, India

1. Introduction

Microgrid cyber-physical structure can be represented by incorporating communication network, power electronic devices, and software-intensive close loop control. The software in closed loop controls as well as communication networks enhances susceptibility for cyber conciliations in the microgrid [1,2]. Because of the weak distribution grid, absence of generational inertia, and dynamic source-load profiles, the cyber threat becomes more noticeable in the inverter-based microgrid configurations. Microgrids are progressively initiated to renovate into cyber-physical microgrids including distributed and decentralized multi-agent features using advanced power electronics [3]. A cyber-physical microgrid structure comprises both cyber and physical layer as illustrated in Fig. 1. The physical layer signifies an interlink electric power system, whereas cyber layer deals with the communication medium for exchange of data between microgrid agents. These agents are interconnected by means of physical layer, that is, power lines, and ahead of physical nodes, cyber communication and information network connects the cyber components. Cyber safety in microgrids have a dominant importance due to the occurrences of adverse, destructive, and undiminished cyber-attacks, particularly at the distribution grid.

The AC microgrid distributed cooperative control [4–6] has now arisen as an alternative to centralize control as it provides improved robustness and smooth control capabilities. In these control strategies, inverters are considered as nodes in a sparse communication digraph. Through the communication digraph, it is ensured that all nodes will reach a common consensus as per measures delivered by the reference node. In AC microgrids, the consensus control is primarily adopted to attain voltage and frequency regulation by information exchange

Microgrid Cyberphysical Systems. https://doi.org/10.1016/B978-0-323-99910-6.00011-6

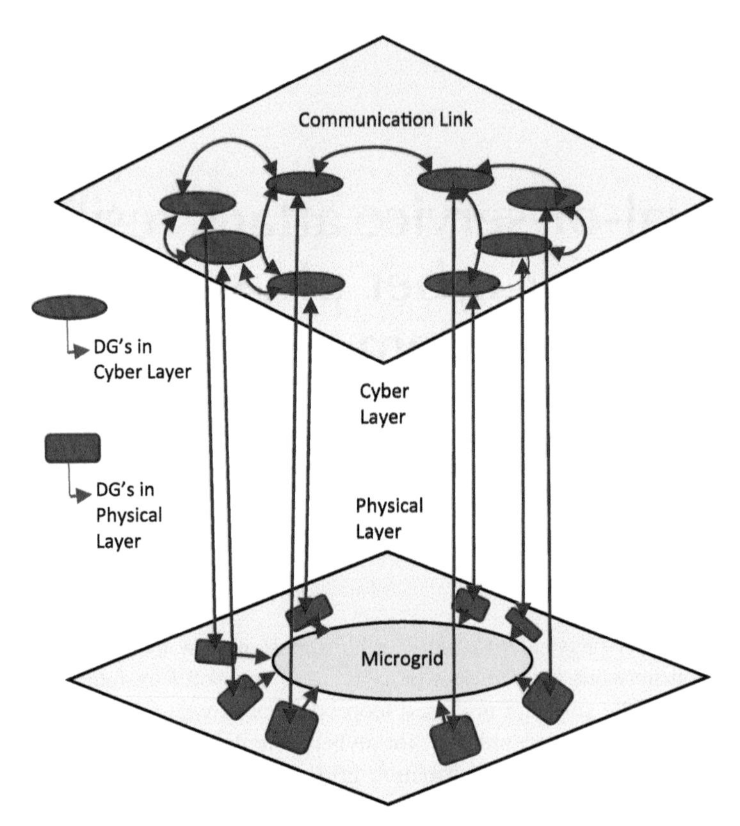

FIG. 1 Multi-DG-based microgridcyber-physical structure.

between local neighboring agents. In a multi-agent distributed AC microgrid, different attacks may deteriorate the system performance as represented in Fig. 2. In these categories, attackers can influence the signals from the leader node through the communication channel between nodes. The attacker may interrupt the neighboring node information and propagate false information to local nodes. Also, the attacker can inject an on-off signal to disrupt the local state-feedback signal. These cyber threat intrusions may be unbounded which can disrupt the synchronization of all agent's mechanisms.

Microgrid control strategies involve three layers of hierarchical structures of control that are primary, secondary, and tertiary in order to achieve smooth and stable operation [7]. In primary control procedure, distributed generations (DGs) have local control loops that are based on the droop control operation. Integrated DGs are the main building component of microgrid that efficiently facilitate the small-scale power system [8,9]. One of the primary control objectives is to take care of frequency and voltage regulation and maintain stability when the microgrid goes into islanded mode. In general, microgrid has two

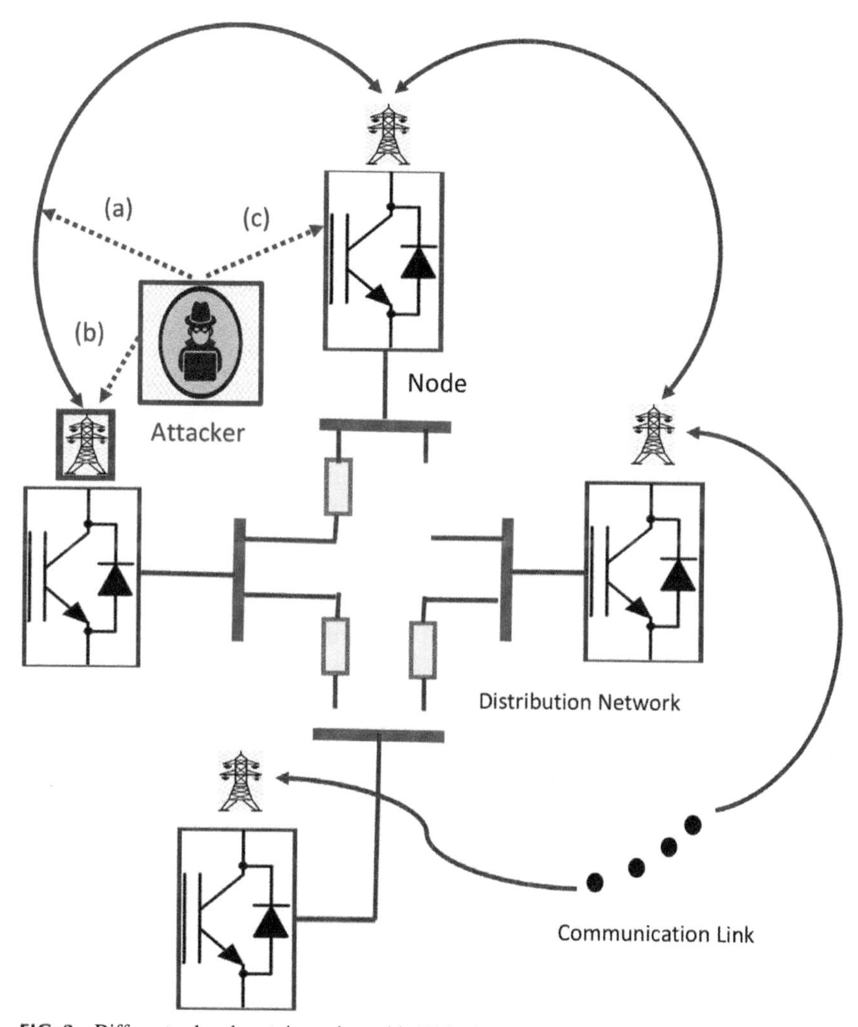

FIG. 2 Different cyber threats in a microgrid: (A) In the communication channel, (B) at the neighboring agent information for corresponding node, and (C) DGs state feedback signal.

modes of operation which are grid connected and islanded mode of operation [7,10]. Secondary control is followed by the primary operation in hierarchical control structure of microgrid to further restore voltage and frequency that may be deviated due to disturbances and system uncertainties [9–11]. It provides reference to all DGs primary control connected through a common bus. Hence, cyber threat in secondary control layer in practical circuits may be introduced due to involvement of communication network in between the bus interconnections. Cyber-physical networks (CPNs) in the previous decade have the center of attention to research community that is evolved from the integration of networking, computation, and physical procedures [12]. Due to the interconnection of cyber components of microgrid with its physical properties under CPNs environment may introduce various aspect to analysis such as fault [13], observer design [14], security issues [15,16], and stability bounds [17,18].

In most of the existing control, concern is focused on the uncertainty and fault compensation which causes failure of communication in the network control tactics [19,20]. These approaches do not effectively cover the malicious cyber adversary, especially in CPNs cases, where physical parameters are interacted with cyber component and hence vulnerable to cyber intimidations. The main classifications of attacks in microgrid communication link are deception and DOS attacks. Deception attack is concerned with the data reliability by influencing transmitted packets over the communication network [21–25], whereas DOS attacks are primarily planned to loss packets in a way to disrupt the exchange of information timeline [26,27]. Investigations on DOS attacks that interrupt the communication availability in microgrid networks are limited. The effect of DOS attack in the study by Srikantha and Kundur [28] is investigated on microgrid communication network by game theory approach using phasor measurement units. Similarly, Liu et al. [29] describe the impact of DOS attack on frequency in microgrid secondary control operation. In both these studies [28,29], it is shown that the DOS attack upsets microgrid stability by avoiding restoration process of secondary controller. An auxiliary communication interface of power line is utilized in network reconfiguration method in Danzi et al. [30] to overcome the impact of DOS attack. To extent the resiliency of microgrid in islanded mode of operation against DOS attack, a fallback control in Chlela et al. [31] is applied for energy storage. Although these algorithms against DOS attack in microgrid are much impactful and promising, it also lacks to present solid microgrid stability assessment during DOS attack which is a key role for an algorithm against DOS [32,33]. In this chapter, a control scheme based on sample data is considered for DOS attack in microgrid communication network. The DOS attacker intended to introduce instability in the controller by preventing communication on control signal to actuator or on the measurement of sensor data to the controller channels. In open loop, the process develops the control as per the last transmitted sample during DOS attacks, whereas in close loop microgrid configuration, the point of interest is to determine close loop bounds to preserve stability in some predefined sense. Pertaining to this DOS attack, modeling is an issue of concern. In previously reported work, it is difficult to explain the problematic packet drop stimulation for a DOS attacker. This chapter follows a simple model of DOS attack where attacker can target the time constraints only in the form of DOS attack duration and its frequency. This configuration of DOS attack can cover various DOS attacks like periodic, trivial, random, and jamming attacks [34–36].

In this chapter, microgrid stability under DOS attack is preserved by characterizing the duration and frequency of DOS attack in the sense of relating stability with the jamming on-off periods [37]. To avoid DOS impact, transmission times are designated to achieve closed loop stability by satisfying suitable condition at the time of communication under the DOS attack. The method is advantageous in a manner to achieve ISS subjected to disturbances under DOS attack through the general Lyapunov-based stability analysis during on-off timings of DOS attack.

The main contribution of the chapter is addressed as:

1. The norms of data transmission event are determined by satisfying the stability condition through the Lyapunov stability criteria for on-off DOS attack periods. In this way, ISS of microgrid under DOS attack is ensured.
2. Robustness of multi-DGs-based microgrid system against disturbances in secondary control design is maintained by siding mode control design.

The remaining part of this chapter is systematized as follows.

In Section 2, multi-DG-based microgrid oriented framework for control formulation for DOS attack is presented. In Section 3, the class of event logics to find ISS of microgrid under DOS attack is explained. The simulation is carried out to present the efficacy of proposed method in Section 4. In Section 5, discussion based on obtained results is summarized in concluding remarks.

2. Microgrid control design for DOS attack

The multi-DG-based microgrid structure is shown in Fig. 3. In this configuration, power electronic DC/AC converters are used to integrate distributed generators (DG's) to AC buses. The interconnected DGs include primary power source and inner current-voltage control loops. This primary control provides output voltage-frequency regulation [38], with their reference *d–q* reference frame. To achieve tracking synchronization with the reference DG, other DGs' primary

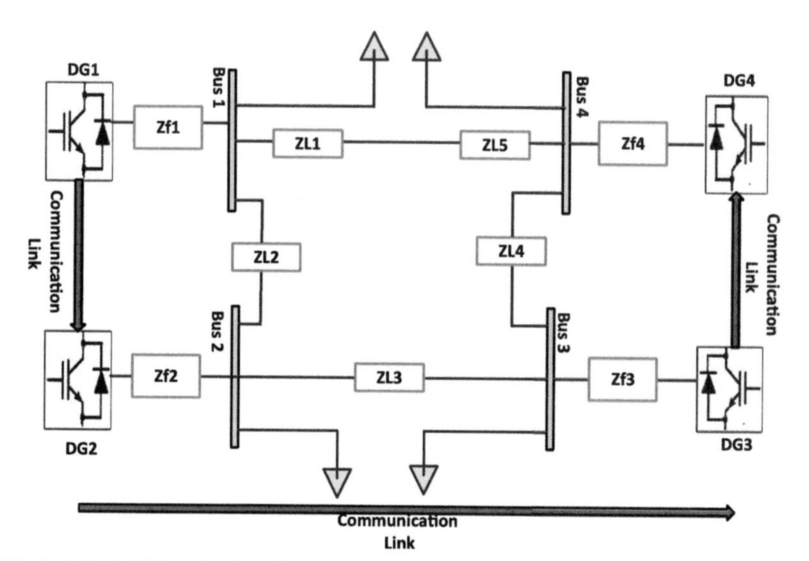

FIG. 3 Multi-DG-based microgrid structure.

control reference must follow the consensus-based secondary control. To fulfill this objective, each DGs voltage magnitude is linked with d-axis of the reference value. Therefore, ith DGs d-axis voltage component can be written as [39]:

$$\begin{cases} v_{odi}^{*} = V_{ni} - n_{Qi}Q_i \\ v_{oqi}^{*} = 0 \end{cases} \tag{1}$$

The V_{ni} represents secondary voltage control so that ($v_{omag} \rightarrow v_{ref}$) [39]. Whereas terminal output magnitude (v_{omag}) of DG voltageis written as:

$$v_{omag} = \sqrt{v_{odi}^2 + v_{oqi}^2} \tag{2}$$

Similarly, the ith DGs' frequency has to follow the secondary frequency as ($\omega_i \rightarrow \omega_{ref}$). The dynamics of ith DG can be given as [39]:

$$\begin{cases} \dot{z}_i = f_i(z_i) + k_i(z_i)\sigma_i + g_i(z_i)u_i \\ y_i = h_i(z_i) + d_iu_i \end{cases} \tag{3}$$

with, $z_i = [\delta_i P_i Q_i \phi_{di} \phi_{qi} \Upsilon_{di} \Upsilon_{qi} i_{Ldi} i_{Lqi} v_{odi} v_{oqi} i_{odi} i_{oqi}]^T$, δ_i represents common angle reference of ith DG. P_i denotes active and Q_i denotes reactive output power at filter terminal. ϕ_{di} and ϕ_{qi} denote auxiliary voltage components, whereas Υ_{di} and Υ_{qi} denote the auxiliary current components. The other parameters like i_{Ldi}, i_{Lqi}, v_{odi}, v_{oqi}, i_{odi}, and i_{oqi} are, respectively, the inverters current, filter output voltage, and output current components of ith DG. The σ_i in Eq. (4) is the vector of $\sigma_i = [v_{com} v_{bdi} v_{bqi}]^T$ as internal AC bus component variables. The value of $y_i = v_{odi}$ and $u_i = V_{ni}$ represents the secondary voltage control. Secondary frequency control is designed as $u_i = \omega_{ni}$ and $y_i = \omega_i$. Rest of the other functions $f_i(z_i)$, $g_i(z_i)$, $h_i(z_i)$ and $k_i(z_i)$ are considered as per Ali Bidram et al. [39].

2.1 Control objective under DOS attack

Here, the DOS is considered as a phenomenon that intended to avoid control signal propagated at a desired event. In this chapter, it is considered that both the control (actuator signal) and measurement (sensor signal) channels are influenced by the DOS attack simultaneously. Therefore, in this case, no exchange of data is executed from the channels. The DOS sequence of off to on is expressed as $\langle h_n \rangle$ with $h_0 \geq 0$. In this, DOS sequencing zero means the communication between channels is possible, whereas one sequencing represents that no communication is possible between channels. Therefore, nth time duration of DOS over which no possibility of communication with $\tau_n \in \mathcal{R}$ length is denoted as:

$$H_n = \langle h_n \rangle \cup [h_n, h_n + \tau_n] \tag{4}$$

The actuator under DoS attack produces the input signal based on previously updated control signal. The time intervals, $\Xi(\tau, t)$ and $\Theta(\tau, t)$, expressed time set of denial and allowance of communication, respectively.

$$\Xi(\tau,t) = \bigcup_{n\epsilon N_0} H_n \bigcup [\tau,t] \tag{5}$$

$$\Theta(\tau,t) = [\tau,t]/\Xi(\tau,t) \tag{6}$$

Here, the relative complement of $\Xi(\tau,t)$ in $[\tau,t]$ is represented as given in Eq. (6). Here, one need to find an event logic that provides robust action under DOS attack. While designing such scheme, one should ensure a positive away from zero inter-execution time of the event-based control logic that is applicable in a physical microgrid network. Therefore, to satisfy it, following definitions are adopted:

Definition 1 ([40]) Any control $\mathcal{C}$ obtaining from Eq. (3) will be ISS if β function based on KL-function and γ function based on a K_∞-function exist in a way that,

$$\|x(t)\| = \beta(\|x(0)\|,t) + \gamma\left(\|\mathcal{D}_{it}\|_\infty\right) \tag{7}$$

for $\mathcal{D}_i \in \mathcal{L}_\infty$ and microgrid state $x(0) \in \mathcal{R}^n$. Here, the values of β and γ are calculated for multi-DG-based microgrid as $\beta = \sqrt{\frac{\alpha_2}{\alpha_1}} e^{-\left(\frac{\omega_1}{2}\right)t}$ and $\gamma = \sqrt{\frac{\gamma_4}{\alpha_1 \mathcal{L}_1}}$.

Definition 2 The sequence control update $\langle t_k \rangle$ will have property of finite sampling rate if inter-execution sampling lower bound rate $\underline{\Delta} \in \mathcal{R}_{(>0)}$ exist in a way that:

$$\Delta_k = t_{k+1} - t_k \geq \underline{\Delta} \tag{8}$$

Having $k \in N_0$.

2.2 Secondary control

The main objective of the chapter is to achieve tracking synchronization with ISS of all DGs voltage and frequency to the leader reference value through secondary control under DOS attack. To achieve this objective, information exchange among neighboring DGs is required by means of sparse communication, which can be exhibited using graph theory approach. The preliminaries of directed graph theory are addressed in this section.

2.3 Preliminaries of graph theory

The edges from jth to ith node is expressed by (v_j, v_i) in the graph theory, where $G = (V, \mathcal{E}, A)$ is the communication link between nodes. The N is the number of finite nodes, whereas vertices of graph are represented by $V = \{v_1, v_2 \ldots v_N\}$. The set of edges are $(\mathcal{E} \subset V \times V)$. The term of adjacency matrix describing graph pattern is expressed as $A = [a_{ij}] \in \mathcal{R}^{N \times N}$ (with a_{ij} as edge (v_j, v_i) weights, a_{ij} is positive if $(v_j, v_i) \in \mathcal{E}$, and $a_{ij} = 0$ is zero otherwise). "D" as in degree matrix is denoted as $D = diag\{d_i\} \in \mathcal{R}^{N \times N}$ with $d_i = \sum_{j \in N_i} a_{ij}$. Another term in graph theory is the Laplacian matrix, where "L" is mathematically represented as $L = D - A$. Spanning tree in graph theory is the direct path between i_l^{th} leader node to other nodes.

2.4 Consensus-based voltage error

Distributed cooperative control is designed to synchronize terminal voltage amplitudes v_{omag} of each DG with reference voltage v_{ref}. By differentiating Eq. (1):

$$\dot{v}_{odi} = \dot{V}_{Ni} - n_{Qi}\dot{Q}_i \equiv u_{vi} \tag{9}$$

Here, u_{vi} is the auxiliary voltage control that calculates V_{Ni} as per Eq. (9). Now the secondary voltage control of N DG microgrid is featured as first-order linear multi-agent tracking synchronization problem.

$$(\dot{v}_{odN} = u_{vN}) \tag{10}$$

$$e_{vi} = \sum_{j \in N_i} a_{ij}\left(v_{odi} - v_{odj}\right) + g_i\left(v_{odi} - v_{ref}\right) \tag{11}$$

The auxiliary control signal u_{vi} are designed based on DG's own information and its neighbor's information through communication graph G, where local neighborhood tracking error is represented as e_{vi}.

2.5 Consensus-based frequency error

The DG's frequency ω_i synchronizes with the leader node frequency ω_{ref} using control input ω_{Ni}. By differentiating frequency droop characteristic $(\omega = \omega_N - m_p P)$ under nonlinear dynamics (Eq. 3) of ith DG yields,

$$\dot{\omega}_i = \dot{\omega}_{Ni} - m_{Pi}\dot{P}_i \equiv u_{\omega i} \tag{12}$$

where $u_{\omega i}$ denotes the auxiliary frequency control that facilitates to calculate ω_{Ni} as per Eq. (12). Secondary frequency control of N DG microgrid is featured as first-order linear multi-agent tracking synchronization problem.

$$(\dot{\omega}_N = u_{\omega N}) \tag{13}$$

The auxiliary controls $u_{\omega i}$ are designed based on DG's own information and with its neighbor's information through communication digraph G where local neighbourhood tracking error $e_{\omega i}$ is represented as

$$e_{\omega i} = \sum_{j \in N_i} a_{ij}\left(\omega_i - \omega_j\right) + g_i\left(\omega_i - \omega_{ref}\right) \tag{14}$$

It is stated that all DG's output powers are to be assigned to meet the following equality.

$$m_{P1}P_1 = \ldots = m_{PN}P_N \tag{15}$$

where droop coefficients m_{Pi} are chosen as per DGs active power rating. Hence, the secondary frequency control must satisfy Eq. (15). To meet this obligation, an additional control is considered for $m_{Pi}\dot{P}_i$ to regulator synchronization for linear and first-order multi-agent system as Eq. (16) using communication digraph G.

$$\left(m_{PN}\dot{P}_N = u_{PN}\right) \tag{16}$$

The auxiliary active power control u_{Pi} is calculated based on own DG information as well as neighbors DG information through the graph with local neighborhood tracking error (e_{Pi}).

$$e_{Pi} = \sum\nolimits_{j\in N_i} a_{ij}\left(m_{Pi}P_i - m_{Pj}P_j\right) \tag{17}$$

The consensus error calculated in Eqs. (11), (14), and (17) are used to achieve auxiliary control u_{vN}, $u_{\omega N}$ and u_{PN}, respectively, using Eq. (18).

$$\begin{cases} V_{Ni} = \displaystyle\int \left(u_{vi} + n_{Qi}\dot{Q}_i\right)dt \\ \omega_{Ni} = \displaystyle\int \left(u_{\omega i} + m_{Pi}\dot{P}_i\right)dt \end{cases} \tag{18}$$

2.6 Sliding mode-based control strategy

The multi-DG-based microgrid dynamics is written as [39],

$$\begin{cases} \dot{x}_i(t) = u_i(t) + \mathcal{D}_i(t) \\ \dot{x}_l(t) = u_l(t) \end{cases} \tag{19}$$

where $x_i(t)\in\mathcal{R}^n$ is the representation of microgrid states of ith DG as frequency, active power, and direct terminal voltage [39]. The auxiliary control input is $u_i(t)\in\mathcal{R}^n$. The unmodeled bounded disturbance of ith DG is denoted as $\mathcal{D}_i(t)\in\mathcal{R}^n$ having $\|\mathcal{D}_i\| \leq D_i$. The leader state vector is $\dot{x}_l(t)\in\mathcal{R}^n$ and $u_l(t)\in\mathcal{R}^n$ is the leader control input. The aim of the chapter is to model dynamic event-based consensus control scheme to mitigate the impact of DOS attack. Let us take $\widetilde{x}_i(t) = x_i(t) - x_l(t) + \beth_i$, and $\widetilde{u}_i(t) = u_i(t) - u_l(t)$, with $\beth_i$ is the desired state deviation.

$$\dot{\widetilde{x}}_i(t) = \widetilde{u}_i(t) + \mathcal{D}_i \tag{20}$$

Lemma 1 [41]: A control system may achieve synchronous tracking for all agents without any disturbances, if the process is considered as:

$$\begin{aligned} \widetilde{u}_i(t) &= X_i^{\eta}(t) \\ \chi_i(t) &= -\sum_{j\in N_i} a_{ij}\left(\widetilde{x}_i(t) - \widetilde{x}_j(t)\right) + g_i\left(\widetilde{x}_i(t)\right) \end{aligned} \tag{21}$$

With $0<\gamma<1$ and $1\leq n_i\leq N$, having N number of DGs as agents.

Lemma 2 [41]: According to graph theory concept, $(L+G)$ matrix has $\mathcal{R}e\left(\lambda_{eign}(L+G)\right) > 0$ in multi-DG-based microgrid with the possible directed spanning tree. The sliding surfaces will converge if Lemma 3 holds.

Lemma 3 [41,42]: If any system represented as $\dot{b} = f(b), f(0) = 0, b \in U_0 \subset \mathcal{R}^n$ and $\mathcal{V}(b)$ is in the vicinity of equilibrium point in a way that $\dot{\mathcal{V}}(b) + C_1\mathcal{V}^{\alpha}(b) \leq 0$, with $C_1 > 0$ and $0 < \alpha < 1$ so that $\mathcal{V}(b)$ reaches to equilibrium point in a finite duration. This finite duration can be obtained as $\left(\mathcal{V}^{1-\alpha}(b(0))/C_1(1-\alpha)\right)$.

Assumption 1 For multi-DG microgrid dynamics (Eq. 20), an integral surface in Eq. (22) is taken in a way to achieve finite time convergence.

$$S_i(t) = \tilde{x}_i(t) - \int_0^t \mathcal{X}_i^{\eta}(t)dt \tag{22}$$

As $S_i(t) = [S_1(t), S_2(t) \ldots S_N(t)]^T$ for $i = 1 \ldots N$. Therefore, in the reaching phase, Eq. (23) is obtained due to $S_i(t) = 0$ and $\dot{S}_i(t) = 0$.

$$\dot{\tilde{x}}_i(t) = \mathcal{X}_i^{\eta}(t) \tag{23}$$

A reaching law to get fast convergence is considered as:

$$\dot{S}_i(t) = -k_1|S_i(t)|^{\frac{1}{2}}sign(S_i(t)) - k_2S_i(t) \tag{24}$$

With sliding mode gains $k_1 = diag\{k_{11}, k_{12}, \ldots, k_{1N}\}$ and $k_2 = diag\{k_{21}, k_{22}, \ldots, k_{2N}\}$, $k_{ij} > 0$ for $\forall i, j$. The mathematic function $sign(S_i(t)) = [sign(S_1(t)), sign(S_2(t)), \ldots, sign(S_N(t))]^T$. Here, the minimum eigen values as $\lambda_{min}(k_1)$ and $\lambda_{min}(k_2)$ can be taken instead of K_1, K_2. The consensus secondary frequency and voltage control is addressed as:

$$\tilde{u}_{yi}(t) = \mathcal{X}_i^{\eta}(t) - k_1|S_i(t)|^{\frac{1}{2}}sign(S_i(t)) - k_2S_i(t) - D_i \tag{25}$$

with $\tilde{u}_{yi} \rightarrow u_{vN}, u_{\omega N} or u_{PN}$.

2.7 Event-based SMC scheme in the absence of DOS attack

The triggering norms in the absence of DOS is calculated in a way that control update occurs when there is violation in the state error bounds, otherwise same input is propagated. The control (Eq. 26) is retained in between inter-execution events $t \in [t^i$ and $t_a^i]$.

$$\tilde{u}_{yi}\left(t_a^i\right) = \mathcal{X}_i^{\eta}\left(t_a^i\right) - k_1\left|S_i\left(t_a^i\right)\right|^{\frac{1}{2}}sign\left(S_i\left(t_a^i\right)\right) - k_2S_i\left(t_a^i\right) \tag{26}$$

The triggering error $\xi_i(t)$ is represented as:

$$\begin{aligned}\xi_i(t) = \Big[&\mathcal{X}_i^{\eta}\left(t_a^i\right) - k_1\left|S_i\left(t_a^i\right)\right|^{\frac{1}{2}}sign\left(S_i\left(t_a^i\right)\right) - k_2S_i\left(t_a^i\right)\\ &-\left(\mathcal{X}_i^{\eta}\left(t^i\right) - k_1\left|S_i\left(t^i\right)\right|^{\frac{1}{2}}sign\left(S_i\left(t^i\right)\right) - k_2S_i\left(t^i\right)\right)\Big]\end{aligned} \tag{27}$$

The triggering error $\xi_i(t)$ is zero at the instant $t_a^i = t^i$. Control input is updated if the triggering error $\xi_i(t)$ contravenes a defined threshold, otherwise previously updated control is utilized [43].

Theorem 1 If triggering rule (Eq. 28) is violated, then the control is updated as per Eq. (26).

$$\|\xi_i(t)\| \le \text{б} \tag{28}$$

With $\text{б} < \left(\lambda_{min}(k_1)\|S_i(t)\|^{\frac{1}{2}} + \lambda_{min}(k_2)\|S_i(t)\| - \mathcal{D}_i\right)$.

Proof 1 A Lyapunov candidate is taken to derive triggering condition by the analysis of system stability in the sense of ISS.

$$\mathcal{V}_1 = \frac{1}{2}S_i^T(t)S_i(t) \tag{29}$$

$$\dot{\mathcal{V}}_1 = S_i^T(t)\dot{S}_i(t),$$

with $\dot{S}_i(t) = \dot{\tilde{x}}_i(t) - \mathcal{X}_i^\eta(t)$, now placing Eqs. (20) and (26),

$$\begin{aligned}
\dot{S}_i(t) &= \mathcal{X}_i^\eta(t_a^i) - k_1|S_i(t_a^i)|^{\frac{1}{2}}sign(S_i(t_a^i)) - k_2S_i(t_a^i) + D_i - X_i^\eta(t)\\
\dot{S}_i(t) &= \xi_i(t) - k_1|S_i(t)|^{\frac{1}{2}}sign(S_i(t)) - k_2S_i(t) + D_i\\
\dot{S}_i(t) &= \xi_i(t) - k_1|S_i(t)|^{\frac{1}{2}}sign(S_i(t)) - k_2S_i(t) + D_i
\end{aligned} \tag{30}$$

Hence,

$$\dot{\mathcal{V}}_1 = S_i^T(t)\left[\xi_i(t) - k_1|S_i(t)|^{\frac{1}{2}}sign(S_i(t)) - k_2S_i(t) + D_i\right]$$

$$\dot{\mathcal{V}}_1 \le \left\|S_i^T(t)\left[\xi_i(t) - k_1|S_i(t)|^{\frac{1}{2}}sign(S_i(t)) - k_2S_i(t) + D_i\right]\right\|$$

It can be written as $sign(S_i(t)) = \frac{S_i(t)}{\|S_i(t)\|}$.

$$\dot{\mathcal{V}}_1 \le \|S_i(t)\|\|\xi_i(t)\| - \lambda_{min}(k_2)\|S_i(t)\|^2 + \mathcal{D}_i\|S_i(t)\| \tag{31}$$

Hence according to Lemma 3, the triggering rule is obtained in Eq. (18).

3. Event-based control in presence of DOS attack

Now, the microgrid close loop control as shown in Fig. 4 relies on the update rule via triggering error $\xi_i(t)$. The control update sequence stabilizes the microgrid system states, if $\xi_i(t)$ satisfies the following norms:

$$\|\xi_i(t)\| \le \sigma\mathcal{L}_1\|x(t)\| + \sigma\mathcal{L}_1\|\mathcal{D}_i\|_\infty \tag{32}$$

With an adequate design parameter, $\sigma \in \mathcal{R}_{(>0)}$.

Here, the parameter σ is designed such that the sequence of control update taking place at finite triggering rate that satisfies Eq. (32) can be written as:

$$\lambda_{min}(k_2) - \sigma > 0 \tag{33}$$

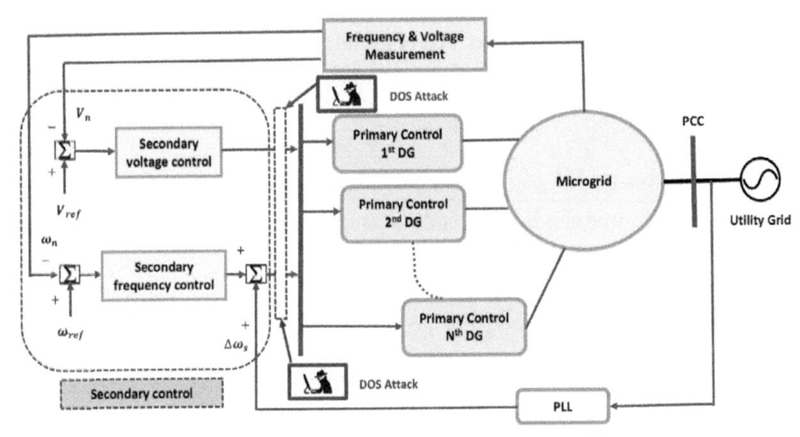

FIG. 4 DOS attack in secondary control of Multi-DG-based microgrid.

Then, $\mathcal{C}$ is input to state stable.

Lemma 4 In absence of DOS, control update rule having sampling rate is less than or equal to:

$$\overline{\Delta}_{\sigma} :=\frac{1}{\vartheta_1}\frac{\sigma\mathcal{L}_1}{(1+\sigma)} \tag{34}$$

will satisfy Eq. (32).

3.1 DOS attack preliminaries time limitations

Preliminary objective is to obtain the DOS constrains that the control system is capable to tolerate before violating the stability norms. In this regard, an adequate constraint should be employed on the duration and frequency of DOS.

(1) Frequency constraint of DOS:

Frequency in terms of occurrence of DOS is determined first by defining elapsed time in between two events of DOS as $\mathcal{J}_n = h_{n+1} - h_n$, $n \in N_0$. From this, it is to be observed that if $\mathcal{J}_n < \underline{\Delta}$ for $\forall n \in N_0$, (where $\underline{\Delta}$ is the lower bound of sampling rate), the stability of the closed system is preserved under DOS attack. In simple words, stability subjected to DOS attack frequency of DOS occurrence is to be much lesser then the minimum inter-sampling rate. Now, the total number of DOS transitions in a duration $[\tau, t[$, (where $\tau \leq t$) is $n(\tau, t)$ expressed as:

$$n(\tau,t) \leq \mathfrak{z} + \frac{t-\tau}{\partial_D} \tag{35}$$

With $\mathfrak{z} \in \mathcal{R}_{(>0)}$ and $\partial_D \in \mathcal{R}_{(>\underline{\Delta})}$.

(2) Duration constraint of DOS:

The duration of DOS is concern with the time interval over which the communication in microgrid system is denied. The length of this time duration of DOS is expressed as:

$$\Xi(\tau, t) \leq g + \frac{t-\tau}{T} \tag{36}$$

With $g \in \mathcal{R}_{(>0)}$ and $T \in \mathcal{R}_{(>1)}$ also $\tau, t \in \mathcal{R}_{(>0)}$ with $t \geq \tau$.

To understand the above DOS attack features, the following example is taken and respective features are described. In the example depicted in Fig. 5, the transitions of DOS attacks are shown. Here, the transitions correspond off to on (↑) and on to off transitions (↓). The ↑ transitions in this example are occurred at 4 s, 9 s, and 14 s. The duration of these DOS attacks is, respectively, 2 s, 3.5 s, and 1 s. The number of attacks in particular durations can be written in this example as: $n(0,3) = 0$, $n(3,10) = 2$ and $n(5,20) = 3$. The duration of DOS defined in the section can be obtained in this example as $\Xi(0,3) = 0$,$\Xi(2,11) = [4,6[\cup[9,11[$, and $\Xi(11,20) = [11,12.5[\cup[14,15[$.

Remark 1 The DOS characteristics discussed above are taken in the chapter for analysis in the microgrid system. Also, in this chapter, it has no impact about the available information to the attacker. Therefore, no assumption in this regard is considered on the probable attacker information such as state feedback, process of dynamics, and control triggering logic. These kinds of assumptions do not impact widely on the designed control configuration, but the assumption of DOS attack duration and frequency is to be the focus of concern as discussed in this section.

Theorem 2 With consideration of the microgrid control system $\mathcal{C}$ constituted by Eq. (19) and having control signal (Eq. 25) under the Lyapunov candidate $\mathcal{V}_1 = \frac{1}{2} S_i^T(t) S_i(t)$ serve as stability analyser. From this, the update sequence of the control input with the finite rate of sampling and satisfying the constraint $\mathcal{J}_n < \underline{\Delta}$, and σ as per Eq. (33). Then, any of the attack sequence of DOS that

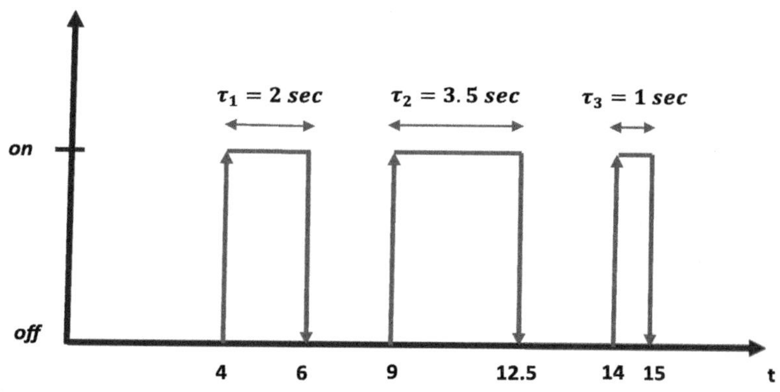

FIG. 5 Duration and frequency transitions in a DOS signal.

satisfies Eqs. (35) and (36) with arbitrary parameters $\mathfrak{z}, g, \partial_D$, and T such that following inequality holds:

$$\frac{\Delta_*}{\partial_D} + \frac{1}{T} < \frac{\omega_1}{\omega_1 + \omega_2} \tag{37}$$

with Δ_* as a strictly positive value that satisfies the following inequality:

$$\sup_{k \in \varsigma} \Delta_k \leq \Delta_* \tag{38}$$

The value of Δ_k is provided in Eq. (8). $\omega_1$is the parameter related to rate of convergence of the close loop microgrid trajectory when the DOS attack is not occurred, whereas ω_2 represents the related parameter to the divergence rate of close loop microgrid trajectory when the DOS attack is occurred. Now the sequence of sampling is taken as $\langle t_k \rangle_{k \in N_0}$ with sequence DOS attack $\langle h_n \rangle_{n \in N_0}$. Then

$$\varsigma := \left\{ k \epsilon N_0 | t_k \epsilon \bigcup_{n \epsilon N_0} H_n \right\} \tag{39}$$

represents the integer set of microgrid control signal update in presence of DOS attack.

3.2 Parameter design for the convergence rate in absence of DOS

Assumption 2 To obtain triggering control logic under DOS attack for the microgrid network system (Eq. 20), it is assumed that function $f(x_i)$ to its argument is Lipschitz and satisfy the following norm Lipschitz coefficient $\mathcal{L}_1$.

$$\| f(x_i(t)) - \mathfrak{z} f(x_i(t_i)) \| < \mathcal{L}_1 \| x_i(t) - x_i(t_i) \| = \mathcal{L}_1 \| \xi_i \| \tag{40}$$

Microgrid control dynamics in terms of sliding surface is represented in relation to the microgrid state as per the inequality (7) is written as:

$$\left\| x_i(t) - \int_0^t \mathcal{X}_i^\eta(t) dt \right\| \leq \| x_i(t) \| \leq \beta(\| x(0) \|, t) + \gamma(\| \mathcal{D}_{ti} \|_\infty)$$

Therefore, the inequality (7) in terms of sliding surface can be expressed as:

$$\| S_i(t) \| \leq \beta(\| x(0) \|, t) + \gamma(\| \mathcal{D}_{ti} \|_\infty)$$

Similarly, the control triggering rule designed in Eq. (28) can be expressed in terms of state error per inter execution event by considering assumption in Eq. (40).

$$\begin{aligned} \xi_i(t) = & \left[\mathcal{X}_i^\eta(t_a^i) - k_1 |S_i(t_a^i)|^{\frac{1}{2}} sign(S_i(t_a^i)) \right. \\ & -k_2 S_i(t_a^i) - \left(\mathcal{X}_i^\eta(t^i) - k_1 |S_i(t^i)|^{\frac{1}{2}} sign(S_i(t^i)) \right. \\ & \left. -k_2 S_i(t^i)) \right] \leq \mathcal{L}_1(x_i(t_a^i)) - \mathcal{L}_1(x_i(t^i)) \end{aligned} \tag{41}$$

or

$$\xi_i(t) \leq \mathcal{L}_1\left(x_i\left(t_a^i\right)\right) - \mathcal{L}_1\left(x_i\left(t^i\right)\right)$$

Therefore, the condition obtained to stabilize microgrid closed loop plant under DOS attack as per Eq. (32). From Eq. (31), it can be written as:

$$\dot{\mathcal{V}}_1 \leq \|S_i(t)\|\|\xi_i(t)\| - \lambda_{min}(k_2)\|S_i(t)\|^2 + \mathcal{D}_i\|S_i(t)\|$$

Now placing the value of Eq. (32) in Eq. (31) and taking $S_i(t) = \mathcal{L}_1 x(t)$:

$$\dot{\mathcal{V}}_1 \leq \mathcal{L}_1\|x(t)\|\left(\sigma\mathcal{L}_1\|x(t)\| + \sigma\mathcal{L}_1\|\mathcal{D}_i\|_\infty\right) - \lambda_{min}(k_2)\mathcal{L}_1{}^2\|x(t)\|^2 + \mathcal{D}_i\|x(t)\|$$

or

$$\dot{\mathcal{V}}_1 \leq -(\lambda_{min}(k_2) - \sigma)\mathcal{L}_1{}^2\|x(t)\|^2 + \sigma\mathcal{L}_1{}^2\|\mathcal{D}_i\|_\infty\|x(t)\| + \mathcal{D}_i\|x(t)\|$$

or

$$\dot{\mathcal{V}}_1 \leq -(\lambda_{min}(k_2) - \sigma)\mathcal{L}_1{}^2\|x(t)\|^2 + \left(1 + \sigma\mathcal{L}_1{}^2\right)\|x(t)\|a(t)$$

where, $a(t) := \sup\{\mathcal{D}_i, \|\mathcal{D}_i\|_\infty\}$, further the above equation can be written as:

$$\dot{\mathcal{V}}_1 \leq -\gamma_1\|x(t)\|^2 + \gamma_2\|x(t)\|a(t) \tag{42}$$

With $\gamma_1 = (\lambda_{min}(k_2) - \sigma)\mathcal{L}_1{}^2$ and $\gamma_2 = \left(1 + \sigma\mathcal{L}_1{}^2\right)$.
Now applying young's inequality with δ as positive real quantity.

$$2\|x(t)\|a(t) = \frac{1}{\delta}\|x(t)\|^2 + \delta a^2(t) \tag{43}$$

By putting $\delta = \frac{\left(1 + \sigma\mathcal{L}_1{}^2\right)}{(\lambda_{min}(k_2) - \sigma)\mathcal{L}_1{}^2}$ *or* $\frac{\gamma_2}{\gamma_1}$ in (42), it can be calculated as:

$$\dot{\mathcal{V}}_1 \leq -\frac{\gamma_1}{2}\|x(t)\|^2 + \frac{\gamma_2{}^2}{2\gamma_1}a^2(t) \tag{44}$$

The strictly positive Lyapunov function $\mathcal{V}_1$ in terms of microgrid states can be expressed with the state inequality as:

$$\alpha_1\mathcal{L}_1\|x(t)\|^2 \leq \mathcal{V}_1(t) \leq \alpha_2\mathcal{L}_1\|x(t)\|^2 \tag{45}$$

From Eqs. (45) and (44), it can be written as:

$$\dot{\mathcal{V}}_1 \leq -\omega_1\mathcal{V}_1(x(t)) + \gamma_3 a^2(t) \tag{46}$$

where, $\gamma_3 = \frac{\gamma_2{}^2}{2\gamma_1}$ and ω_1 as:

$$\omega_1 = \frac{(\lambda_{min}(k_2) - \sigma)\mathcal{L}_1{}^2}{2\alpha_2\mathcal{L}_1} \tag{47}$$

Here $\|a_t\|_\infty = \|\mathcal{D}_{it}\|_\infty$ for $t \in R_{>0}$. Thus, from Eq. (46), the differential inequalities will result in the expression as:

$$\mathcal{V}_1(x(t)) \leq e^{-\omega_1 t}\mathcal{V}_1(x(0)) + \gamma_4\|\mathcal{D}_{it}\|_\infty^2 \tag{48}$$

where, $_4 = _3/_1$, now from Eqs. (45) and (48), it can be written as:

$$\|x(t)\|^2 \leq \frac{\alpha_2}{\alpha_1} e^{-\omega_1 t}\|x(0)\|^2 + \frac{\gamma_4}{\alpha_1 \mathcal{L}_1}\|\mathcal{D}_{it}\|_\infty^2 \tag{49}$$

Since it is well known that $a^2+b^2 \leq (a+b)^2$ for the positive real values of a and b, it can be finally written as:

$$\|x(t)\| \leq \sqrt{\frac{\alpha_2}{\alpha_1}} e^{-\left(\frac{\omega_1}{2}\right)t}\|x(0)\| + \sqrt{\frac{\gamma_4}{\alpha_1 \mathcal{L}_1}}\|\mathcal{D}_{it}\|_\infty \tag{50}$$

That Eq. (50) is the result in the stability bounds as expressed in Eq. (7).

3.3 Parameter design for the divergence rate in presence of DOS

The control update in absence of the DOS will be successful. Therefore, according to Eq. (41), the error dynamics will satisfy the following inequality:

$$\frac{d}{dx}\|\xi_i(t)\| \leq \left\|\frac{d}{dx}\xi_i(t)\right\| \leq \left\|\frac{d}{dx}\left[\mathcal{X}_i^\eta(t_a^i) - k_1|S_i(t_a^i)|^{\frac{1}{2}} sign(S_i(t_a^i)) - k_2 S_i(t_a^i)\right]\right\| \tag{51}$$

As the previous control remains same during the consecutive events, Eq. (51) is obtained.

$$\frac{d}{dx}\|\xi_i(t)\| \leq \left\|\frac{d}{dt}\left[\mathcal{X}_i^\eta(t)\right]\right\| - \left\|\frac{d}{dt}\left[k_1|S_i(t)|^{\frac{1}{2}} sign(S_i(t))\right]\right\| - \left\|\frac{d}{dt}\left[k_2 S_i(t)\right]\right\|$$

$$\begin{aligned}\frac{d}{dx}\|\xi_i(t)\| &\leq \eta n^{3-2\eta}\|(L+G)\otimes I_n\|\|\mathcal{X}(t)\|^{2\eta-1}\\ &-\lambda_{min}(k_1)\left\|\left(tanh\,(\beta S_i(t))\frac{d}{dt}\left[|S_i(t)|^{\frac{1}{2}}\right] + |S_i(t)|^{\frac{1}{2}}\frac{d}{dt}\left[sign(S_i(t))\right]\right)\right\|\\ &-\lambda_{min}(k_2)\|\dot{S}_i(t)\|\end{aligned}$$

$$\begin{aligned}\frac{d}{dx}\|\xi_i(t)\| &\leq \eta n^{3-2\eta}\|(L+G)\otimes I_n\|\|\mathcal{X}(t)\|^{2\eta-1}\\ &-\lambda_{min}(k_1)\left(\left\|tanh\,(\beta S_i(t))\frac{d}{dt}|S_i(t)|^{\frac{1}{2}}\right\| + \|S_i(t)\|^{\frac{1}{2}}\|(1 - tanh^2(\beta S_i(t)))\beta\dot{S}_i(t)\|\right)\\ &-\lambda_{min}(k_2)\|\dot{S}_i(t)\|\end{aligned}$$

Now taking $\|(1_{n\times n} - tanh^2(\beta S_i(t)))\| \leq \|1_{n\times n}\| = n$

$$\begin{aligned}\frac{d}{dx}\|\xi_i(t)\| &\leq \eta n^{3-2\eta}\|(L+G)\otimes I_n\|\|\mathcal{X}(t)\|^{2\eta-1}\\ &-\lambda_{min}(k_1)\left(\left\|tanh\,(\beta S_i(t))\frac{d}{dt}|S_i(t)|^{\frac{1}{2}}\right\| + n\beta\|S_i(t)\|^{\frac{1}{2}}\|\dot{S}_i(t)\|\right)\\ &-\lambda_{min}(k_2)\|\dot{S}_i(t)\|\end{aligned}$$

$$\frac{d}{dx}\|\xi_i(t)\| \leq \eta n^{3-2\eta}\|(L+G)\otimes I_n\|\|\mathcal{X}(t)\|^{2\eta-1}$$
$$-\lambda_{min}(k_1)\left(\frac{1}{2}\|S_i(t)\|^{\frac{1}{2}}\|\dot{S}_i(t)\|\|\tanh(\text{ß}S_i(t))\| + n\text{ß}\|S_i(t)\|^{\frac{1}{2}}\|\dot{S}_i(t)\|\right)$$
$$-\lambda_{min}(k_2)\|\dot{S}_i(t)\|$$

$$\frac{d}{dx}\|\xi_i(t)\| \leq \eta n^{3-2\eta}\|(L+G)\otimes I_n\|\|\mathcal{X}(t)\|^{2\eta-1}$$
$$-\left(\lambda_{min}(k_1)\|S_i(t)\|^{\frac{1}{2}}\left(\frac{1}{2}\|\tanh(\text{ß}S_i(t))\| + n\text{ß}\right) + \lambda_{min}(k_2)\right)\|\dot{S}_i(t)\|$$

By using Assumption 2 and putting the value of $\dot{S}_i(t)$ from Eq. (30):

$$\frac{d}{dx}\|\xi_i(t)\| \leq \eta n^{3-2\eta}\mathcal{L}_2\|(L+G)\otimes I_n\|\|\xi_i(t)\|$$
$$-\left(\lambda_{min}(k_1)\|S_i(t)\|^{\frac{1}{2}}\left(\frac{1}{2}\|\tanh(\text{ß}S_i(t))\| + n\text{ß}\right) + \lambda_{min}(k_2)\right)\|\xi_i(t)$$
$$-k_1\|S_i(t)\|^{\frac{1}{2}}sign(S_i(t)) - k_2S_i(t) + D_i\| \tag{52}$$

$$\frac{d}{dx}\|\xi_i(t)\| \leq \gamma_5\|\xi_i(t)\|$$
$$-\gamma_6\left[\|\xi_i(t)\| - \lambda_{min}(k_1)\|S_i(t)\|^{\frac{1}{2}} - \lambda_{min}(k_2)\|S_i(t)\| + \|\mathcal{D}_{it}\|_\infty\right] \tag{53}$$

where $\gamma_5 = \eta n^{3-2\eta}\mathcal{L}_2\|(L+G)\otimes I_n\|$ and $\gamma_6 = \left(\lambda_{min}(k_1)\|S_i(t)\|^{\frac{1}{2}}\left(\frac{1}{2}\|\tanh(\text{ß}S_i(t))\| + n\text{ß}\right) + \lambda_{min}(k_2)\right)$. By taking Assumption 2 in consideration, Eq. (53) can be written as:

$$\frac{d}{dx}\|\xi_i(t)\| \leq (\gamma_5-\gamma_6)\|\xi_i(t)\| + \gamma_6\mathcal{L}_1[\lambda_{min}(k_1) + \lambda_{min}(k_2)]\|x(t)\| - \gamma_6\|\mathcal{D}_{it}\|_\infty$$

or

$$\frac{d}{dx}\|\xi_i(t)\| \leq \gamma_7\|\xi_i(t)\| + \gamma_8\|x(t)\| - \gamma_9\|\mathcal{D}_{it}\|_\infty \tag{54}$$

where $\gamma_7 = (\gamma_5-\gamma_6)$, $\gamma_8 = \gamma_6\mathcal{L}_1[\lambda_{min}(k_1) + \lambda_{min}(k_2)]$ and $\gamma_9 = \gamma_6$.

Solving Eq. (54),

$$\|\xi_i(t)\| \leq \vartheta_1\int_{t_k}^{t} e^{\mu_{\gamma_7}(t-s)}\left[\|x(t)\| + \|\mathcal{D}_{it}\|_\infty\right]ds$$

where $\vartheta_1 = \max\{\gamma_8, \gamma_9\}$, now by considering $f(t-t_k) = \int_{t_k}^{t} e^{\mu_{\gamma_7}(t-s)}ds$ and using the fact $x(t_k) = x(t) + \frac{1}{\mathcal{L}_1}\xi_i(t)$:

$$\|\xi_i(t)\| \leq \frac{\vartheta_1}{\mathcal{L}_1}f(t-t_k)\|\xi_i(t)\| + \vartheta_1 f(t-t_k)\left(\|x(t)\| + \|\mathcal{D}_{it}\|_\infty\right) \tag{55}$$

Here, we assume $f(0)=0\&f(t-t_k)$ is strictly increasing with respect to t. Now for real positive Δ and from Eq. (32), it can be written as:

$$\|\xi_i(t)\| \leq \frac{\vartheta_1}{\mathcal{L}_1} f(\Delta)\left(\sigma\mathcal{L}_1\|x(t)\| + \sigma\mathcal{L}_1\|\mathcal{D}_i\|_\infty\right) + \vartheta_1 f(\Delta)\left(\|x(t)\| + \|\mathcal{D}_{it}\|_\infty\right) \tag{56}$$

or

$$\|\xi_i(t)\| \leq \vartheta_1 f(\Delta)(1+\sigma)\left(\|x(t)\| + \|\mathcal{D}_i\|_\infty\right) \tag{57}$$

Comparing Eq. (57) with Eq. (32),

$$f(\Delta) \leq \frac{1}{\vartheta_1}\frac{\sigma\mathcal{L}_1}{(1+\sigma)} \tag{58}$$

Therefore, from Eq. (58), any microgrid state control update law such that $\Delta_k \leq \Delta$ that satisfies Eq. (32) for $t \in R_{\geq 0}$.Under the attack of DOS, the main idea is to divide time axis in two intervals, in one-time interval the condition in Eq. (32) is satisfied and in another time interval, condition in Eq. (32) may not hold due to the presence of DOS attack. In this scenario, the closed loop microgrid dynamics is to be analyzed in a way that it changes in between stable and unstable regions.

Lemma 5 [37]: For an interval τ, $t \in R_{\geq 0}$, having $0 \leq \tau \leq t$, the duration $[\tau, t]$ is the sum of $\overline{\Theta}(\tau,t)$ & $\overline{\Xi}(\tau,t)$. Here, $\overline{\Theta}(\tau,t)$ is representing the interval division of $[\tau,t]$ for which Eq. (32) is satisfied; whereas, $\overline{\Xi}(\tau,t)$ representing the subinterval of $[\tau,t]$ for which Eq. (32) does not hold. Now, the two real positive sequence of numbers $\langle\zeta_m\rangle_{m\in N_0}$, $\langle v_m\rangle_{m\in N_0}$are exist in such a way that

$$\overline{\Xi}(\tau,t) = \bigcup_{m \in N_0} X_m \cap [\tau,t] \tag{59}$$

$$\overline{\Theta}(\tau,t) = \bigcup_{m \in N_0} Y_{m-1} \cap [\tau,t] \tag{60}$$

with

$$X_m = \langle\zeta_m\rangle \cup [\zeta_m, \zeta_m + v_m] \tag{61}$$

$$X_m = \langle\zeta_m + v_m\rangle \cup [\zeta_m + v_m, \zeta_{m+1}] \tag{62}$$

and where $\zeta_{-1} = v_{-1} = 0$.

It is noted that in $[\tau,t]$, incorporate the sum of interval ranges $\overline{\Xi}(\tau,t)$ and $\overline{\Theta}(\tau,t)$. Also, the intersection set is empty. Therefore, any successful update of control essentially arises at $\zeta_m + v_m$ by the calculation for each $m \in N_0$, also over the range of Y_m, no DOS is present.

In the duration of Y_m, $m \in N_0$, for which the condition in Eq. (32) satisfies. From Eq. (48), it is obtained as:

$$\mathcal{V}_1(x(t)) \leq e^{-\omega_1(t-\zeta_m-v_m)}\mathcal{V}_1(x(\zeta_m + v_m)) + \gamma_4\|\mathcal{D}_{it}\|_\infty^2 \tag{63}$$

for all time interval $t \in Y_m$ and for all values of $m \in N_0$. The intervals X_m, $m \in N_0$, where the inequality (32) does not essentially hold true. To design a condition on the rate of enhancement of stability in terms of $\mathcal{V}_1(x(t))$, some transitional steps are desired. So, it can be written for every $m \in N_0$

$$\|\xi_i(t)\| = \|x_{t+k} - x_t\| = \|x_{t+k} - x(\zeta_m) + x(\zeta_m) - x_t\|$$

$$\|\xi_i(t)\| \leq \|x_{t+k} - x(\zeta_m)\| + \|x(\zeta_m)\| + \|x_t\|$$

$$\|\xi_i(t)\| \leq \|\xi_i(\zeta_m)\| + \|x(\zeta_m)\| + \|x_t\|$$

From Eq. (32),

$$\|\xi_i(t)\| \leq \sigma\mathcal{L}_1\|x(\zeta_m)\| + \sigma\mathcal{L}_1\|\mathcal{D}_i(\zeta_m)\|_\infty + \|x(\zeta_m)\| + \|x_t\|$$

or

$$\|\xi_i(t)\| \leq (1 + \sigma\mathcal{L}_1)\|x(\zeta_m)\| + \|x_t\| + \sigma\mathcal{L}_1\|\mathcal{D}_i\|_\infty \tag{64}$$

for all $t \in X_m$. Recall that

$$\xi_i(t) \leq \mathcal{L}_1(x_i(\zeta_m)) - \mathcal{L}_1(x_i(\mathrm{t}))$$

Here, for all time interval of $t \in X_m$, $x(\zeta_m)$ denotes process state value of the microgrid with the latest successful update of control signal till ζ_m. If suppose $\zeta_0 = 0$, then $x(\zeta_m(0)) = 0$ in a way (64) is effective. (32) satisfies for all $t \in Y_m$. therefore, for the microgrid state continuity it can be written as:

$$\|\xi_i(\zeta_m)\| \leq \sigma\mathcal{L}_1\|x(\zeta_m)\| + \sigma\mathcal{L}_1\|\mathcal{D}_i(\zeta_m)\|_\infty \tag{65}$$

And from the Lyapunov candidate analysis it can be written as:

$$\dot{\mathcal{V}}_1 \leq \mathcal{L}_1\|x(t)\|\|\xi_i(t)\| - \lambda_{min}(k_2)\mathcal{L}_1^2\|x(t)\|^2 + \mathcal{D}_i\mathcal{L}_1\|x(t)\| \tag{66a}$$

Placing Eq. (64) into Eq. (66a) results

$$\begin{aligned}\dot{\mathcal{V}}_1 \leq \mathcal{L}_1\|x(t)\|((1 + \sigma\mathcal{L}_1)\|x(\zeta_m)\| + \|x_t\| + \sigma\mathcal{L}_1\|\mathcal{D}_i\|_\infty)\\ - \lambda_{min}(k_2)\mathcal{L}_1^2\|x(t)\|^2 + \mathcal{D}_i\mathcal{L}_1\|x(t)\|\end{aligned}$$

$$\begin{aligned}\dot{\mathcal{V}}_1 \leq (1 + \sigma\mathcal{L}_1)\mathcal{L}_1\|x(\zeta_m)\|\|x(t)\| + \left(\mathcal{L}_1 - \lambda_{min}(k_2)\mathcal{L}_1^2\right)\|x(t)\|^2\\ + \left(\sigma\mathcal{L}_1^2 + \mathcal{L}_1\right)\|x(t)\|a(t)\end{aligned}$$

$$\dot{\mathcal{V}}_1 \leq (1 + \sigma\mathcal{L}_1)\mathcal{L}_1\|x(\zeta_m)\|\|x(t)\| + \gamma_{10}\|x(t)\|^2 + \gamma_{11}\|x(t)\|a(t) \tag{66b}$$

where, $a(t) := \sup\{\mathcal{D}_i, \|\mathcal{D}_i\|_\infty\}$, $\gamma_{10} = \left(\mathcal{L}_1 - \lambda_{min}(k_2)\mathcal{L}_1^2\right)$ and $\gamma_{11} = \left(\sigma\mathcal{L}_1^2 + \mathcal{L}_1\right)$.

Now further applying the young's inequality with δ_1 as positive real quantity.

$$2\|x(t)\|a(t) = \frac{1}{\delta_1}\|x(t)\|^2 + \delta_1 a^2(t) \tag{67}$$

By putting $\delta_1 = \frac{(\sigma\mathcal{L}_1^2 + \mathcal{L}_1)}{(\mathcal{L}_1 - \lambda_{min}(k_2)\mathcal{L}_1^2)}$ *or* $\frac{\gamma_{11}}{\gamma_{10}}$ in (66a), it can be calculated as:

$$\dot{\mathcal{V}}_1 \leq (1 + \sigma\mathcal{L}_1)\mathcal{L}_1 \|x(\zeta_m)\| \|x(t)\| + \frac{3\gamma_{10}}{2} \|x(t)\|^2 + \frac{\gamma_{11}^2}{2\gamma_{10}} a^2(t) \tag{68a}$$

or

$$\dot{\mathcal{V}}_1 \leq \gamma_{12} \|x(\zeta_m)\| \|x(t)\| + \frac{3\gamma_{10}}{2} \|x(t)\|^2 + \frac{\gamma_{11}^2}{2\gamma_{10}} a^2(t) \tag{68b}$$

where $\gamma_{12} = (1 + \sigma\mathcal{L}_1)\mathcal{L}_1$.

Now again repeating the sequence of applying young's inequality with δ_2 as positive real quantity.

$$2\|x(\zeta_m)\| \|x(t)\| = \frac{1}{\delta_2} \|x(t)\|^2 + \delta_2 \|x(\zeta_m)\|^2 \tag{69}$$

By putting $\delta_2 = \frac{(1 + \sigma\mathcal{L}_1)\mathcal{L}_1}{(\mathcal{L}_1 - \lambda_{min}(k_2)\mathcal{L}_1^2)}$ *or* $\frac{\gamma_{12}}{\gamma_{10}}$ in Eq. (66a), it can be calculated as:

$$\dot{\mathcal{V}}_1 \leq 2\gamma_{10} \|x(t)\|^2 + \frac{\gamma_{12}^2}{2\gamma_{10}} \|x(\zeta_m)\|^2 + \frac{\gamma_{11}^2}{2\gamma_{10}} a^2(t) \tag{70}$$

as $\|a_t\|_\infty = \|\mathcal{D}_{it}\|_\infty$ for $t \in R_{\geq 0}$, and $\gamma_{13} = \frac{\gamma_{11}^2}{2\gamma_{10}}$ then it can be written as:

$$\dot{\mathcal{V}}_1 \leq 2\gamma_{10} \|x(t)\|^2 + \frac{\gamma_{12}^2}{2\gamma_{10}} \|x(\zeta_m)\|^2 + \gamma_{13} \|\mathcal{D}_{it}\|_\infty^2 \tag{71}$$

Now from Eq. (45) and Eq. (71), it can be written as:

$$\dot{\mathcal{V}}_1 \leq \omega_2 \max\{\mathcal{V}_1(x(t)), \mathcal{V}_1(x(\zeta_m))\} + \gamma_{13} \|\mathcal{D}_{it}\|_\infty^2 \tag{72}$$

where, the value of parameter related to divergence rate of system stability in presence of DOS ω_2 is obtained as:

$$\omega_2 = \frac{2\gamma_{10}}{\alpha_1 \mathcal{L}_1} \tag{73}$$

and

$$\dot{\mathcal{V}}_1 \leq e^{\omega_2(t-\zeta_m)} \mathcal{V}_1(x(\zeta_m)) + \gamma_{14} e^{\omega_2(t-\zeta_m)} \|\mathcal{D}_{it}\|_\infty^2 \tag{74}$$

for $t \in X_m$, with $\gamma_{14} = \frac{\gamma_{13}}{\omega_2}$.

4. Verification and result analysis of the proposed scheme

The resilient control design for the multi-DG-based microgrid network under DOS attack in Section 3 is validated in the MATLAB/Simulink platform. For the validation of the control strategy, four DG-based microgrid structure shown in Fig. 3 is considered for the simulation purpose. The respective parameters describing the microgrid specification are presented in Table 1. This

TABLE 1 System properties and specifications.

	Symbols	DG1	DG2	DG3	DG4
DGs	m_P	0.000078	0.000067	0.000067	0.000054
	n_Q	0.0015	0.0013	0.0013	0.0011
Lines	Z_1, Z_3	$0.15+j0.1$			
	Z_2	$0.70+j0.84$			
	Z_4, Z_5	$0.35+j0.42$			
RL-load	Load-bus 1,2	$320+j0.48$			
	Load-bus 3,4	$250+j0.95$			
Gains	k_1, k_2	1.43, 2.21	1.18, 3.10	1.22, 2.09	1.8, 3.31
	λ	1.2			

control scheme concern with the cooperative secondary voltage frequency control in islanded mode of operation of the microgrid under the threat of cyber-attack. DOS attack sequence is scheduled from the source block set taken as random interrupt signal that is incorporated through the programming under the communication graph theory.

4.1 Simulation results obtained from the proposed scheme and discussion

Four DG-based microgrid structure shown in Fig. 3 is simulated in MATLAB/Simulink with the specification given in Table 1 under the DOS attack. The DOS sequences as generated randomly through the communication graph theory on the secondary control channel connected with all DGs of the network microgrid. The DOS attack duration is represented by the gray color bars plots in Figs. 6 and 7. Under the impact of DOS, different results with respect to microgrid parameter state variations are obtained that are shown from Figs. 6–12. The variation in the frequency (ω) of each DG connected to the microgrid is depicted in Fig. 6. The variation in the direct bus voltage of each DG is shown in the Fig 7.

Figs. 6 and 7 incorporate the sequence of DOS. After the initial transient, these secondary microgrid states are stabilized under the sliding mode-based consensus control, but due to the occurrence of DOS, these states are perturbed from their stabilized equilibrium at the duration of DOS beyond the stability

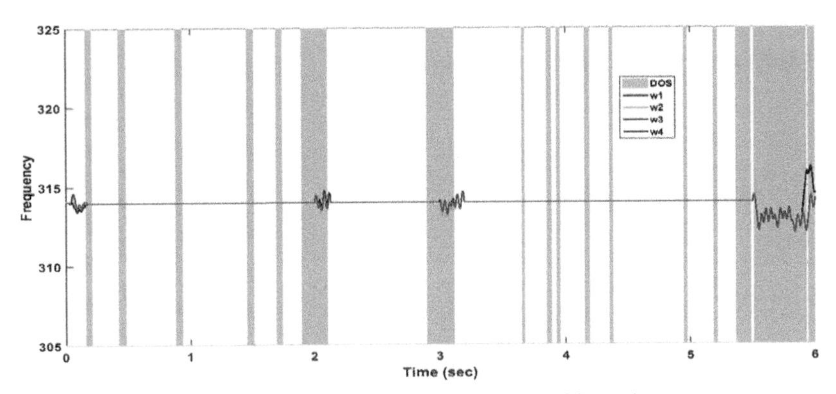

FIG. 6 Impact on the bus frequencies of the microgrid under DOS attack.

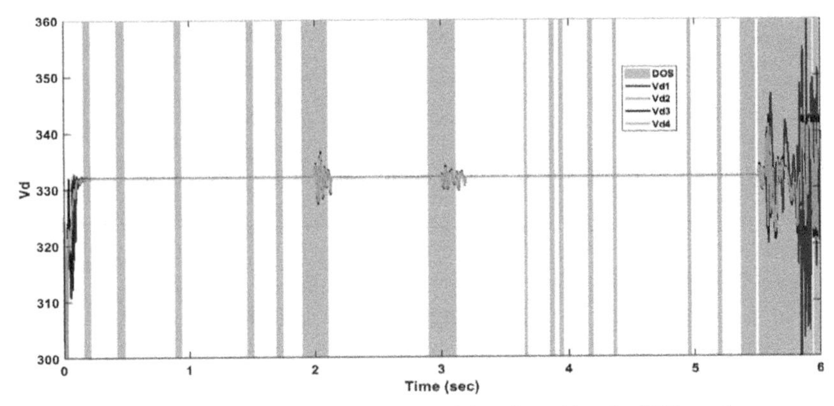

FIG. 7 Impact on the direct voltages at each bus of the microgrid under DOS attack.

bound. At the time of high DOS duration at 2 and 3 s, the states are deviated but the impact of DOS is in the tolerable limit. These states at 2 and 3 s are stabilized after the DOS active duration. Small duration of the DOS is tolerable to an extent. As it can be seen from the results, when the duration of DOS is highly increased, all microgrid states under secondary operation gets deviated from the stable trajectory as seen after 5.5 s in Figs. 6 and 7.

Fig. 8 shows the consensus error plot of the frequency corresponding to all DGs connected to the microgrid. At each DOS event, the deviation in the respective error between DG configuration is observed under violation of stability bounds at 2 and 3 s. After the active duration of DOS, error states are stabilized at 2 and 3 s. In case of large DOS duration, the deviation of error is abruptly deteriorated. The similar results are obtained for the consensus-based direct voltage error of the network microgrid as illustrated in Fig. 9. The impact of DOS sequence can be analyzed on the quadrature component of the bus voltages from Fig. 10. All DGs quadrature components are shown in the Fig. 10.

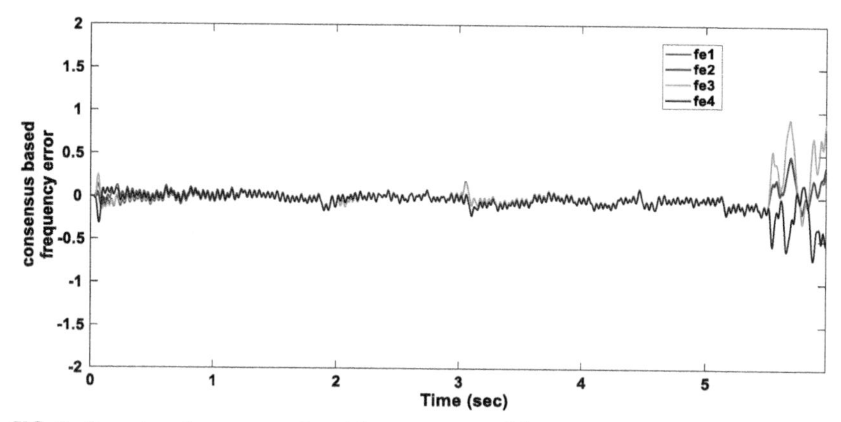

FIG. 8 Impact on the consensus-based frequency error of the microgrid under DOS attack.

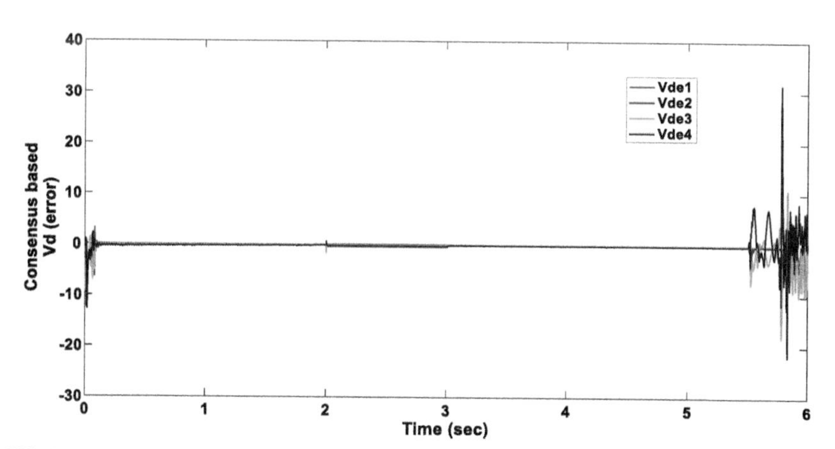

FIG. 9 Impact on the consensus-based voltage error of the microgrid under DOS attack.

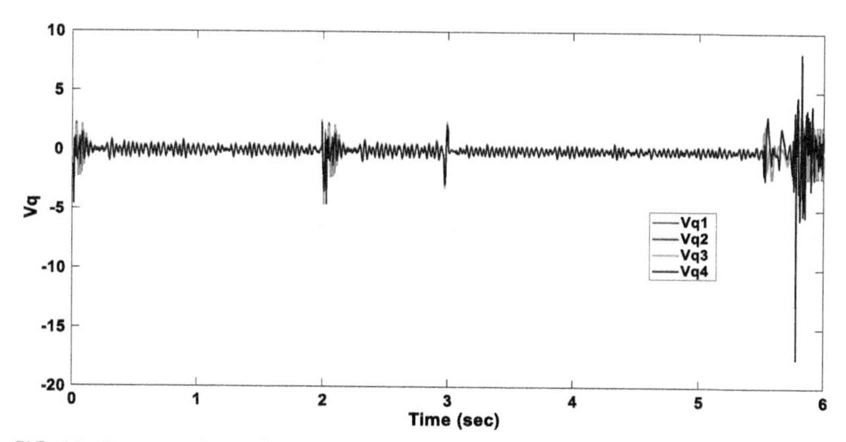

FIG. 10 Impact on the quadrature component of the bus voltages of the microgrid under DOS attack.

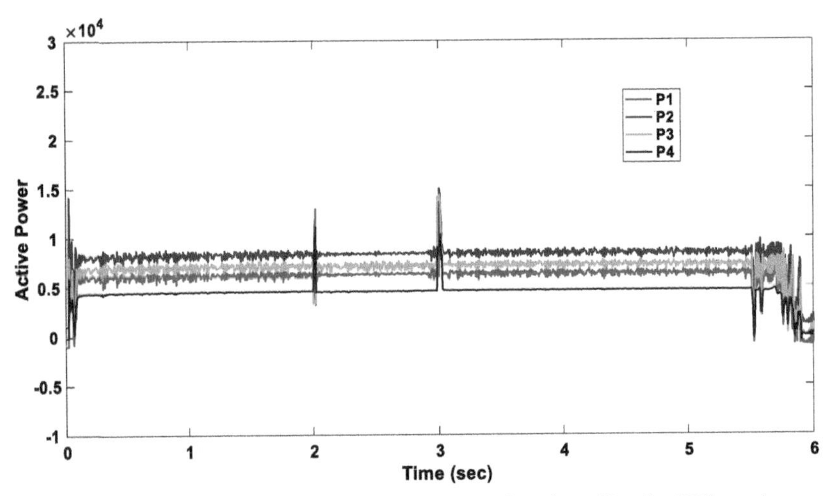

FIG. 11 Impact on the bus active powers of each DG in the microgrid under DOS attack.

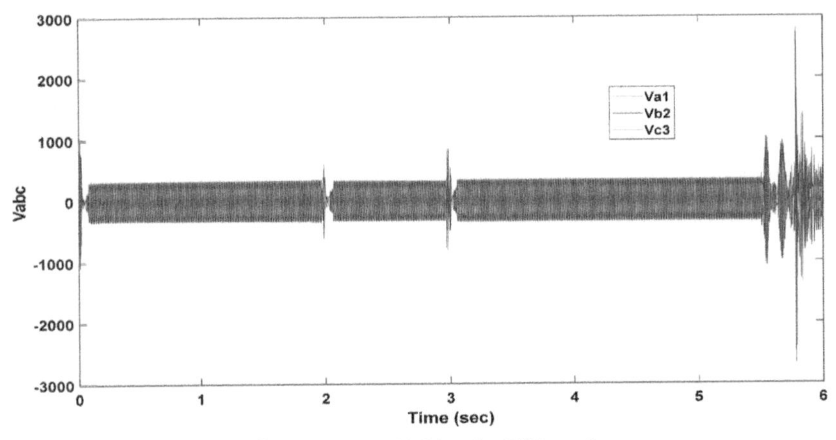

FIG. 12 Impact on the $3-\varnothing$ bus voltage of DG2 under DOS attack.

The corresponding deviation due to DOS is reflected in the quadratic components of the bus voltages at 2 and 3 s. Also, after the 5.5 s, the quadratic voltage components deteriorate. DOS impact on the corresponding rated DG powers in the droop ratio with the frequency is shown in the Fig. 11. Deviations at 2 and 3 s are observed that are further stabilized as per the deactivated DOS duration. Unstable power states region is seen after 5.5 s due to high impact of DOS. The three-phase voltage at bus 2 under DOS attack is shown in the Fig. 12. Here, the similarity with the other microgrid states is observed as per the sequence of DOS attack. Stability region, deviation of three phase voltage at 2 and 3 s, and deterioration of the voltage after 5 s can be seen from the result shown in Fig. 12.

5. Conclusion

The proposed control scheme shows robust performance with minimal control effort in the microgrid structure under DOS attack. The stability in terms of Lyapunov candidature is proved with the analysis of DOS frequency and duration for which closed loop microgrid stability remain preserved. Under DOS, attack stability in terms of converging and diverging parameters is discussed by division of sub intervals of the DOS attack sequence with the on-off jamming ratio. The multi-DG-based microgrid control strategy is governed by sliding mode control design that provides the sampling rule for which ISS of the microgrid system is achieved and its robustness against system disturbance is ensured. Further, the DOS attack resilient control algorithm is applied on the secondary voltage/frequency control of the microgrid. From the obtained results, it is observed that all microgrid states under attack are stabilized to an extent of stability bounds over which the deterioration of the microgrid states is started. DOS tolerance limit is determined by the relevant triggering-based control design using sliding mode. Pertaining to this control design for microgrid effect of delay in information exchange can be the future scope of the work. As a matter of fact, results verify the objective of proposed technique for autonomous AC microgrid operating in islanded mode of operation.

References

[1] X. Liu, M. Shahidehpour, Y. Cao, L. Wu, W. Wei, X. Liu, Microgrid risk analysis considering the impact of cyber-attacks on solar PV and ESS control systems, IEEE Trans. Smart Grid 8 (3) (2017) 1330–1339.

[2] H. He, J. Yan, Cyber-physical attacks and defenses in the smart grid: a survey, IET Cyber Phys. Syst. Theory Appl. 1 (1) (2016) 13–27.

[3] A. Vasilakis, et al., The evolution of research in microgridscontrol, IEEE OAJPE 7 (2020) 331–343.

[4] L. Che, M. Shahidehpour, A. Alabdulwahab, Y. Al-Turki, Hierarchical coordination of a community microgrid with AC and DC microgrids, IEEE Trans. Smart Grid 6 (6) (2015) 3042–3051.

[5] S. Zuo, A. Davoudi, Y. Song, F.L. Lewis, Distributed finite-time voltage and frequency restoration in islanded AC microgrids, IEEE Trans. Ind. Electron. 63 (10) (2016) 5988–5997.

[6] V. Kumar, S.R. Mohanty, Dynamic event driven robust control design with uncertainty compensator for agent misbehave in autonomous AC microgrid, in: 2021 IEEE International Conference on Systems, Man, and Cybernetics (SMC), 2021, pp. 1406–1411.

[7] Bidram, A. Davoudi, Hierarchical structure of microgrids control system, IEEE Trans. Smart Grid 3 (4) (2012) 1963–1976.

[8] B. Fahimi, A. Kwasinski, A. Davoudi, R.S. Balog, M. Kiani, Charge it, Power Energy Mag. 9 (4) (2011) 54–64.

[9] J.M. Guerrero, J.C. Vásquez, J. Matas, M. Castilla, Vicuña, L.G.d., Castilla, M., Hierarchical control of droop-controlled AC and DC microgrids – a general approach toward standardization, IEEE Trans. Ind. Electron. 58 (1) (2011) 158–172.

[10] C.K. Sao, W. Lehn, Control and power management of converter fed microgrids, IEEE Trans. Power Syst. 23 (3) (2008) 1088–1098.

[11] R. Majumder, Some aspects of stability in microgrids, IEEE Trans. Power Syst. 28 (3) (2013) 3243–3252.

[12] X. Guan, B. Yang, C. Chen, W. Dai, A comprehensive overview of cyber-physical systems: From perspective of feedback system, IEEE/CAA J. Autom. Sinica 3 (1) (2016) 1–14.

[13] X. Jin, Fault tolerant finite-time leader-follower formation control for autonomous surface vessels with LOS range and angle constraints, Automatica 68 (2016) 228–236.

[14] Y. Shoukry, et al., SMT-based observer design for cyber-physical systems under sensor attacks, in: Proc. ACM/IEEE 7th Int. Conf. Cyber-Phys. Syst., Vienna, Austria, 2016, pp. 1–10.

[15] S. Sridhar, A. Hahn, M. Govindarasu, Cyber-physical system security for the electric power grid, Proc. IEEE 100 (1) (2012) 210–224.

[16] B. Zheng, P. Deng, R. Anguluri, Q. Zhu, F. Pasqualetti, Cross-layer codesign for secure cyber-physical systems, IEEE Trans. Comput.-Aided Des. Integr. Circuits Syst. 35 (5) (2016) 699–711.

[17] A. Farraj, E. Hammad, D. Kundur, A cyber-physical control framework for transient stability in smart grids, IEEE Trans. Smart Grid 9 (2) (2018) 1205–1215.

[18] H.S. Li, L.F. Lai, H.V. Poor, Multicast routing for decentralized control of cyber physical systems with an application in smart grid, IEEE J. Sel. Areas Commun. 30 (6) (2012) 1097–1107.

[19] L. Schenato, B. Sinopoli, M. Franceschetti, K. Poolla, S. Sastry, Foundations of control and estimation over lossy networks, Proc. IEEE 95 (1) (2007) 163–185.

[20] J. Hespanha, P. Naghshtabrizi, Y. Xu, A survey of recent results in networked control systems, Proc. IEEE 95 (1) (2007) 138–162.

[21] Y. Liu, M. Reiter, P. Ning, False data injection attacks against state estimation in electric power grids, in: Proc. ACM Conf. Computer and Commun. Security, Chicago, IL, USA, 2009.

[22] G. Dan, H. Sandberg, Stealth attacks and protection schemes for state estimators in power systems, in: Proc. IEEE Int. Conf. Smart Grid Commun., Gaithersburg, MD, USA, 2010, pp. 214–219.

[23] H. Fawzi, P. Tabuada, S. Diggavi, Secure state-estimation for dynamical systems under active adversaries, in: Proc. Annu. Allerton Conf. Commun., Control, Comput, 2011.

[24] F. Pasqualetti, F. Dorfler, F. Bullo, Attack detection and identification in cyber-physical systems, IEEE Trans. Autom. Control 58 (11) (2013) 2715–2729.

[25] Y. Fujita, T. Namerikawa, K. Uchida, Cyber-attack detection and faults diagnosis in power networks by using state fault diagnosis matrix, in: Proc. Eur. Control Conf., Zurich, Switzerland, 2013.

[26] W. Xu, K. Ma, W. Trappe, Y. Zhang, Jamming sensor networks: Attack and defense strategies, IEEE Netw. 20 (3) (2006) 41–47.

[27] D. Thuente, M. Acharya, Intelligent jamming in wireless networks with applications to 802.11b and other networks, in: Proc. 25th IEEE Commun. Soc. Military Commun. Conf., Washington, DC, USA, 2006.

[28] P. Srikantha, D. Kundur, Denial of service attacks and mitigation for stability in cyber-enabled power grid, in: Proc. IEEE Power Energy Soc. Innovative Smart Grid Technol. Conf, February 2015, pp. 1–5.

[29] S. Liu, P.X. Liu, X. Wang, Effects of cyber-attacks on islanded microgrid frequency control, in: Proc. IEEE 20th Int. Conf. Comput. Supported Cooperative Work Des, May 2016, pp. 461–464.

[30] P. Danzi, M. Angjelichinoski, C. Stefanovic, P. Popovski, Antijamming strategy for distributed microgrid control based on power talk communication, in: Proc. IEEE Int. Conf. Commun. Workshops, May 2017, pp. 911–917.

[31] M. Chlela, D. Mascarella, G. Joos, M. Kassouf, Fallback control for isochronous energy storage systems in autonomous microgrids under denial-of-service cyber-attacks, IEEE Trans. Smart Grid 9 (5) (2018) 4702–4711.

[32] A. Teixeira, K. Paridari, H. Sandberg, K.H. Johansson, Voltage control for interconnected microgrids under adversarial actions, in: Proc. IEEE 20th Conf. Emerg. Technol. Factory Automat, September 2015, pp. 1–8.

[33] R. Fu, X. Huang, J. Sun, Z. Zhou, D. Chen, Y. Wu, Stability analysis of the cyber physical microgrid system under the intermittent dos attacks, Energies 10 (5) (2017) 1–15.

[34] W. Xu, W. Trappe, Y. Zhang, T. Wood, The feasibility of launching and detecting jamming attacks in wireless networks, in: Proc. ACM Int. Symp. Mobile Ad-Hoc Netw. Comput, 2005.

[35] B. DeBruhl, P. Tague, Digital filter design for jamming mitigation in 802.15.4 communication, in: Proc. Int. Conf. Comput. Commun. Netw., Maui, HI, USA, 2011.

[36] P. Tague, M. Li, R. Poovendran, Mitigation of control channel jamming under node capture attacks, IEEE Trans. Mobile Comput. 8 (9) (2009) 1221–1234.

[37] C. De Persis, P. Tesi, Input-to-state stabilizing control under denial-of-service, IEEE Trans. Autom. Control 60 (11) (2015) 2930–2944.

[38] Y. Xu, Q. Guo, H. Sun, Z. Fei, Distributed discrete robust secondary cooperative control for islanded microgrids, IEEE Trans. Smart Grid (2018) 1.

[39] A. Bidram, F.L. Lewis, A. Davoudi, Distributed control systems for small-scale power networks: using multiagent cooperative control theory, IEEE Control Syst. Mag. 34 (6) (2014) 56–77.

[40] E. Sontag, Input to state stability: basic concepts and results, Nonlin. Optim. Control Theory Lect. Notes Math. 1932 (2008) 163–220.

[41] Y. Zhu, X. Guan, X. Luo, S. Li, Finite-time consensus of multiagent system via nonlinear event-triggered control strategy, IET Control Theory Appl. 9 (17) (2015) 2548–2552.

[42] S. Yu, X. Yu, B. Shirinzadeh, Z. Man, Continuous finite time control for robotic manipulators with terminal sliding mode, Automatica 41 (11) (2005) 1957–1964.

[43] V. Kumar, S.R. Mohanty, S. Kumar, Event trigger super twisting sliding mode control for DC micro grid with matched/unmatched disturbance observer, IEEE Trans. Smart Grid 11 (5) (2020) 3837–3849.

Chapter 2

Distributed generating system integration: Operation and control

Priyatosh Mahish, Manas Ranjan Mishra, and Sukumar Mishra
Indian Institute of Technology Delhi, New Delhi, India

1. Introduction

In the recent years, the renewable energy resource (RES); mostly solar photovoltaic (PV) and wind-based distributed generations (DGs) is the prime interest throughout the world to achieve sustainable growth of power system. In this context, microgrid operation is quite beneficial to overcome geographical constraints of RES. Further, the microgrid operation reduces the need of transmission system which helps to avoid congestion of the network [1,2]. The microgrid operation is much beneficial for electrification of rural areas due to nonrequirement of transmission infrastructure. It is important to define characteristics of microgrid to know the nature of its operation. Thereby, the U.S. Department of Energy has made a definition of microgrid, from which it is understood that [3],

- microgrid obtains specific electrical boundary,
- DGs and loads inside a microgrid must take control action in such a way that they behave like a unit,
- total capacity of the microgrid must ensure to satisfy maximum critical load within the defined electrical boundary,
- the flexibility should be achieved with the distribution network, so that the microgrid can switch its operation from grid-connected to islanded mode and vice-versa.

1.1 Research on distributed generating system integration: Focus area

Intermittent and variable natures of DGs raise critical challenges to operate power systems. Different methods are proposed in literatures to operate the system within voltage limits and frequency limits. The energy storage systems, that

Microgrid Cyberphysical Systems. https://doi.org/10.1016/B978-0-323-99910-6.00002-5

is, battery, and vehicle-to-grid (V2G) based technology are also used to maintain power balance against the DGs. In this book chapter, different control techniques of voltage source converters (VSCs) are discussed in detail, addressing the challenges, that is, voltage stability, frequency stability, and reactive power compensation.

The distributed generating system is considered mostly as voltage source inverter (VSI)-dominated microgrid, operating in grid forming (GFM) or grid following (GFL) mode [4]. Due to the intermittent nature of DGs, power balance of the system is maintained properly, which degrades frequency regulation and voltage regulation. Therefore, autonomous operation of microgrid through GFM mode of VSIs is more challenging as compared with grid-connected operation of the microgrid through GFL mode of VSIs. Lack of reactive power compensation in the microgrid directly affects voltage stability of the distributed generating system. To solve this issue, the reactive power capability of the VSIs can be used. Also, the V2G technology has potential to provide reactive power in need.

The transitions from GFL to GFM or vice versa are other challenges in distributed generating system integration. Different control strategies are proposed in the literatures from smooth transition between these modes. The methods for transition from GFL to GFM mode are categorized into mainly two types:

(i) Transition of power/current control of GFL mode to voltage control of GFM mode [5–7].
(ii) Unified control on both the GFL and GFM modes [8–11].

The power/current control mode of GFL mode is usually operated with maximum power point of DGs. On the other hand, the voltage control of GFM mode supplies power to maintain active power balance. The methods proposed in Yao et al. [5] and Wai et al. [12] reduce current to zero, through the line connecting VSI to the grid. However, avoiding the line current to zero is possible through high speed transition from GFL to GFM [6,13]. For unified control, the same controller can operate during GFL mode, GFM mode, and transient period. Therefore, it is challenging to develop such robust controller. In this control operation, the big size DGs are operated in voltage control and small size DGs are operated in current control. However, both the voltage control and current control need to be modified to apply them in both the grid-connected and autonomous operation of distributed generating system. As an example, conventional droop control is modified with combination of virtual impedance-based droop control and PI-based droop control [9,14,15]. The control techniques for transition from GFM to GFL mode are categorized into mainly three types:

(i) Transition of voltage control of GFM mode to power/current control of GFL mode [16–18].
(ii) Unified control on both the GFL and GFM modes [19].
(iii) Active and passive synchronization control [16,19].

The synchronization process is popular to connect VSIs with the utility grid. During the synchronization, the voltage mismatch between the VSIs and grid creates transients. The passive synchronization is useful if the voltage mismatch is low. In this case, the phase angle needs to be the same on microgrid and utility grid. However, in most cases, voltage mismatch is high and phase angle may not be equal during reconnection of microgrid. In such situation, active synchronization is effective. In this process, energy storage devices, that is, batteries, are required to coordinate with the distributed energy resources (DERs). In few cases, an independent source, that is, diesel generator, is utilized to deliver a synchronization signal to the VSIs for connecting microgrid to the utility grid. In GFM mode of operation, the active synchronization process is classified into two different ways. First, few DERs are introducing synchronization signal by an independent voltage source, and remaining DERs in the microgrid are following the signal through their Phase locked loop (PLL). Second, all the DERs are taking responsibility in active synchronization. The control techniques for GFL-GFM mode transition explained here is summarized in the Fig. 1 [4].

1.2 IEEE standards for distributed generating system integration

High penetration of renewable resources in low and medium voltage microgrid is increasing day to day. This results a gradual change from passive to active distribution network. Further, replacement of conventional synchronous generators with the DERs makes change in operating nature of utility grid. It is obvious that the small capacity DERs cannot make significant impact on the grid. However, cumulative effect of large number of such DERs in the form of

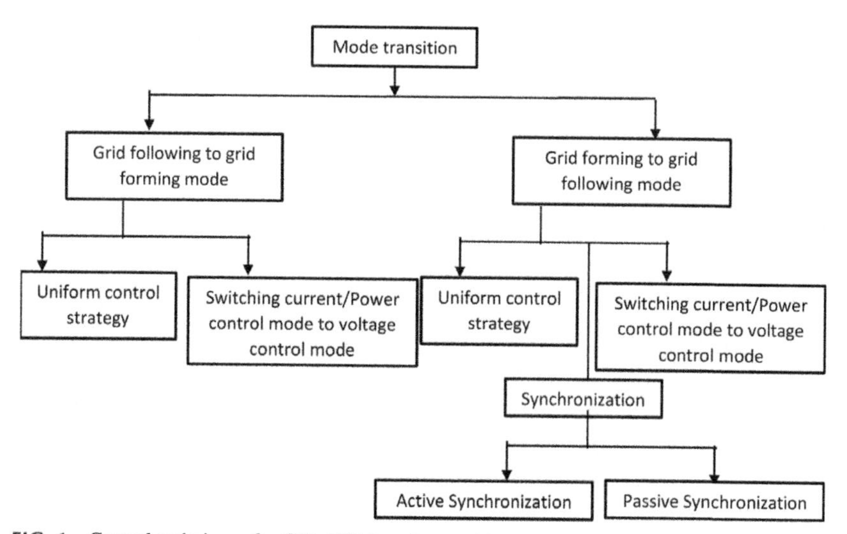

FIG. 1 Control techniques for GFL-GFM mode transition.

grid-connected microgrids can make huge impact on the system, which makes concern on power system stability and reliability. To cope up with this situation, IEEE has provided new guidelines, that is, IEEE-Std 1547-2003 [20]. This standard is further revised to IEEE-Std 1547-2018 [20–22]. Based on voltage frequency regulation and voltage regulation, IEEE-Std 1547-2018 classifies DERs into two categories. Based on low-voltage-ride-through (LVRT), DERs are classified into three categories. More details about the standards are elaborated later in this chapter.

2. Grid supporting mode of VSI in distributed generating system

2.1 Standards for distributed generating system integrations

With continuous growth of DERs, the importance of proper guidelines and standards is realized in many parts of the world. Tables 1–6 provide a brief summarization of popular standards for distributed generating system integration to the distribution grid.

Table 1 illustrates important standards for DG integration to the distribution grid. For large-scale DG integration, IEEE has provided the guidelines through IEEE std. 1547. Whereas, for medium- and small-scale DG integration, countries like USA, Germany, and Australia provide guidelines through RULE 21, VDE-AR-4105, and AS 4777, respectively. The IEEE std. 929 is applicable for solar PV only. Table 2 illustrates guidelines for allowable power factor, voltage regulation, and synchronization parameters. The voltage regulation is recommended to ±0.05 pu. The guidelines for power factor vary with the countries. RULE 21 and IEEE std. 1547 provide same guideline for synchronization limits. Whereas, VDE-AR-4105, AS 4777, and IEEE std. 929 have no recommendations for synchronization limits. Table 3 illustrates clearing time against variations in voltage at the point of connection. IEEE std. 929 recommends

TABLE 1 Important standards for distributed generating system integrations to the distribution grid.

Standards	Capacity	DG type	Publication
RULE 21 [23]	LV network	Any	California, USA
VDE-AR-4105 [24]	–	Any	Germany
AS 4777 [25–27]	<10kW	Any	Australia
IEEE std. 929 [28]	<10kW	PV	IEEE
IEEE std. 1547 [20]	<10MVA	Any	IEEE

TABLE 2 Allowable power factor, voltage regulation, and synchronization parameters.

Standards		RULE 21	VDE-AR-4105	AS 4777	IEEE std. 929	IEEE std. 1547
Voltage regulation		±0.05 pu	0.05 pu	–	–	±0.05 pu
Power factor		>0.9	>0.9 (for $P_{DER} > 0.2\ P_{Rated}$)	>0.95 (for $I_{DER} < 20A$)	>0.85 (for $P_{DER} > 0.1\ P_{Rated}$)	>0.9
Synchronization limits	Δf	<0.3 Hz	–	–	–	<0.3 Hz
	ΔV	<0.1 pu				<0.1 pu
	$\Delta \theta$	$<20^0$				$<20^0$

TABLE 3 Clearing time against variations in voltage at the point of connection.

Standards	Voltage	Clearing time
RULE 21	$V<0.45$ pu	0.16 s
IEEE std. 1547	$0.45\,\text{pu}<V<0.6$ pu	1 s
	$0.6\,\text{pu}<V<0.88$ pu	2 s
	$1.1\,\text{pu}<V<1.2$ pu	1 s
	$1.2\,\text{pu}<V$	0.16 s
VDE-AR-4105	$V<0.8$ pu	0.1 s
	$V>1.1$ pu	0.1 s
AS 4777	$V<0.2$ kV	2 s
	$V>0.27$ kV	2 s
IEEE Std. 929	$V<0.5$ pu	6 cycles
	$0.5\,\text{pu}\leq V<0.88$ pu	120 cycles
	$1.1\,\text{pu}<V\leq 1.2$ pu	120 cycles
	$1.2\,\text{pu}\leq V$	6 cycles

TABLE 4 Clearing time against variations in grid frequency.

Standards	Frequency (in pu)	Clearing time
RULE 21	$f<0.95$	0.16 s
IEEE Std. 1547	$0.95<f<0.99$	2 s
	$1.01<f<1.03$	2 s
	$1.03<f$	0.16 s
VDE-AR-4105	$f>1.03$	0.1 s
	$f<0.95$	0.1 s
AS 4777	$f<0.9$	2 s
	$f>1.1$	2 s
IEEE Std. 929	$f<0.99$	6 cycles
	$1.01<f$	6 cycles

TABLE 5 Standards for harmonics and DC injection at the point of connection.

Harmonics (h)		RULE 21 IEEE Std. 1547 IEEE Std. 929	AS 4777	VDE-AR-4105	
				Harmonics (h)	Current limit (A/MVA)
Odd harmonics	3	<4%	<4%	3	3
	5	<4%	<4%	5	1.5
	7	<4%	<4%	7	1
	9	<4%	<4%	9	0.7
	11	<2%	<2%	11	0.5
	13	<2%	<2%	13	0.4
	15	<2%	<2%	15	0.35
	17	<1.5%	<1.5%	17	0.3
	19	<1.5%	<1.5%	19	0.25
	21	<1.5%	<1.5%	21	0.2
	$23 \leq h \leq 33$	<0.6%	<0.6%	23	0.2
	$h > 33$	<0.3%	<0.3%	25	0.15
Even harmonics	$2 < h < 8$	<25% of limits of odd harmonics	<1%	$h < 40$	1.5
	$10 < h < 32$		<0.5%	$42 < h < 178$	4.5
DC injection	4	$<0.005 I_{Rated}$	$<0.01 I_{Rated}$	0.2 s clearing time for $I_{DC} > 1$ A	

TABLE 6 Clearing time with respect to average leakage current at grid-integrated PV.

Standards	Average leakage current (A)	Clearing time (ms)
AS 4777	0.03	300
	0.06	150
	0.1	40
	0.3	300

clearing time in cycles, whereas the other standards provide the time in seconds. It is realized from Table 4 that there is significant diversity in the guidelines through different standards, for the variations in grid frequency. Table 5 illustrates the harmonics and DC injection at the point of connection. Table 6 demonstrates clearing time with respect to average leakage current at grid-integrated PV.

2.2 Frequency Support

2.2.1 Fast grid frequency support from distributed energy resources: NREL REPORT, March, 2021

As more nonsynchronous generators, such as solar PV systems and battery energy storage systems are connected to electric power networks, certain synchronous generator-based power plants are partially or fully shut down. Because of synchronous generators' rotational inertia and primary frequency response (PFR) maintaining grid frequency in short-term frequency regulation, replacing synchronous generation with the DERs degrade frequency stability. To confront such grid stability issues with high DER penetration, certain control formulation needs to be done in the VSI control which may be attributed as frequency support [29]. DERs should not only avoid aggravating disruptions but also actively assist in maintaining frequency stability [30]. This is accomplished by DER inverters swiftly adjusting their power output in response to frequency fluctuations on the grid [31,32].

Most PV and storage inverters on the market in the United States have the ability to reduce power output in response to over frequency events using a frequency-watt droop curve, and IEEE 1547-2018, which was recently published, mandates that all DERs provide frequency droop control for both under frequency and over frequency events [21]. Frequency-watt control is a self-operating feature that does not require communication to work. The inverter measures the AC grid frequency at its terminals and responds by modulating its output power to follow a droop curve such as the one shown in Fig. 2. In

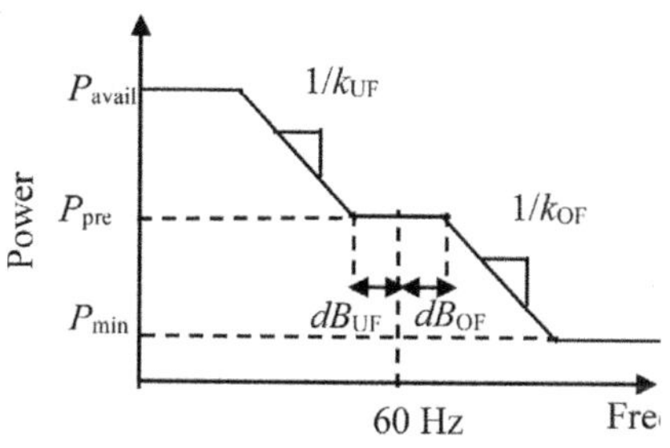

FIG. 2 Frequency-watt droop curve.

the inertial and PFR time scales, frequency response of power grid is depicted in Fig. 3 [33].

The following are examples of DER frequency assistance that have been proposed:

- Fast frequency response (FFR), in which a DER quickly adjusts its output power in response to a trigger such as frequency or rate-of-change-of-frequency (ROCOF) exceeding a preset level or a transmitted signal, using a predetermined step change command.
- Automatic generation control (also referred to as secondary frequency regulation) and other longer time-scale controls. While inertial and PFR time scales (subsecond to a few tens of seconds) are used by droop control, virtual inertia, and FFR, DERs may be designed to function on larger time scales [34,35].

  FFR: While inverters lack inertia, their ability to fast response allows them to maintain grid frequency during a small disturbance [36]. In low-inertia power systems, this capacity to respond rapidly might be advantageous since frequency events occur extreme quickly (i.e., with a high ROCOF); therefore, resources participating in PFR must respond fast. A demonstration for such an event is shown in Fig. 4. In case of an over

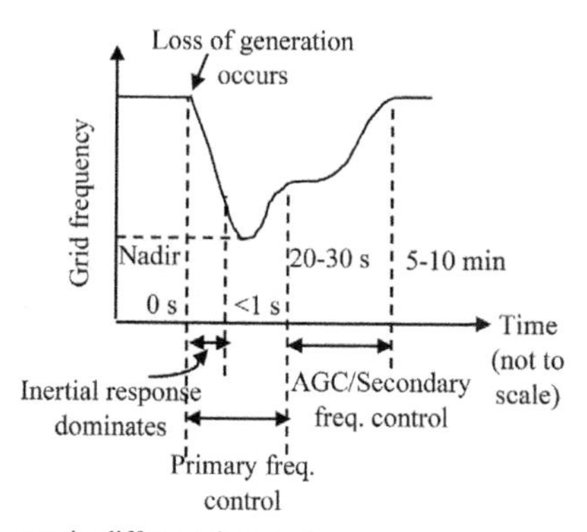

FIG. 3 Frequency response in different time scales.

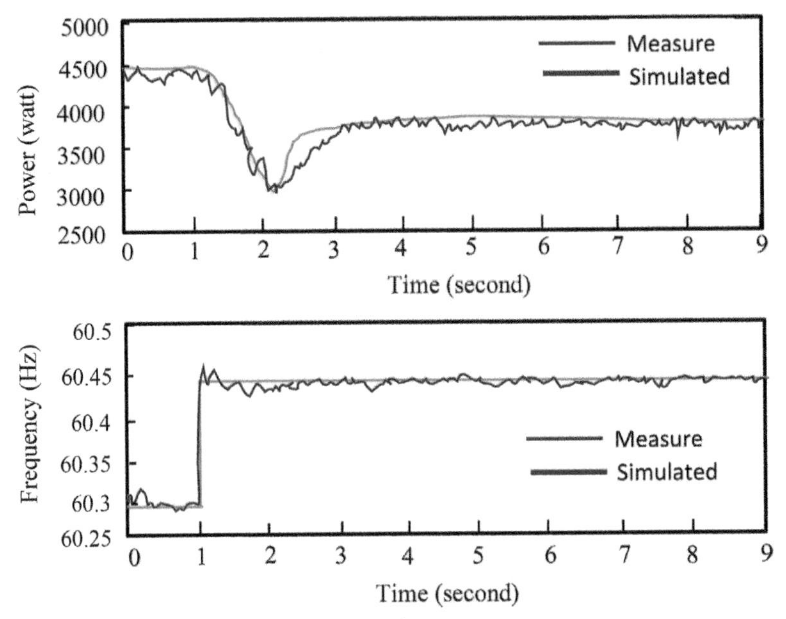

FIG. 4 Example of inverter to frequency-watt time response.

frequency event at 1 s, the active power injection by the DER drops as a first-order response with some delay. As stated in (1), the power response of this inverter was modeled using a first-order time constant and a set delay τ.

$$H(s) = \frac{e^{-\tau s}}{\tau s + 1} \tag{1}$$

2.3 Voltage support

2.3.1 *An overview of issues related to IEEE Std 1547-2018 requirements regarding voltage and reactive power control*

One of the most essential tasks of distribution utilities is to keep voltage within the specified bounds as defined in the grid code. The "specified" voltage level is established by American National Standards Institute, ANSI C84.1, and is commonly referred to in utility electric service standards and interconnection guidelines. Utilities use a range of equipment, including capacitors and voltage regulators, to keep the voltage at certain levels.

2.3.2 *Voltage ranges defined by ANSI C84.1*

The American National Standards Institute proposes C84.1 standard which provides voltage ratings and standards for electrical systems in the United States.

The standard specifies that the voltage ranges must be built into all distribution-connected equipment, including all end-user loads and DERs.

The distribution operator shall keep the service voltage between the minimum and maximum values specified for Range A during typical working conditions. Voltage may violate Range A on rare occasions. However, boundaries of Range B cannot be violated at any condition. Table 7, lists the base voltage ranges, limits, and corresponding per-unit (p.u.) values for residential single-phase systems specified in ANSI C84.1.

Voltage typically decreases in service cables toward the load end from the distribution stations owing to the losses along the length of the cables depending on the power drawn by the connected load. Fig. 5 shows the ANSI voltage limits for service and utilization voltage ranges A and B for a nominal 120 V system. The per-unit voltage at the electric outlet will be less than the voltage at the distribution primary due to voltage drop at distribution feeder, distribution transformer, secondary service wiring, and customer wiring.

Traditional voltage regulation methods

Voltage regulation of distribution system violates the limits for a many reasons, such as daily or seasonal variations in loading circumstances, low load to high load and load density along the feeder, load balancing activities, unbalanced loads and feeders, etc. To address voltage control challenges, utilities have typically used a range of typical distribution equipment. The partial list of such equipment is as follows [37]:

(a) Generator voltage regulators
(b) Voltage-regulating equipment at the distribution substation
(c) Capacitors at the distribution substation
(d) Rebalancing load on the feeder primary
(e) Increasing feeder conductor capacity
(f) Changing feeder sections from single phase to multiphase

TABLE 7 ANSI voltage ranges and limits for residential single-phase systems.

	Nominal system voltage	Service voltage		Utilization voltage	
	V *(p.u.)*	*Minimum* V *(p.u.)*	*Maximum* V *(p.u.)*	*Minimum* V *(p.u.)*	*Maximum* V *(p.u.)*
Voltage range A	120 (1.00)	114 (0.95)	126 (1.05)	108 (0.90)	125 (≈1.04)
Voltage range B	120 (1.00)	110 (≈0.92)	127 (≈1.06)	104 (≈0.87)	127 (≈1.06)

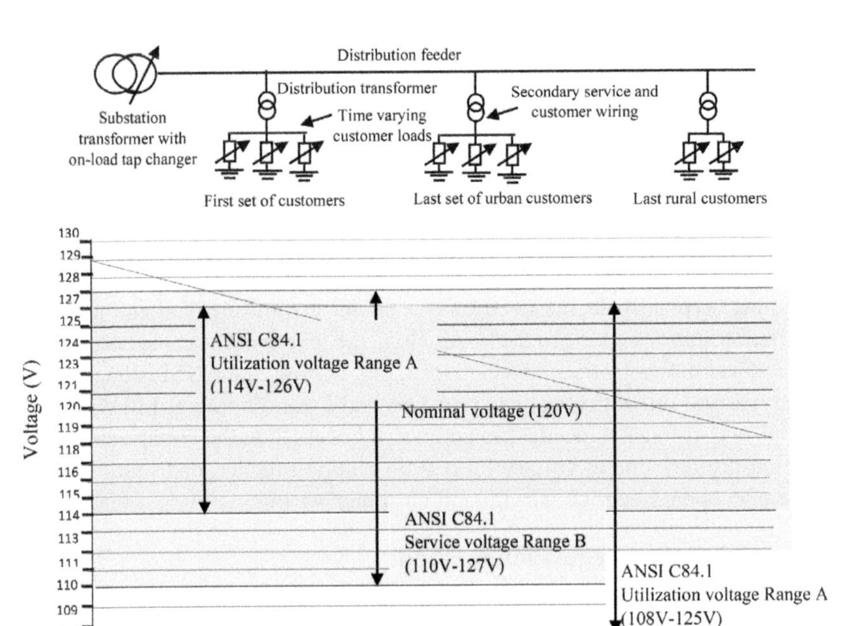

FIG. 5 Example voltage profile along distribution feeder on a 120-V basis.

(g) Transferring load to different feeders
(h) Installing new substation and primary feeder
(i) Increasing primary voltage level
(j) Using voltage regulators on the feeder primary
(k) Using shunt capacitors on the feeder primary

Figs. 6 and 7 depict the effects of voltage control by a shunt capacitor and a voltage regulator.

Voltage regulation using inverters

Fig. 8 shows a block schematic of a Volt-Var control system utilizing a three-phase grid-connected PV system. The reactive power, or Var, of a PV generating system is controlled by the grid-connected PV inverter. Using the Volt-Var control curve, the smart PV-inverter may deliver or absorb Var depending on the inverter terminal voltage (V_g). The Volt-Var control curve's input is the inverter terminal voltage V_g, and the four distinct voltage set points are V_1, V_2, V_3, and V_4. The inverter's reactive power command Q^*_{inv} is calculated using the input voltage and voltage set points and ranges from 0% to 100% of the inverter's rated power, as stated by the following equation [38]:

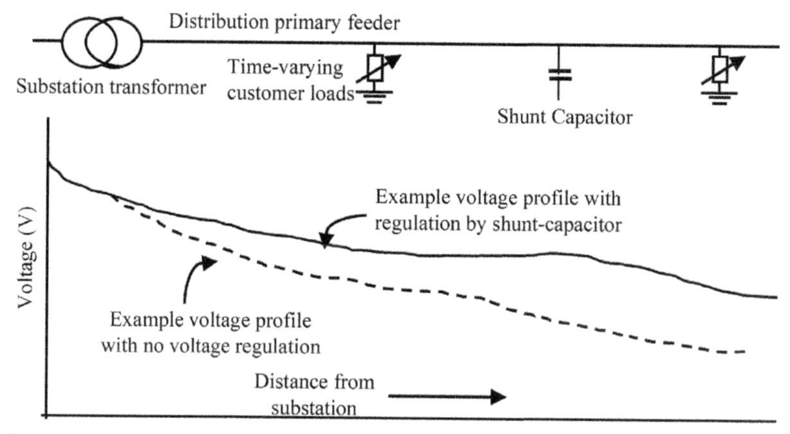

FIG. 6 Example of voltage profile at peak load with regulation by shunt capacitor.

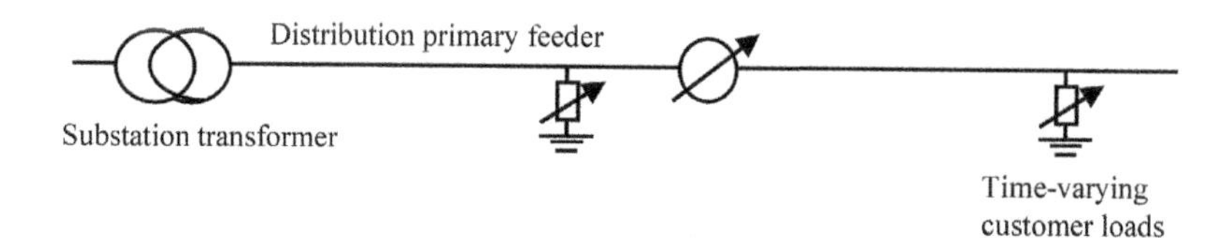

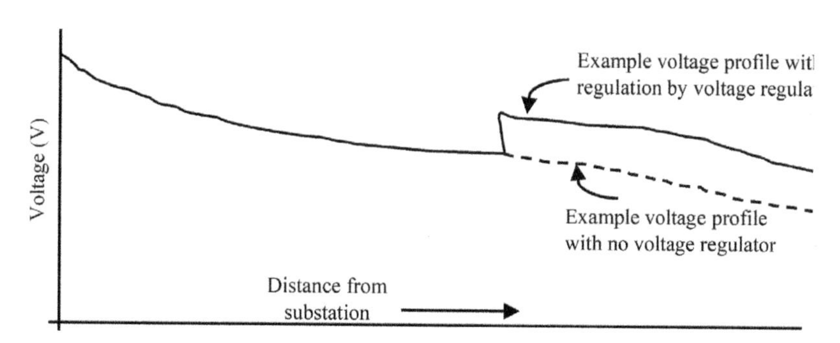

FIG. 7 Example voltage profile at peak load with regulation by voltage regulator.

$$Q_{inv}^{*} = \left\{ \begin{array}{ll} 0-100\%(\text{capacitive region}-1) & \text{if } V_g \leq V_1; \\ 0-100\%(\text{capacitive region}-2) & \text{if } V_1 \leq V_g \leq V_2; \\ 0\ (\text{region}-3\ (\text{zero var})) & \text{if } V_2 \leq V_g \leq V_3; \\ 0-100\%(\text{inductive region}-4) & \text{if } V_3 \leq V_g \leq V_4; \\ 0-100\%(\text{inductive region}-5) & \text{if } V_g \leq V_4; \end{array} \right\} \quad (2)$$

Regions 1 and 2 are the reactive power or Var generation regions which are known as capacitive regions of Var. Region-1 is the constant Var generation

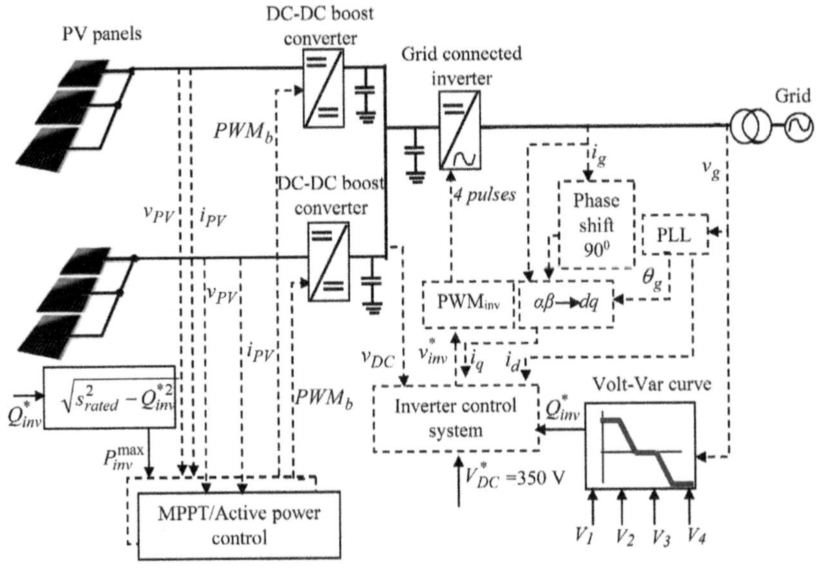

FIG. 8 Volt-Var control system.

region and Region-2 is the variable Var generation region. Region-3 contains the zero Var generation region. Regions 4 and 5 demonstrate the inductive effect. The inverter reactive power control reference Q^*_{inv} is the control input along with V^*_{DC} as shown in Fig. 9. With the inverter's reactive reference set

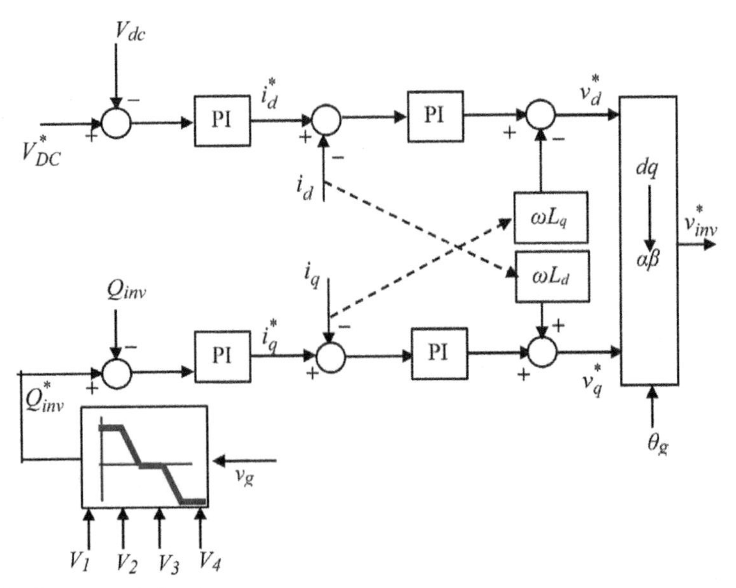

FIG. 9 Inverter control system.

to Q^*_{inv}, the maximum active power output P^{max}_{inv} can be delivered by the PV inverter, which is obtained as,

$$S^2_{rated} = P^2_{inv} + Q^2_{inv} \quad (3)$$

$$P^{max}_{inv} = \sqrt{S^2_{rated} + Q^{*2}_{inv}} \quad (4)$$

where, S^*_{rated}, Q^*_{inv}, and P^*_{inv} is the apparent power, reactive, and active power reference. The actual power output of the PV panels can be formulated as follows:

$$P_{out}(t) = \left\{ \begin{array}{c} P_{pv}(t), \ \text{MPPT (normal operation)} \\ P^{max}_{inv}(t), \ \text{Volt} - \text{Var control operation and } P^{max}_{inv} < P_{pv} \end{array} \right\} \quad (5)$$

where, $P_{out}(t)$ is the instantaneous power output of PV panels, $P_{pv}(t)$ is the instantaneous maximum PV output power at maximum power point tracking (MPPT) mode of operation. $P^{max}_{inv}(t)$ is the instantaneous maximum power when Volt-Var control is applied and $P^{max}_{inv} < P_{pv}$.

In Fig. 10, the MPPT controller generates reference voltage signal Q^*_{inv}, while a PI controller generates a pulse width modulation (PWM_b) gate signal for the boost converter. Also, in the Volt-Var control mode of operation, the PI controller controls the P^{max}_{inv} and generates a PWM_b gate signal for the boost converter.

2.4 Inertial support

The decrease of carbon footprint and increase in clean energy production are the key driving forces for deploying and developing a substantial quantity of renewable energy integrated into power systems. Despite being sought globally, using renewable energies to partial replacement of fossil fuels may require retrofitting the entire power grid and pose a threat to the stability of contemporary power systems. One of the key difficulties, which has already been recognized in

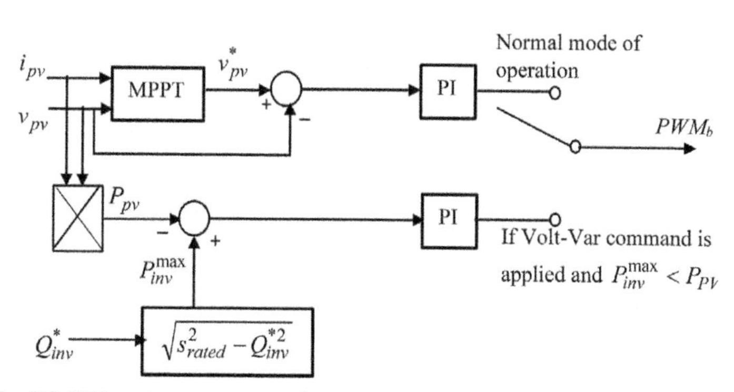

FIG. 10 DC-DC boost converter control system.

small-scale power systems, such as those in Ireland and the United Kingdom, is frequency instability caused by a lack of inertia in the systems with a high penetration level of power-converter-interfaced renewable energy [39,40].

The major grid interfaces in conventional power systems are synchronous generators operating at a speed synchronized with the grid frequency. Synchronous generators automatically slow down or speed up in accordance with the grid frequency when a frequency change occurs owing to a power imbalance between generation and demand [41]. However, in current power systems, this occurrence changes since most renewable energy sources, such as wind and solar PV, are tied to the power grid via power electronic converters [42]. The inertia of power system reduces drastically as more synchronous generators are phased out and replaced by power converters, posing a serious challenge for stability and control. Without enough inertia, the grid frequency and/or RoCoF may be prone to be exceeding the permitted range during extreme events, resulting in generation tripping, unwanted load shedding, or even system collapse [30].

2.4.1 Three-phase grid-connected power converters with virtual inertia control

Fig. 11 shows a three-phase grid connected power converter with virtual inertia control, where the power grid is depicted as a series connection of an ideal voltage source v_{abc} through grid inductor L_g. A three-phase power converter with two levels and an output inductor filter L_c is employed as a coupling inductor. The system's control structure is divided into two sections, as illustrated in Fig. 12, a virtual inertia controller and a traditional cascaded voltage/current controller implemented in the synchronous dq-frame [43,44]. The virtual inertia controller is expected to modify the dc-link voltage reference by supplying a voltage difference Δv_{dc_ref}. To emulate inertia, the virtual inertia controller's voltage adjustment should be proportional to the change in grid angular frequency $\Delta\omega_r$, while the voltage/current controller simply regulates the dc-link voltage v_{dc} to follow the voltage reference adjusted by the virtual inertia

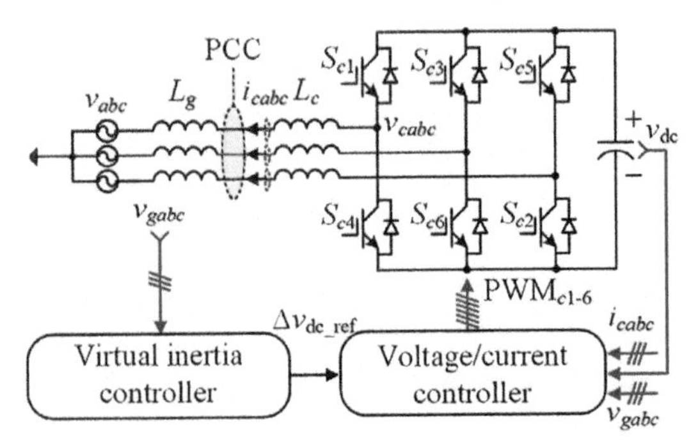

FIG. 11 Schematic diagram of a three-phase grid-connected power converter with the virtual inertia control (PWM stands for pulse width modulation).

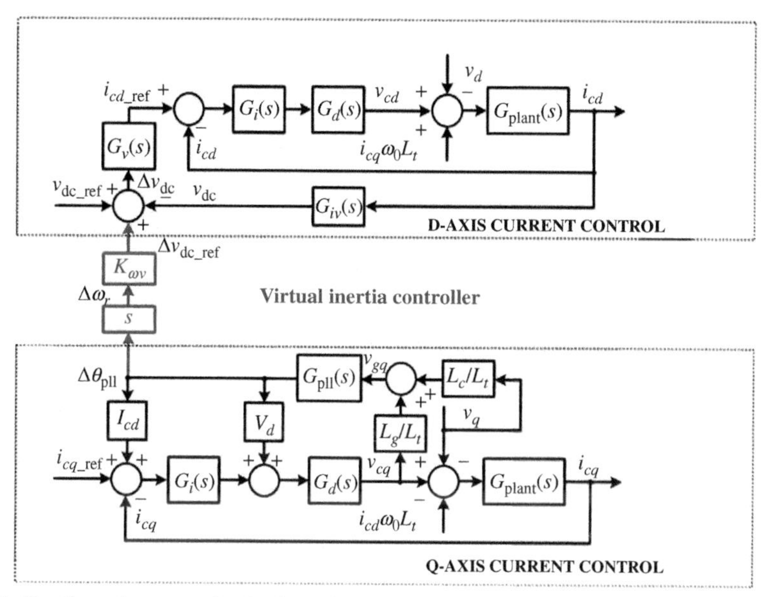

FIG. 12 Control structure for the three-phase power converter with the virtual inertia control.

controller. The mapping between the synchronous generator and the capacitor, where the inertia constant of the synchronous generator H is defined as the ratio of the kinetic energy stored in the rotating masses of the synchronous generator ($J\omega_{Om}^2/2$) to its rated power S_{base}, which can be expressed as

$$H = \frac{J\omega_{om}^2}{2S_{base}} \tag{6}$$

where J denotes the combined moment of inertia of the synchronous generator and turbine. ω_{Om} stands for rated rotor mechanical speed. Similarly, the inertia constant of the capacitor H_{cap} can be defined as the ratio of the electrical energy stored in the capacitor ($C_{dc} \cdot V_{dc}^2/2$) to its rated power S_{base}, which can be expressed as follows:

$$H_{cap} = \frac{C_{dc} \cdot V_{dc\_ref}{}^2}{2S_{base}} \tag{7}$$

where, C_{dc} and V_{dc_ref} represent the capacitance and rated capacitor voltage of dc-link, respectively. The rotor speed ω_r and the capacitor voltage V_{dc} have the same function to determine the inertia constant H_{cap} and H. After the per unit capacitor voltage and the per unit rotor speed or grid frequency are coupled by the proportional gain $K_{\omega v\_pu}$, the virtual inertia with an analogous inertia coefficient $H_{cap}K_{\omega v\_pu}$ may be predicted from the capacitor [43].

$$\Delta v_{dc\_pu} = K_{\omega v\_pu} \cdot \Delta\omega_{r\_pu} \tag{8}$$

2.4.2 Virtual inertia control mechanism

In Fig. 12, $G_i(s)$ and $G_v(s)$ represent proportional integral (PI) controller for the current control and voltage control, respectively, which are expressed as follows:

$$G_i(s) = K_{cp} + K_{ci}/s$$
$$G_v(s) = K_{vp} + K_{vi}/s \tag{9}$$

A temporal delay exhibits in the control structure, which is expressed as follows:

$$G_d(s) = \frac{1}{T_d(s) + 1} \tag{10}$$

in which $T_d = 1.5/f_s$ with f_s being the sampling frequency. The cascaded voltage/current controller's control architecture shown in Fig. 20. The conventional philosophy is being implemented in the grid-tied inverter-based system to regulate power from the DC bus to the grid. V_{gabc} is the point of common coupling (PCC) point voltage sensed by the voltage sensor and passed on to the PLL (to detect the phase angle of V_{gabc}) represented by $G_{pll}(s)$. The d-axis and q-axis components of converter voltages are denoted by v_{cd} and v_{cq}, respectively. The d-axis and q-axis components of grid voltages are denoted by v_d and v_q, respectively. The d-axis and q-axis components of converter currents are denoted by i_{cd} and i_{cq}, respectively. L_t stands for total inductance ($L_t = L_c + L_g$), and the plant transfer function is modeled as

$$G_{plant}(s) = \frac{1}{L_t(s)} \tag{11}$$

while ω_0 represents the fundamental angular frequency. $-i_{cd}\omega_0 L_t$ and $i_{cq}\omega_0 L_t$ are the cross coupling elements between d-axis and q-axis. The extra terms $I_{cd} \cdot \Delta\theta_{pll}$ and $V_d \cdot \Delta\theta_{pll}$ are derived from the PLL output. I_{cd} denotes the rated d-axis current derived from the rated power. The input-output relationship in the PLL can be represented by [45],

$$G_{pll}(s) = \frac{\Delta\theta_{pll}(s)}{v_{gq}(s)} \tag{12}$$

Through control of the d-axis current i_{cd}, the dc-link voltage v_{dc} is regulated to its reference v_{dc_ref}. Variations in the d-axis current i_{cd} makes impact on v_{dc} through a transfer function $G_{iv}(s)$, which is induced by the three-phase power converter's real-time power balancing between the ac-side and dc-side. $G_{iv}(s)$ can be derived as

$$G_{iv}(s) = \frac{-3V_d}{2V_{dc\_ref} C_{dc} s} \tag{13}$$

where, V_d and V_{dc_ref} represents the rated voltage and reference value at dc-link.

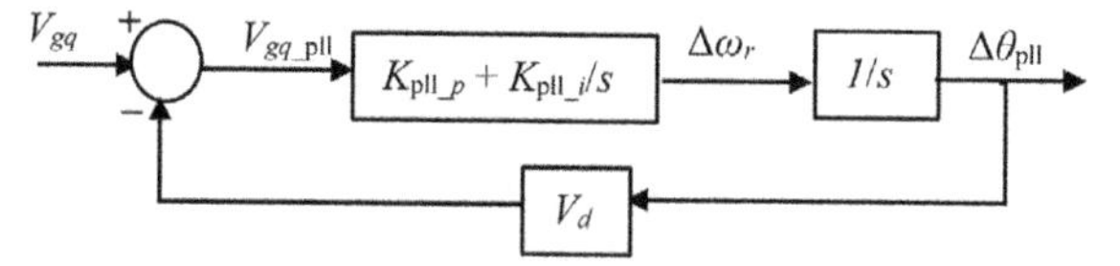

FIG. 13 Linear small-signal model of the PLL.

Virtual inertia can be produced by the three-phase grid-connected converter by directly linking the change of the capacitor voltage reference Δv_{dc_ref} and the change of grid angular frequency $\Delta\omega_r$ through a proportional gain $K_{\omega v}$.

Fig. 13 represents the linear small signal model of PLL. It is observed that $\Delta\omega_r$ is obtained from time derivative of the phase angle, $\Delta\Theta_{pll}$. Mathematically, it can be represented as,

$$\Delta\omega_r(s) = s \cdot \Delta\Theta_{pll}(s) \tag{14}$$

As a result, the virtual inertia control adds a differential operator between $\Delta\Theta_{pll}$ and Δv_{dc_ref} inherently, as shown in Fig. 13, which shows the full control structure for the three-phase power converter with the virtual inertia control.

3. Grid following mode of VSI in distributed generating system

The grid following VSCs are the DERs which are connected to the strong grid and they follow the PCC point voltage and frequency by a synchronization unit known as PLL. To operate the grid following VSC in strong grids, many techniques [46,47] have been discussed in the literature, such as synchronous reference frame (SRF)-control [46], proportional resonant (PR) controller [48], hysteresis controller [47], etc. Among the above-mentioned control SRF-control commonly called as *dq* control is widely accepted in the literature.

3.1 Structure of grid-imposed frequency VSC system

Fig. 14 shows a schematic diagram of a grid-imposed frequency VSC system. The VSC represents a two-level VSC. The VSC is modeled by an equivalent DC-bus capacitor, a current source representing the VSC switching power loss, and series on-state resistances at the AC side representing the VSC conduction power loss. The DC side of the VSC may be interfaced with a DC voltage source or a DC power source. Each phase of the VSC is interfaced with the AC system via a series resistive-inductive (RL) branch.

The AC system is modeled as an infinitely stiff system by an ideal three-phase voltage source, V_{sabc}. It is also assumed that V_{sabc} is balanced, sinusoidal, and of a relatively constant frequency. The VSC system exchanges the real- and reactive-power components $P_s(t)$ and $Q_s(t)$ with the AC system, at the PCC. Depending on the control strategy, the VSC system is used as either a real- or reactive-power controller or a controlled DC-voltage power port.

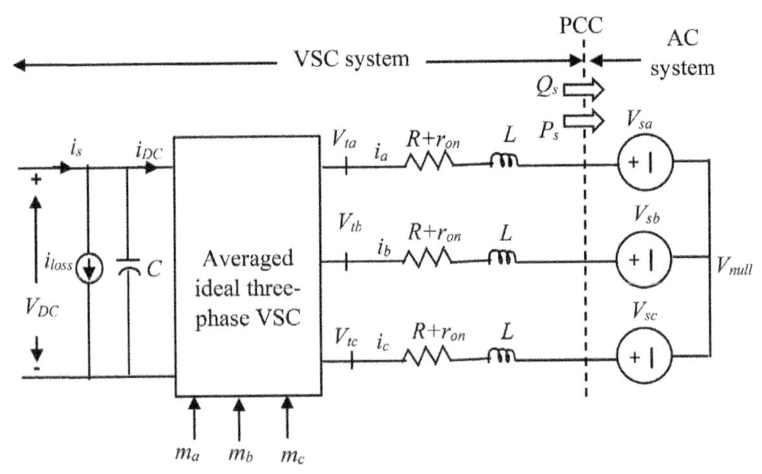

FIG. 14 Schematic diagram of a grid-imposed frequency VSC system.

3.2 Real and reactive power controller

The grid-imposed frequency VSC system of Fig. 15 can be employed as a real- or reactive-power controller. As such, the VSC DC side is connected in parallel with a DC voltage source. The objective is to control the instantaneous real and reactive power that the VSC system exchanges with the AC system, that is, $P_s(t)$ and $Q_s(t)$.

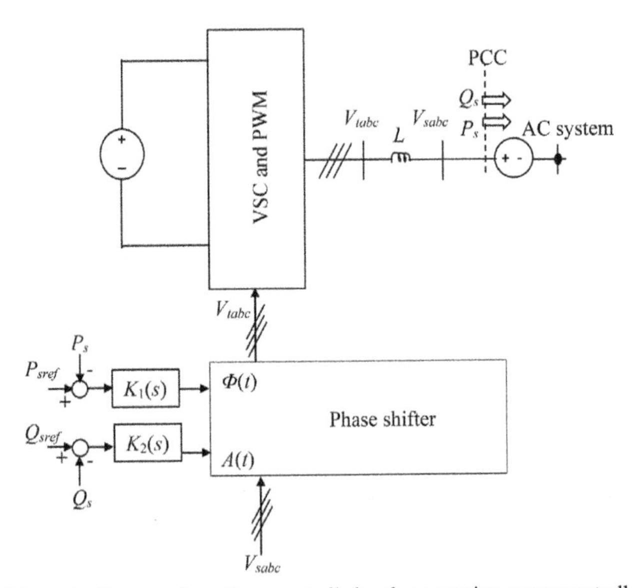

FIG. 15 Schematic diagram of a voltage-controlled real- or reactive-power controller.

In a voltage-controlled VSC system, the phase angle and the amplitude of the VSC AC-side terminal voltage relative to the PCC voltage control the real and reactive power, respectively, as depicted in Fig. 15 [49]. If the amplitude and phase angle of V_{tabc} are close to those of V_{sabc}, the real and reactive power are almost decoupled and two independent compensators can be employed for their control. The voltage-mode control is simple and has a low number of control loops. However, the main shortcoming of the voltage-mode control is that there is no control loop closed on the VSC line current. Consequently, the VSC is not protected against overcurrent, and the current may undergo large excursions if the power commands are rapidly changed, or faults take place in the AC system.

3.3 VSC current control-SRF (DQ-control)

Fig. 16 shows a schematic diagram of a current-controlled real- or reactive-power controller, illustrating that the control is performed in dq-frame. The second approach to control the real and reactive power in the VSC system is referred to as the current-mode control. In this approach, the VSC line current is tightly regulated by a dedicated current-control scheme, through the VSC AC-side terminal voltage. Then, the real and reactive power are controlled by the phase angle and the amplitude of the VSC line current with respect to the PCC voltage. Thus, due to the current regulation scheme, the VSC is protected against overcurrent conditions. Other advantages of the current-mode control include robustness against variations in parameters of the VSC system

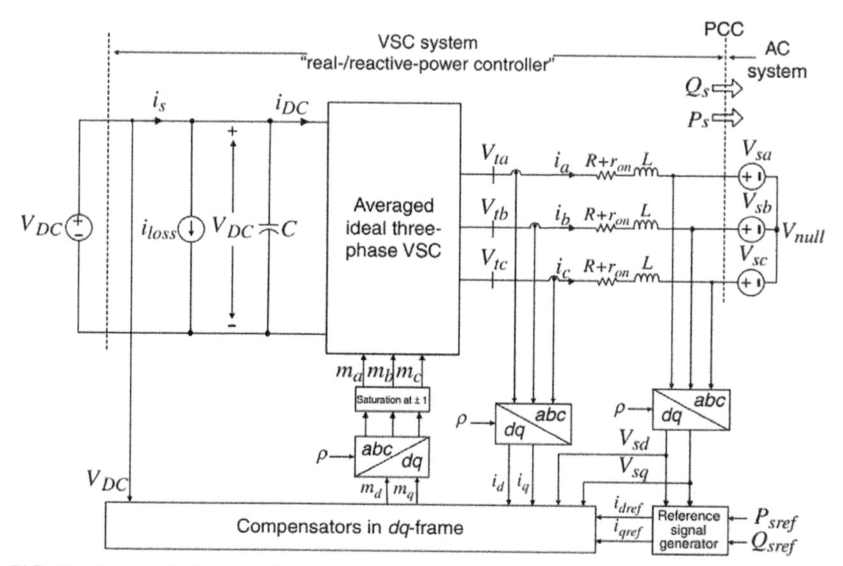

FIG. 16 Schematic diagram of a current-controlled real- or reactive-power controller in dq-frame.

and the AC system, superior dynamic performance, and higher control precision [50]. P_s and Q_s are controlled by the line current components *id* and *iq*. The feedback and feed-forward signals are first transformed to the *dq*-frame and then processed by compensators to produce the control signals in *dq*-frame. Finally, the control signals are transformed to the *abc*-frame and fed to the VSC. To protect the VSC, the reference commands i_{dref} and i_{qref} are limited by the corresponding saturation blocks (not shown in the figure).

The *dq*-frame control of the real- or reactive-power controller is based on equations (1) and (2). Assuming a steady-state operating condition and substituting for $\omega(t) = \omega_0$ in (1) and (2), we deduce

$$L\frac{dI_d}{dt} = L\omega_0 i_q - (R + r_{on})i_d + V_{td} - V_{sd} \tag{15}$$

$$L\frac{di_q}{dt} = -L\omega_0 i_d - (R + r_{on})i_d + V_{td} - V_q \tag{16}$$

where, V_{td} and V_{tq} are

$$V_{td}(t) = \frac{V_{DC}}{2} m_d(t) \tag{17}$$

$$V_{tq}(t) = \frac{V_{DC}}{2} m_d(t) \tag{18}$$

Eqs. (17) and (18) represent the VSC model in *dq*-frame. The model is applicable to both the two-level VSC and the three-level NPC. The i_d and i_q are state variables, V_{td} and V_{tq} are control inputs, and V_{sd} and V_{sq} are disturbance inputs. Due to the presence of $L\omega_0$ terms in (15) and (16), dynamics of i_d and i_q are coupled. To decouple the dynamics, we determine m_d and m_q as,

$$m_d = \frac{2}{V_{dc}}(u_d - L\omega_0 i_q + V_{sd}) \tag{19}$$

$$m_q = \frac{2}{V_{dc}}(u_d - L\omega_0 i_q + V_{sd}) \tag{20}$$

where u_d and u_q are two new control inputs [51] which is passed on to PWM to generate the switching signals for the switches.

3.4 Phase locked loop (PLL)

PLL is an integral dedicated unit of any grid tied system which helps in synchronization of any inverter-based distributed energy source to the utility [52]. The basic structure of the PLL is shown in Fig. 17. To analyze the stability margin and tune its parameter, accurate modeling of the PLL is required. The purpose of the SRF-PLL is to estimate the phase of the supply voltage that is stated as follows:

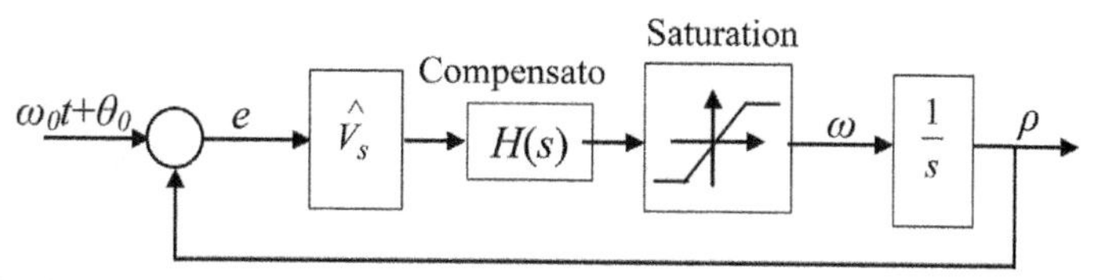

FIG. 17 Control diagram of the PLL.

$$v(t) = V_{PCC} \cos(\omega t + \theta_0) \tag{21}$$

where V_{pcc}, ω, and θ_0 are the voltage at the PCC point, angular frequency, and phase angle of the grid voltage. It should be mentioned that the system variables are time dependent due to the integrator in the PI controller. The transfer function of the $H(s)$ is

$$H(s) = k_p + \frac{k_i}{s} \tag{22}$$

3.5 Experimental results

In Fig. 18, the dynamic response of the PLL to sudden change in the grid voltage waveform V_{sabc} at 0.05 s is shown. Due to the voltage imbalance at 0.05 s, a transient in the frequency at 0.05 s is observed and then it reaches to steady state following the dynamics of the PLL. It can be further observed that post disturbance at 0.05 s, the V_{sdq} oscillates at double frequency of the fundamental frequency due to the imbalance.

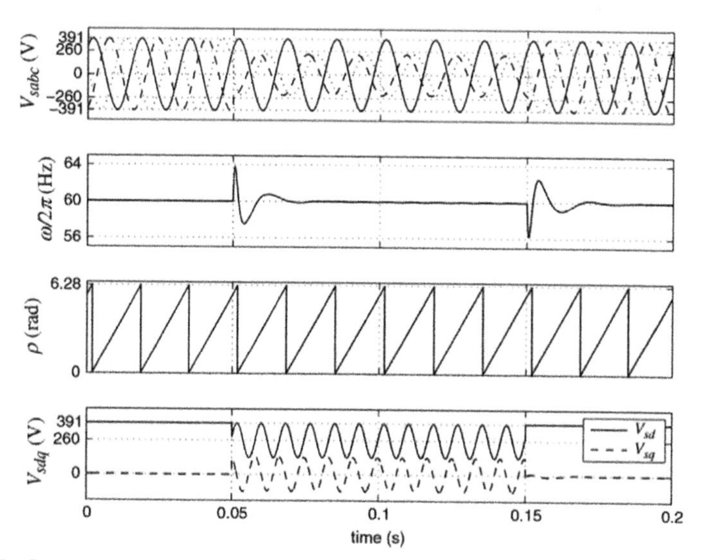

FIG. 18 Response of the PLL to a sudden AC system voltage imbalance.

4. Grid forming mode of VSI in distributed generating system

The concept of microgrid is effective in increasing power generation from DERs. These DERs are integrated to the system through VSIs. In general, for grid connected microgrid, the VSI control attempts to follow both frequency and voltage of the grid through phase locked loop. The autonomous or islanded operation of the microgrid needs to balance the local loads and DER generations by maintaining frequency and voltage of the system within limits, which requires the VSIs to be operated in GFM mode. Further, there is no inertial support from the grid, resulting more challenge against GFM operation of VSIs. Addressing these problems, the following sections will discuss about different GFM control techniques.

4.1 Conventional droop control approach

The GFM control structure of a VSI can be explained in three main parts: power controller, voltage controller, and current controller, as shown in Fig. 19 [53]. The power controller provides a reference voltage magnitude and phase for the VSI output voltage fundamental component. The droop characteristic is implemented for active and reactive power with consideration of fundamental component of voltage output. The harmonic components of power sharing can be considered through different control approach [54].

4.1.1 Power controller

The main idea of power controller is to operate a droop controller which is pretty similar of the primary control of a conventional synchronous generator. An active droop controller in VSI decreases reference frequency within allowable limit for an increase in load or vice versa. Similarly, the voltage reference is changed within a limit by reactive power variation through droop characteristic.

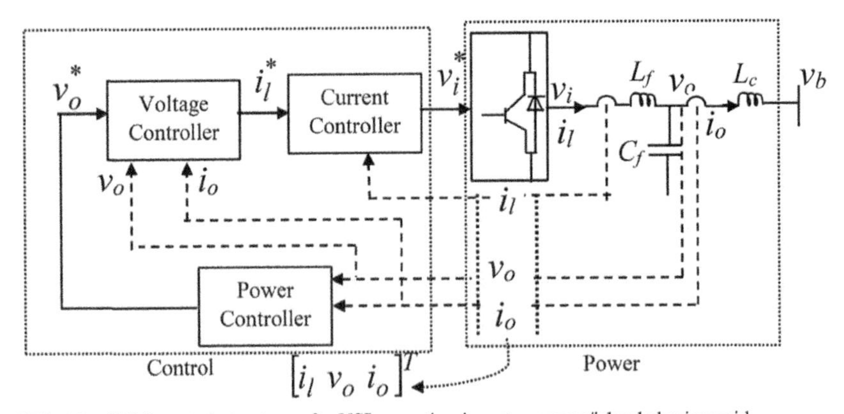

FIG. 19 GFM control structure of a VSI operation in autonomous/islanded microgrid.

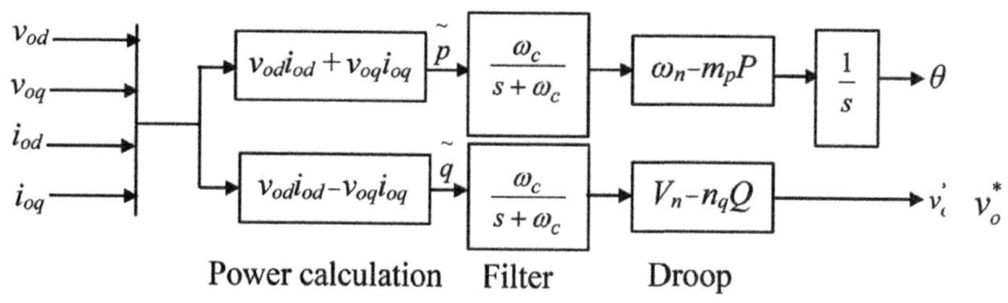

FIG. 20 Block diagram of power controller.

Fig. 20 depicts the block diagram of power controller. The instantaneous active power ($\widetilde{p}$) and reactive power ($\widetilde{q}$) output of the VSI is obtained from voltage measured at the output terminal, as in (23).

$$\begin{aligned} \widetilde{p} &= v_{od}i_{od} - v_{oq}i_{oq} \\ \widetilde{q} &= v_{od}i_{oq} - v_{oq}i_{od} \end{aligned} \tag{23}$$

The low pass filters are designed with cut-off frequency of ω_c to obtain fundamental component of active power (P) and reactive power (Q). The linear droop characteristic with slope of $-m_p$ and $-n_q$ are used to obtain operating frequency and voltage of the VSI, as in (24) and (25), respectively.

$$\left.\begin{aligned} \omega &= \omega_n - m_pP \\ \theta &= \omega_nt - \int m_pPdt \\ \alpha &= -\int m_pPdt \\ \alpha &= -m_pP \end{aligned}\right\} \tag{24}$$

$$v_{od}^* = v_n - n_qQ; \quad v_{oq}^* = 0 \tag{25}$$

The constant droop gains m_p and n_q are obtained from operating voltage limits and frequency limits, as in (26).

$$\begin{aligned} m_p &= \frac{\omega_{\max} - \omega_{\min}}{P_{\max}} \\ n_q &= \frac{V_{od\max} - V_{od\min}}{Q_{\max}} \end{aligned} \tag{26}$$

It is clear from the above discussion that each VSI may operate different frequency according to their droop characteristic. However, for a common system, all the VSIs should operate at same frequency. To overcome this issue, the frequency of a particular VSI is considered as reference, and other VSIs share their power according to difference in their operating frequencies with respect to the reference. The frequency difference of a VSI (other than reference VSI) can be expressed in terms of δ, as in (27).

$$\delta = \int \left(\omega - \omega_{ref}\right) dt \tag{27}$$

4.1.2 Voltage controller

Voltage controller output is obtained from PI controller. The block diagram of voltage controller is depicted in Fig. 21. The control model can be realized through set of differential equations (28) and algebraic equation (29).

$$\begin{aligned} \phi_d &= v_{od}^* - v_{od} \\ \phi_q &= v_{oq}^* - v_{oq} \end{aligned} \tag{28}$$

$$\begin{aligned} i_{ld}^* &= Fi_{od} - \omega_n C_f v_{oq} + K_{pv}\left(v_{od}^* - v_{od}\right) + K_{iv}\phi_d \\ i_{lq}^* &= Fi_{oq} - \omega_n C_f v_{od} + K_{pv}\left(v_{oq}^* - v_{oq}\right) + K_{iv}\phi_q \end{aligned} \tag{29}$$

4.1.3 Current controller

Similar to voltage controller, current controller output is obtained from PI controller. The block diagram of current controller is shown in Fig. 22. The control model can be realized through set of differential equations (30) and algebraic equation (31).

$$\begin{aligned} \dot{\gamma}_d &= i_{ld}^* - i_{ld} \\ \dot{\gamma}_q &= i_{lq}^* - i_{lq} \end{aligned} \tag{30}$$

$$\begin{aligned} v_{id}^* &= -\omega_n L_f i_{lq} + K_{pc}\left(i_{lq}^* - i_{lq}\right) + K_{ic}\gamma_d \\ v_{id}^* &= -\omega_n L_f i_{ld} + K_{pc}\left(i_{ld}^* - i_{ld}\right) + K_{ic}\gamma_q \end{aligned} \tag{31}$$

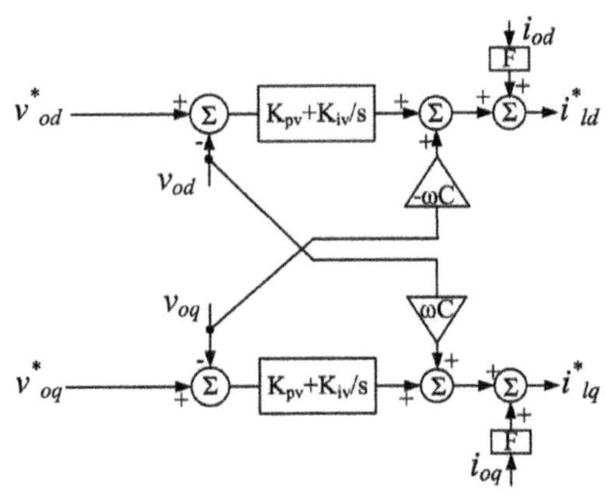

FIG. 21 Block diagram of voltage controller.

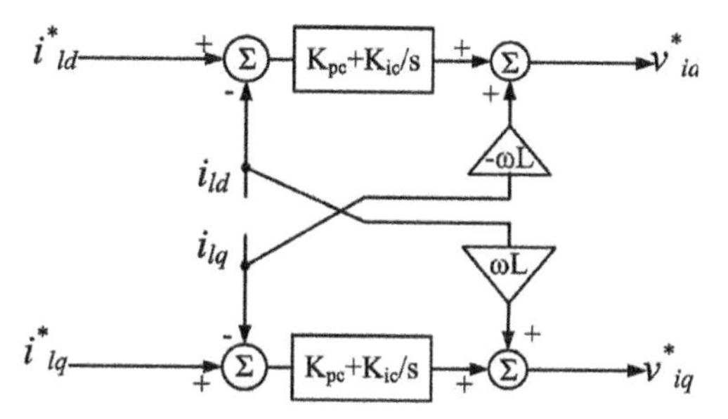

FIG. 22 Block diagram of current controller.

4.2 Virtual synchronous machine (VSM) based control approach

In autonomous microgrid, the VSIs may suffer due to lack of inertia in the system. Implementation of VSM-based control has the potential to improve the situation [55,56]. In the absence of PLL in GFM mode of operation, the VSM approach relies on similar generation-load balance technique as in the conventional SM-based system. An overview of VSM-based control structure for autonomous microgrid is depicted in Fig. 23. There are two important parts in this control approach, which follow the SM control technique: (1) voltage control mimics the automatic voltage regulator (AVR) control of SM and (2) frequency control integrated with virtual inertia model which mimics the inertia and governor control of SM.

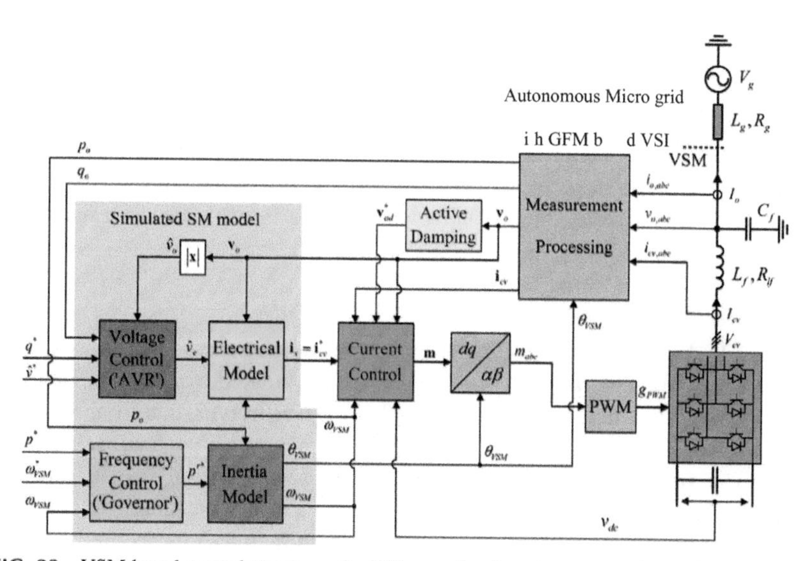

FIG. 23 VSM based control structure of a VSI operating in autonomous micro grid.

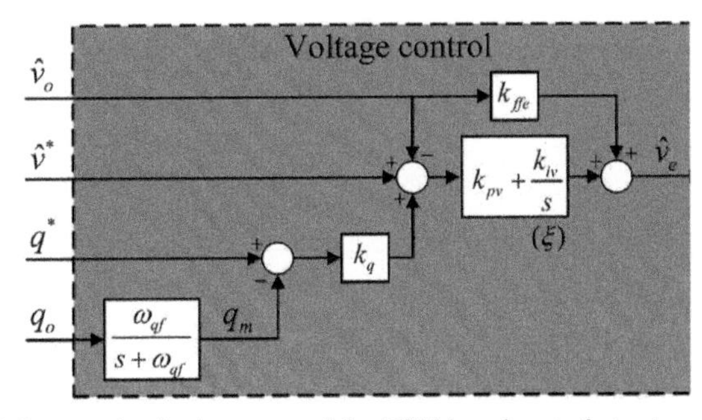

FIG. 24 signal flow graph of voltage control for VSM based control structure.

4.2.1 Voltage controller

The output of voltage controller $\widehat{v}_e$ is similar to induced emf at the armature of SM, which is regulated by an AVR to control the magnitude of $\widehat{v}_o$ at the filter capacitor C_f. Fig. 24 shows the signal flow graph of voltage control of VSM-based control structure. A low pass filter is designed to obtain fundamental component of reactive power output of the VSI q_m, which is used to calculate necessary change in voltage through a droop gain k_q. A PI control takes responsibility to regulate the voltage considering a reference voltage of $\widehat{v}^*$. The output of voltage controller $\widehat{v}_e$ is expressed as in (32).

$$\widehat{v}_e = k_{pv}\left(\widehat{v}^* - \widehat{v}_e\right) + k_{pv}k_q\left(q^* - q_m\right) + k_{iv}\xi + k_{ffe}\widehat{v}_0 \tag{32}$$

ξ is the integrator state of the PI controller [55].

4.2.2 Frequency controller

The frequency controller is implemented as an ideal power generator (high speed governor) to deliver the desired power with fast response. The power reference is obtained from a frequency regulator with a droop gain of k_w as in (33).

$$p^{r*} = p^r + k_w\left(\omega^*_{VSM} - \omega_{VSM}\right) \tag{33}$$

Fig. 25 depicts the signal flow graph of inertia control for VSM-based control structure. In general, the damping torque is emulated from the VSI inertia control using grid frequency through a PLL [57]. However, due to absence of PLL in GFM mode of operation, the VSI uses a negative feedback signal through a low-pass filter with crossover frequency of ω_d, as shown in Fig. 25. The damping torque is proportional to a constant gain k_d. However, a large k_d may create angle instability (monitored by θ_{VSM}) for a small change in ω_{VSM}. Therefore, optimum value of k_d is desired.

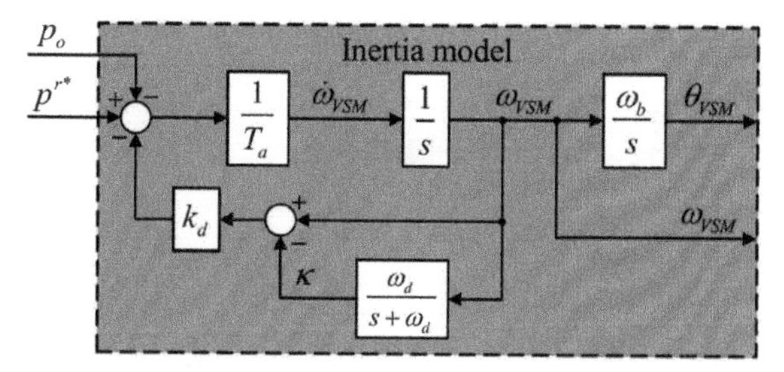

FIG. 25 Signal flow graph of inertia control for VSM based control structure

5. Converter control in EV-based distributed generating system

The intermittent nature of DERs and lack of inertia, due to operating more non-rotating components, in the distributed generating system degrades regulation of both the frequency and voltage of the system. Addressing these issues, different control techniques for VSIs in the distributed generation side is discussed in the previous sections. However, these control approaches may not be sufficient due to unpredictable availability of DERs. The controllable load, that is, EV has proven their effectiveness in power balancing against the DERs. This section elaborates different control techniques for the converters involved in such applications.

5.1 Modified droop control for frequency regulation of EV-based system

As discussed in the earlier section, the conventional droop control approach changes reference power through power controller, presuming the power is available at the DER. However, this may not be true for all the time. To solve this problem, modified droop control considers addition signals from EVs available for V2G control. Fig. 26 shows the system structure of the modified droop controller (MDC) [58].

The MDC communicates with the EVs through an aggregator to know their available powers. DER management system (DERMS) is there to monitor overall system performance. Fig. 27 depicts the control block diagram of the aggregator. The state of charge of each EV is communicated to the aggregator in centralized manner to obtain $P_{tot,EV}$ which is further used in V2G controller and MDC. The MDC sends signal $P_{ref,AG}$ to the V2G controller to control charging and discharging status of the EVs. The performance curves are available in DERMS, which are shown in Fig. 28.

It is realized from the Fig. 28A–D is that the DER, that is, wind and PV sources availability is difficult to match with their references. Therefore, the power mismatch is balanced through DGs and EVs. Fig. 28E shows the

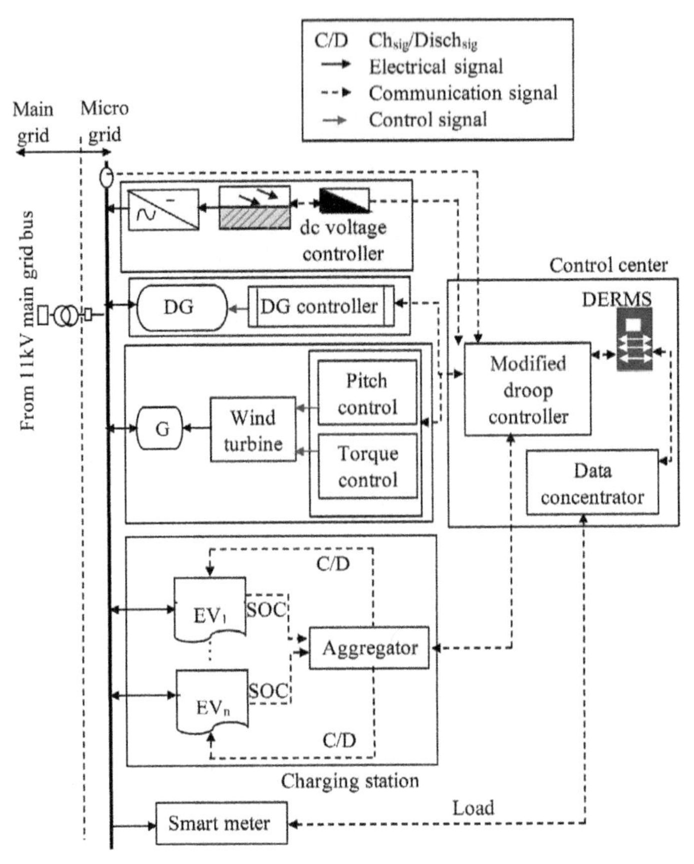

FIG. 26 System structure of the MDC.

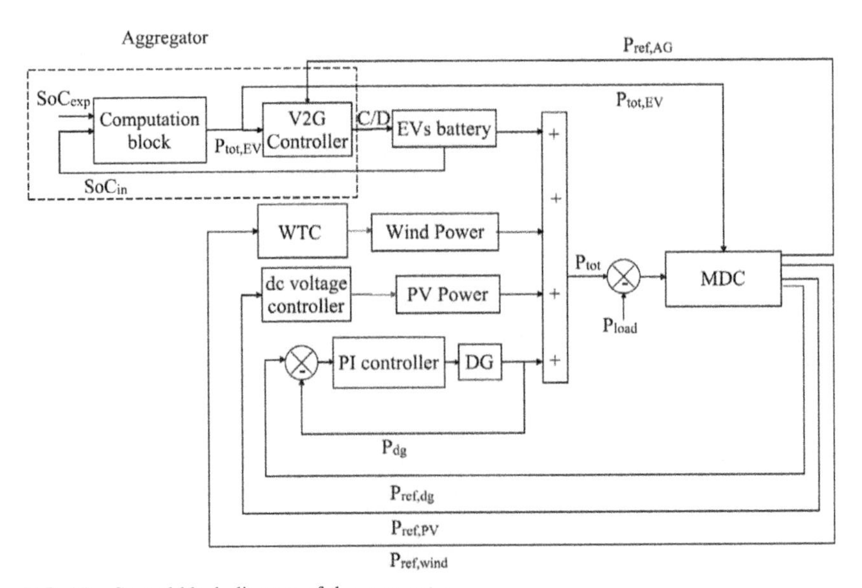

FIG. 27 Control block diagram of the aggregator.

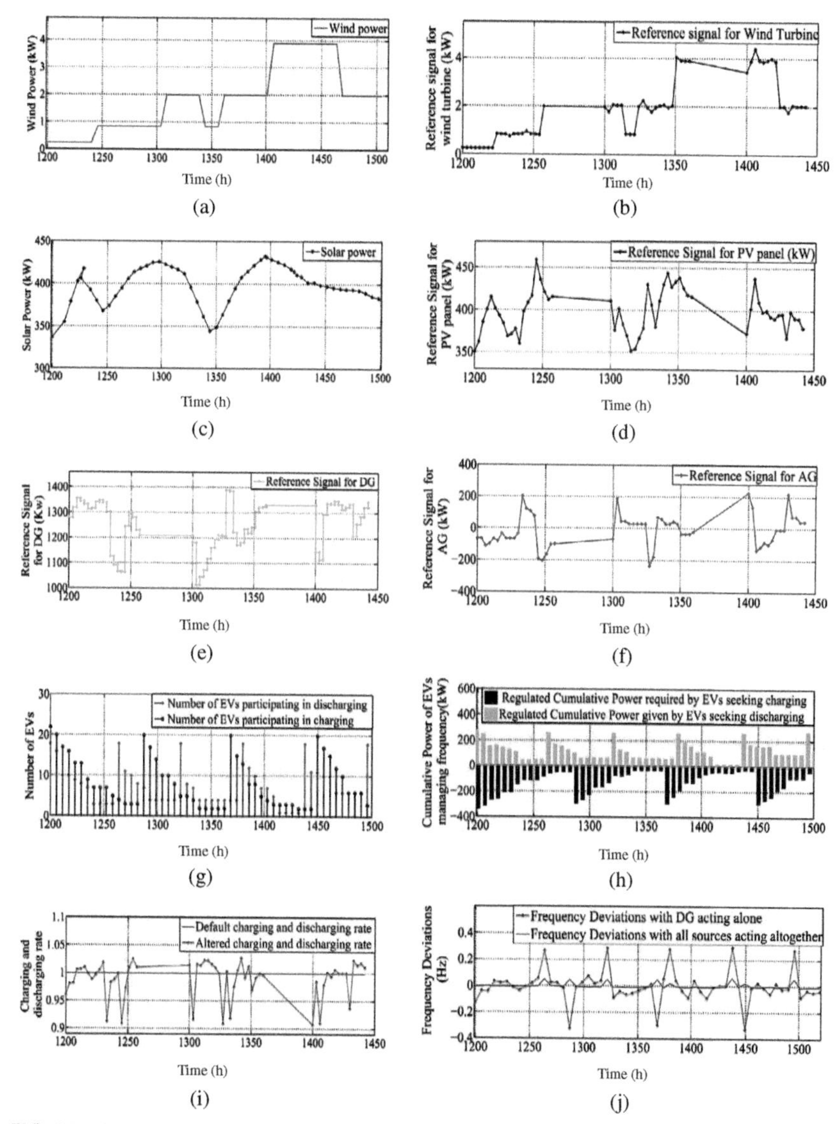

FIG. 28 (A) Available wind power, (B) reference signal for wind turbine, (C) available solar PV power, (D) reference signal for solar PV panel, (E) reference signal for DG, (f) reference signal for aggregator, (G) number of available EVs participating in V2G control, (H) Cumulative power of EVs, (I) charging and discharging rate of EVs, (J) frequency deviation at 11 kV substation.

reference signal for DG. The V2G controller controls charging and discharging rate of EVs, which is depicted in Fig. 28I. The frequency regulation of the grid is improved significantly with participation of EVs along with DGs as evident from Fig. 28J.

5.2 Distributed control for voltage regulation of EV-based system

Apart from the frequency regulation, in the literatures, EV chargers have proven their ability in improving voltage regulation of the system as well [59,60]. Topological variety of different single-phase EV chargers is represented in [59]. The reactive power compensation capability of different three-phase bidirectional EV chargers are analyzed and demonstrated in Buja et al. [61]. The experimental studies are also conducted to reduce power conversion loss in V2G technology [60]. Hence, it is to be believed that EV chargers have huge potential in reactive power support to the grid. This needs an unified model of EV chargers which can be applicable widely into the grid.

5.2.1 *Model of EV chargers*

Fig. 22 shows the V2G enabled EV charger model and its range of operation. The main component of this model is a three-phase converter with controlled operating voltage v_o. v_{RL} represents voltage drop at the line connecting EV to the grid. The grid voltage v_G is constant. Hence, the bidirectional current I between EV charger and the grid is limited by v_{RL}. Operating range of EV charger is demonstrated through a P-Q diagram, which is circle and divided in four quadrants. Thus, the P-Q diagram is expressed as,

$$P^2 + Q^2 \leq S_{\max}^2 \tag{34}$$

$S_{\max}$ is the rated capacity of the EV charger.

Quadrant I and IV are significant for EV is in charging mode. Quadrant II and III are significant for EV is in discharging mode. Positive and negative Q signify inductive and capacitive power delivered from the grid. From Figs. 29 and 26, it is evident that reactive power margin improves at lower active power operation of the EV charger. Therefore, in improving voltage regulation of the grid, the EV chargers are effective during their idle operation.

5.2.2 *Distributed control framework*

Fig. 30 shows the framework of distributed control of EV chargers participating in V2G technology. The framework is divided into three parts: cyber layer, V2G layer, and grid layer. Each EV charger in V2G layer interacts among each other through dedicated point-to-point (P2P) communication medium to obtain their optimum operating voltage to improve voltage regulation of the grid layer. Based on the operating voltage of each EV charger, the deliverable reactive power from the chargers is controlled with distributed model predictive controller (DMPC). Fig. 31 depicts the control block diagram of DMPC in V2G technology. A sensitivity calculation module precalculates the change in voltage at charger output terminal, which further helps to obtain optimum operating voltages of the chargers. A detailed prediction model of DMPC is available in Hu [62].

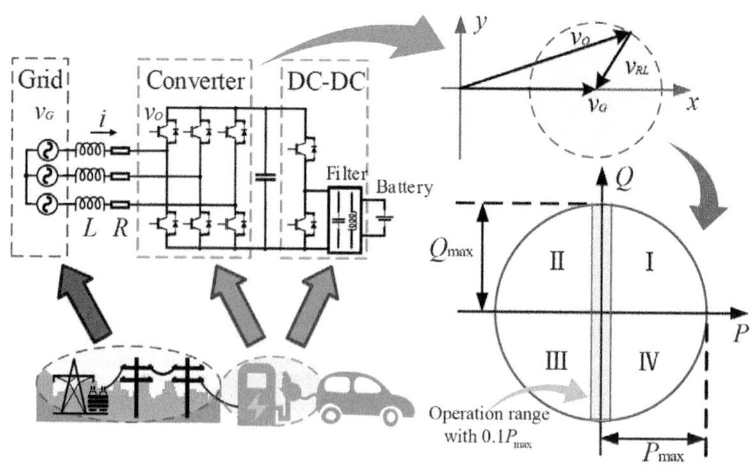

FIG. 29 V2G enabled EV charger model and its range of operation.

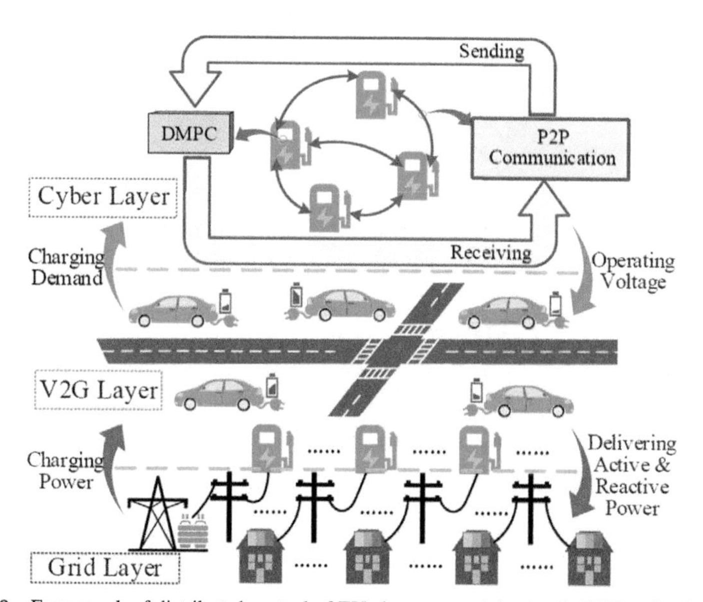

FIG. 30 Framework of distributed control of EV chargers participating in V2G technology.

5.2.3 Case study

The DMPC is verified in modified IEEE European low voltage feeder system [63]. The system is depicted in Fig. 32. There are 55 feeders and 28 EV chargers distributed throughout the system.

Fig. 33A and B shows voltage of the feeders throughout a day without and with participation of EV chargers, respectively. The improvement of voltage

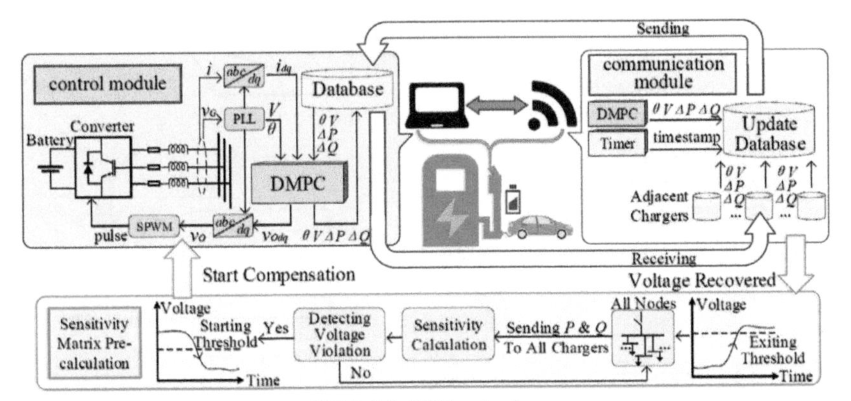

FIG. 31 Control block diagram of DMPC in V2G technology.

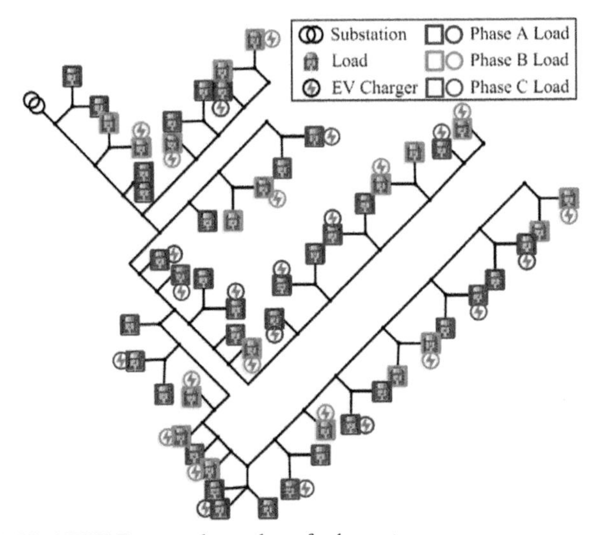

FIG. 32 Modified IEEE European low voltage feeder system.

regulation throughout a day using EV chargers is evident by comparing these two figures.

6. Conclusions

As per the above discussions, there are many challenges against the integration of distributed generating systems to the grid. There are certain standards and regulation which are to be maintained for effective operation of power systems with high penetration level of DERs, which is not a easy job. However, different control schemes for VSCs have attempted to make the operation as simple as possible. The main intention of these schemes is to make the power system

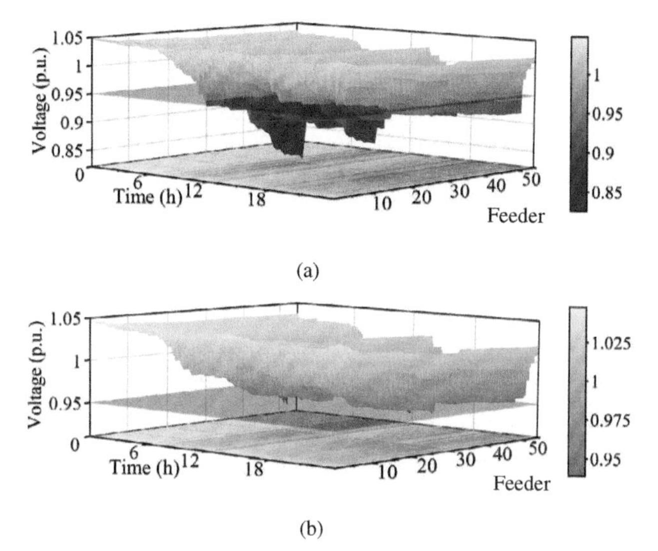

FIG. 33 Voltage of the feeders throughout a day (A) without participation of EV chargers (B) with participation of EV chargers.

adaptive with the changing behavior of DERs. Few of these control schemes have already implemented in different countries. However, the other schemes are still under research, which may be required to be modified in future with more and more integration distributed generating systems.

References

[1] A. Hirsch, Y. Parag, J. Guerrero, Microgrids: a review of technologies, key drivers, and outstanding issues, Renew. Sustain. Energy Rev. 90 (2018) 402–411.

[2] R. Singh, C. Bansal, A.R. Singh, R. Naidoo, Multi-objective optimization of hybrid renewable energy system using reformed electric system cascade analysis for islanding and grid connected modes of operation, IEEE Access 6 (2018) 47332–47354.

[3] M. Smith, D. Ton, Key connections: the U.S. Department of Energy's Microgrid initiative, IEEE Power Energy Mag. 11 (4) (2013) 22–27.

[4] M. Ahmed, L. Meegahapola, A. Vahidnia, M. Dutta, Stability and control aspects of microgrid architectures-a comprehensive review, IEEE Access 8 (2020) 144730–144766.

[5] Z. Yao, L. Xiao, Y. Yan, Seamless transfer of single-phase gridinteractive inverters between grid-connected and stand-alone modes, IEEE Trans. Power Electron. 25 (6) (2010) 1597–1603.

[6] M. Karimi-Ghartemani, Universal integrated synchronization and control for single-phase DC/AC converters, IEEE Trans. Power Electron. 30 (3) (2015) 1544–1557.

[7] H. Karimi, A. Yazdani, R. Iravani, Negative-sequence current injection for fast islanding detection of a distributed resource unit, IEEE Trans. Power Electron. 23 (1) (2008) 298–307.

[8] Y.A.R.I. Mohamed, A.A. Radwan, Hierarchical control system for robust microgrid operation and seamless mode transfer in active distribution systems, IEEE Trans. Smart Grid 2 (2) (2011) 352–362.

[9] F. Gao, M.R. Iravani, A control strategy for a distributed generation unit in grid-connected and autonomous modes of operation, IEEE Trans. Power Del. 23 (2) (2008) 850–859.
[10] J. Matas, M. Castilla, L.G.D. Vicuña, J. Miret, J.C. Vasquez, Virtual impedance loop for droop-controlled single-phase parallel inverters using a second-order general-integrator scheme, IEEE Trans. Power Electron. 25 (12) (2010) 2993–3002.
[11] J.M. Guerrero, J.C. Vasquez, J. Matas, M. Castilla, et al., Control strategy for flexible microgrid based on parallel line interactive UPS systems, IEEE Trans. Ind. Electron. 56 (3) (2009) 726–736.
[12] R.J. Wai, C.Y. Lin, Y.R. Huang, Design of highperformance stand-alone and grid-connected inverter for distributed generation applications, IEEE Trans. Ind. Electron. 60 (4) (2013) 1542–1555.
[13] I.J. Balguer, Q. Lei, S. Yang, U. Supatti, et al., Control for grid-connected and intentional islanding operations of distributed power generation, IEEE Trans. Ind. Electron. 58 (1) (2011) 147–157.
[14] J. Kim, M. Guerrero, P. Rodriguez, R. Teodorescu, et al., Mode adaptive droop control with virtual output impedances for an inverter based flexible AC microgrid, IEEE Trans. Power Electron. 26 (3) (2011) 689–701.
[15] J. He, Y.W. Li, Hybrid voltage and current control approach for DG-grid interfacing converters with LCL filters, IEEE Trans. Ind. Electron. 60 (5) (2013) 1797–1809.
[16] C. Cho, J.H. Jeon, J.Y. Kim, S. Kwon, et al., Active synchronizing control of a microgrid, IEEE Trans. Power Electron. 26 (12) (2011) 3707–3719.
[17] T.L. Vandoorn, B. Meersman, D. Kooning, Vandevelde., Transition from islanded to grid-connected mode of microgrids with voltage-based droop control, IEEE Trans. Power Syst. 28 (3) (2013) 2545–2553.
[18] H. Xu, X. Zhang, F. Liu, D. Zhu, et al., Synchronization strategy of microgrid from islanded to grid-connected mode seamless transfer, IEEE Int. Conf. IEEE Region (TENCON), China (2013) 1–4.
[19] Y. Jia, D. Liu, J. Liu, A novel seamless transfer method for a microgrid based on droop characteristic adjustment, in: 7th Int. Power Electron. Motion Control Conf., Harbin. China, 2012, pp. 362–367.
[20] IEEE-Std 1547-2003, Standard for Interconnecting Distributed Resources With Electric Power Systems, IEEE Standard 1547-2003, 2003.
[21] IEEE Standard 1547-2018, IEEE standard for interconnection and interoperability of distributed energy resources with associated electric power systems interfaces, in: IEEE Standard 1547-2018, 2018, pp. 1–138.
[22] IEEE Standards Association 1547.4-2011, IEEE Guide for Design, Operation, and Integration of Distributed Resource Island Systems With Electric Power Systems, IEEE Standards Association, NY, USA, 2011.
[23] Rule 21 Generating, Rule 21 Generating Facility Interconnections, California Public Utilities Commission (CPUC), San Francisco, CA, USA, 2014.
[24] Power Generation, Power generation systems connected to the low-voltage distribution network—technical minimum requirements for the connection to and parallel operation with low-voltage distribution networks, VDE-ARN 4105-2011, 2011.
[25] AS 4777.1-2005, Grid Connection of Energy Systems via Inverters Part 1: Installation Requirements, Council of Standards Australia, AS 4777.1-2005, 2005.
[26] AS 4777.3-2005, Grid Connection of Energy Systems via Inverters Part 3: Installation Requirements, Council of Standards Australia, AS 4777.3-2005, 2005.

[27] AS 4777.2-2005, Grid Connection of Energy Systems via Inverters Part 2: Installation Requirements, Council of Standards Australia AS 4777.2-2005, 2005.

[28] IEEE Std 929-2000, Recommended Practice for Utility Interface of Photovoltaic Systems, IEEE Std 929-2000, 2000.

[29] NREL Report, Fast Grid Frequency Support From Distributed Energy Resources, NREL Report, 2021.

[30] Y. Liu, S. You, J. Tan, Y. Zhang, et al., Frequency response assessment and enhancement of the U.S. power grids toward extra-high photovoltaic generation penetrations—an industry perspective, IEEE Trans. Power Syst. 33 (3) (2018) 3438–3449.

[31] A. Hoke, D. Maksimovic, Active power control of photovoltaic power systems, in: 1st IEEE Conference on Technologies for Sustainability (SusTech), Portland, USA, 2013, pp. 70–77.

[32] J. Neely, J. Johnson, J. Delhotal, S. Gonzalez, et al., Evaluation of PV frequency-watt function for fast frequency reserves, in: Applied Power Electronics Conference, CA, USA, 2016.

[33] B. Neill, V. Gevorgian, PV controls utility-scale demonstration project, in: UVIG Fall Technical Workshop, San Diego, 2015 (Online) https://www.nrel.gov/docs/fy16osti/65284.pdf.

[34] A. Hoke, S. Chakraborty, M. Shirazi, E. Muljadi, et al., Rapid active power control of photovoltaic systems for grid frequency support, IEEE J. Emerg. Sel. Top. Power Electron. 5 (3) (2017) 1154–1163.

[35] B.I. Crăciun, T. Kerekes, D. Sera, R. Teodorescu, Frequency support functions in large PV power plants with active power reserves, IEEE J. Emerg. Sel. Top. Power Electron. 2 (4) (2014) 849–858.

[36] J. Aho, et al., A tutorial of wind turbine control for supporting grid frequency through active power control, in: American Control Conference (ACC), Montreal, QC, Canada, 2012, pp. 3120–3131.

[37] T. Gönen, Electric Power Distribution System Engineering, second ed., CRC Press Taylor & Francis Group, 2008.

[38] A.M. Howlader, S. Sadoyama, L.R. Roose, S. Sepasi, Distributed voltage regulation using volt-var controls of a smart PV inverter in a smart grid: an experimental study, Renew. Energy 127 (2018) 145–157.

[39] F. Blaabjerg, R. Teodorescu, M. Liserre, A.V. Timbus, Overview of control and grid synchronization for distributed power generation systems, IEEE Trans. Ind. Electron. 53 (5) (2006) 1398–1409.

[40] International Review of Frequency Control Adaptation, Australia Energy Market Operator, Melbourne, 2016. http://www.aemo.com.au. (Online).

[41] P. Kundur, Power System Stability and Control, McGraw-Hill, New York, NY, USA, 1994.

[42] J.M. Carrasco, et al., Power electronic systems for the grid integration of renewable energy sources: a survey, IEEE Trans. Ind. Electron. 53 (4) (2006) 1002–1016.

[43] J. Fang, H. Li, Y. Tang, F. Blaabjerg, Distributed power system virtual inertia implemented by grid-connected power converters, IEEE Trans. Power Electron. 33 (10) (2018) 8488–8499.

[44] K. Guo, J. Fang, Y. Tang, Autonomous dc-link voltage restoration for grid-connected power converters providing virtual inertia, in: IEEE energy convers. Congr. Expo., Portland, OR, USA, 2018.

[45] B. Wen, D. Boroyevich, R. Burgos, P. Mattavelli, et al., Analysis of d-q small-signal impedance of grid-tied inverters, IEEE Trans. Power Electron. 31 (1) (2016) 675–687.

[46] P. Rodriguez, J. Pou, J. Bergas, I. Candela, et al., Double synchronous reference frame PLL for power converters control, in: IEEE 36th Power Electronics Specialists Conference, Dresden, Germany, 2005, pp. 1415–1421.

[47] S. Prakash, S. Mishra, A three-sample-based PLL-less hysteresis current control and stability analysis of a single-phase active distribution system, IEEE Trans. Ind. Electron. 68 (8) (2021) 7045–7060.

[48] M. Ebad, B. Song, Improved design and control of proportional resonant controller for three-phase voltage source inverter, in: IEEE Power Electronics and Machines in Wind Applications, Denver, CO, USA, 2012, pp. 1–5.

[49] A.R. Bergen, V. Vittal, Power System Analysis, Prentice-Hall, New Jersey, USA, 2000.

[50] M.P. Kazmierkowski, L. Malesani, Current-control techniques for three-phase voltage-source PWM converters: a survey, IEEE Trans. Ind. Electron. 45 (5) (1998) 691–703.

[51] C.D. Schauder, R. Caddy, Current control of voltage-source inverters for fast four-quadrant drive performance, IEEE Trans. Ind. Electron. IA-18 (2) (1982) 163–171.

[52] T. Tran, T. Chun, H. Lee, H. Kim, et al., PLL-based seamless transfer control between grid-connected and islanding modes in grid-connected inverters, IEEE Trans. Power Electron. 29 (10) (2014) 5218–5228.

[53] N. Pogaku, M. Prodanovik, T.C. Green, Modeling, analysis and testing of autonomous operation of an inverter-based microgrid, IEEE Trans. Power Electron. 22 (2) (2007) 613–625.

[54] M.N. Marwali, J. Jung, Keyhani., Control of distributed generation systems—part II: load sharing control, IEEE Trans. Power Electron. 19 (6) (2004) 1551–1561.

[55] O. Mo, S. Arco, A. Suul, Evaluation of virtual synchronous machines with dynamic or quasi-stationary machine models, IEEE Trans. Power Electron. 64 (7) (2017) 5952–5962.

[56] Y. Chen, R. Hesse, Turschner, H.P. Beck, Investigation of the virtual synchronous machine in the island mode, in: 3rd IEEE Innov. Smart Grid Technol. Eur. (ISGT Europe), Berlin, Germany, 2012, pp. 1–6.

[57] H. Wu, et al., Small-signal modeling and parameters design for virtual synchronous generators, IEEE Trans. Ind. Electron. 63 (7) (2016) 4292–4303.

[58] R. Rana, M. Singh, S. Mishra, Design of modified droop controller for frequency support in microgrid using fleet of electric vehicles, IEEE Trans. Power Syst. 32 (5) (2017) 3627–3636.

[59] M.C. Kisacikoglu, B. Ozpineci, L.M. Tolbert, EV/PHEV bidirectional charger assessment for V2G reactive power operation, IEEE Trans. Power Electron. 28 (12) (2013) 5717–5727.

[60] D.B.W. Abeywardana, P. Acuna, B. Hredzak, R.P. Aguilera, et al., Single-phase boost inverter-based electric vehicle charger with integrated vehicle to grid reactive power compensation, IEEE Trans. Power Electron. 33 (4) (2018) 3462–3471.

[61] G. Buja, Bertoluzzo, C. Fontana, Reactive power compensation capabilities of V2G-enabled electric vehicles, IEEE Trans. Power Electron. 32 (12) (2017) 9447–9459.

[62] J. Hu, C. Ye, Y. Ding, J. Tang, et al., A distributed MPC to exploit reactive power V2G for real-time voltage regulation in distribution networks, IEEE Trans. Smart Grid 13 (1) (2022) 576–588.

[63] The IEEE European Low Voltage Test, The IEEE European low voltage test feeder, in: IEEE Power & Energy Society, Tech. Rep, 2015. https://site.ieee.org/pes-testfeeders/resources/. (Online).

Chapter 3

Short-term solar irradiance forecasting using ground-based sky images

Pravat Kumar Ray[a] and Hoay Beng Gooi[b]
[a]*Department of Electrical Engineering, National Institute of Technology, Rourkela, India,* [b]*School of Electrical and Electronic Engineering, Nanyang Technological University, Singapore, Singapore*

1. Introduction

Due to the increasing rate of climatic changes and global warming, more and more concerns arise about the environmental impacts of generating electricity using conventional fossil fuels. Uncertainties of solar PV generation result in the need of larger energy storage or reserve capacities to regulate the power output for meeting ancillary service requirements. The power output of solar PV generation is heavily dependent on the instantaneous solar irradiance, and the stability of power generation is very much influenced by the cloud condition. A widespread change in solar irradiance caused by large clouds covering the sunlight results in the power output of solar PVs to drop drastically, affecting the system frequency and voltage and possibly leading to system breakdown or blackout [1]. Energy storage alone might not be possible to feed the electricity grid and would require the ancillary services to ramp up or down [2] to meet the change in electricity supply while maintaining the system stability and power quality.

In general, the solar irradiance forecasting can be done through two different approaches, which are numerical prediction or deterministic measurement-based forecasting. The Numerical Weather Prediction (NWP) method forecasts the solar irradiance by building the weather model and forecast solar irradiance for a rather longer period of time. Artificial intelligence and machine learning techniques are often applied in constructing a more dynamic mathematical model to represent the solar irradiance trend at specific location. NWP forecasts are able to perform better than forecast based on macro-scale satellite imaging for Forecast Horizon (FH) longer than a few hours [3]. However, since NWP

Microgrid Cyberphysical Systems. https://doi.org/10.1016/B978-0-323-99910-6.00009-8

only forecasts on the mathematical value of solar irradiance, it cannot predict the position of the cloud cover and so their effect on solar irradiance over a specific location is neglected.

Another approach is based on real-time measurements, be it macro-scale or micro-scale. Macro-scale measurements refer to satellite images which are used in several published literatures to predict long-term solar irradiance [4,5]. Micro-scale measurements on the other hand refer to sky imagers such as Total Sky Imagers (TSIs). These micro-scale measurements enable detail forecast to be done regarding the position of cloud cover at next time step and determine the blockage of solar beam radiation, thus forecasting the solar irradiance. These features allow micro-scale deterministic approaches to achieve a relatively higher accuracy and better spatial and temporal resolutions for solar irradiance prediction with FH of a shorter period of time ranging from a couple of minutes to an hour.

Solar irradiance forecast typically gives either global horizontal irradiance (GHI) or direct normal irradiance (DNI) [6], depending on the technique applied. GHI refers to the total amount of solar irradiance that will be received by a surface horizontal to the ground. It is also the summation of DNI and diffuse horizontal irradiance (DHI). On the other hand, DNI refers to the solar irradiance received by a surface that is held perpendicular to the sunlight ray. Out of these two values, the one that will affect the solar photovoltaic generation output is GHI. Therefore, in this chapter, the solar irradiance prediction has been done in terms of GHI.

However, since the effects of cloud distribution and the optical depth on DHI are typically small and insignificant, the variation of GHI is somehow correlated to the variation of DNI, when forecasting is done based on cloud coverage [6,7]. To determine the GHI value, the presence of cloud in between the ground location and the sun is detected, as it will block the solar beam from penetrating the ground, thus causing the GHI value to differ from its supposed clear sky value. Hence, in the forecasting of solar irradiance, Kt plays a vital role. As far as accuracy in forecasting is concerned, Kt values have been updated using the RLS [8] and VLLMS [9,10] algorithms.

This chapter has been organized as: Section 2 describes the methodology of solar irradiance forecasting using sky images. Section 3 demonstrates the updating of clearness index Kt using the RLS algorithm for forecasting of solar irradiance. Section 4 illustrates the updating of Kt using the VLLMS algorithm for solar irradiance forecasting. Section 5 concludes the chapter.

2. Methodology adopted for forecasting of solar irradiance using sky images

There are mainly four steps, such as (i) preprocess sky images, (ii) cloud pixel determination, (iii) cloud movement determination, and (iv) forecast solar irradiance, to be followed, for forecasting of solar irradiance using sky images. These steps of implementation have been discussed in the following.

2.1 Preprocess sky images

TSI captures sky images [11] through fish-eye lens, which causes nearby objects in the horizon such as poles and buildings to be captured in the images unintentionally. The pixels containing these disturbances are removed from processing as well. To remove these false pixels, three levels of masks, such as circular mask, masking of poles, and building and shadow band, are created to mask out these obstructions. Fig. 1A shows the original sky images captured by the TSI and Fig. 1B shows the sky images after preprocessing/masking.

2.2 Cloud pixel determination

Each image pixel of the cloud has been classified into three categories, such as "Thin Cloud," "Thick Cloud," and "Clear Sky." The red channel and blue channel for each image pixel have been extracted and their respective R/B ratio is calculated to classify it into one of the three categories. The R/B ratio has been compared with certain threshold value in order to classify into certain category. The threshold value has been computed on creating a clear sky library. Fig. 2 shows the flow chart for the cloud determination algorithm.

2.2.1 Clear sky library

To increase the accuracy of cloud determination, Clear Sky Library (CSL) has been compiled to create different threshold values for different pixels in the image. For creating the CSL, a completely clear day within the data set has been

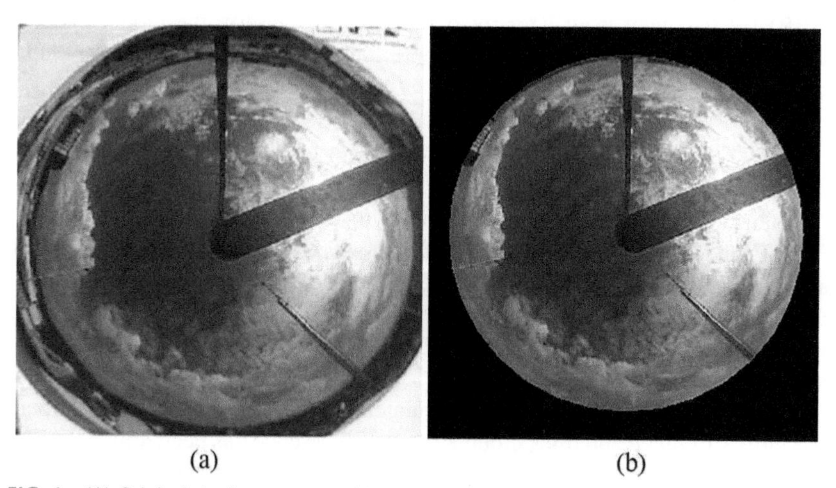

FIG. 1 (A) Original sky image captured by TSI, (B) sky image after masking.

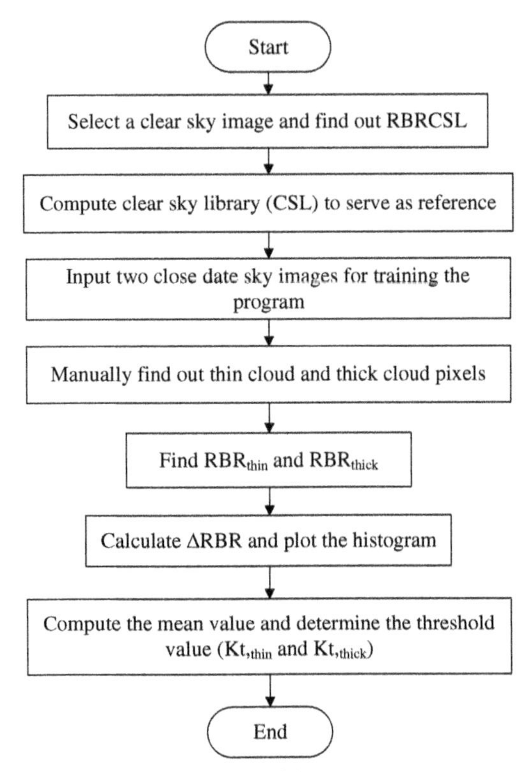

FIG. 2 Flow chart for cloud determination algorithm.

chosen to be the reference. The R/B ratio of every pixel of this image has been calculated and stored in a variable RBRcsl. This reference red to blue ratio (RBR) [6] has been used later for comparison to classify the cloud. Fig 3A shows the image that is used to compute the CSL and Fig. 3B displays the R/B ratio value of the CSL.

2.2.2 Computation of threshold value

To determine the threshold value for classifying each pixel into "Thick Cloud," "Thin Cloud," or "Clear Sky," two different sets of training data are collected and compared with the R/B ratio of CSL. In each of the training image, the thin cloud and thick cloud have been manually drawn. The region of pixels containing the clouds has been recorded and the R/B ratio values of these pixels are extracted. These R/B ratio values have been compared with the R/B ratio from CSL and the differences have been calculated. The histogram for the thin cloud R/B ratio difference and thick cloud R/B ratio difference has been drawn. Based on the histogram, the mean value is computed and treated as the threshold value for classifying the image pixel as "Clear Sky," "Thin Cloud," or "Thick Cloud."

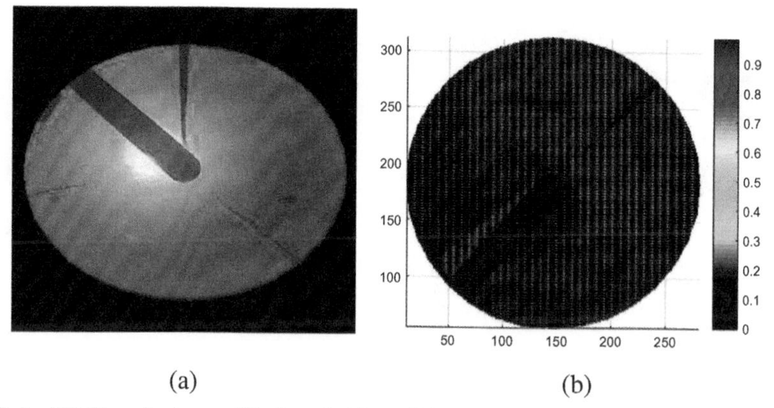

FIG. 3 (A) Clear sky image, (B) clear sky library R/B ratio.

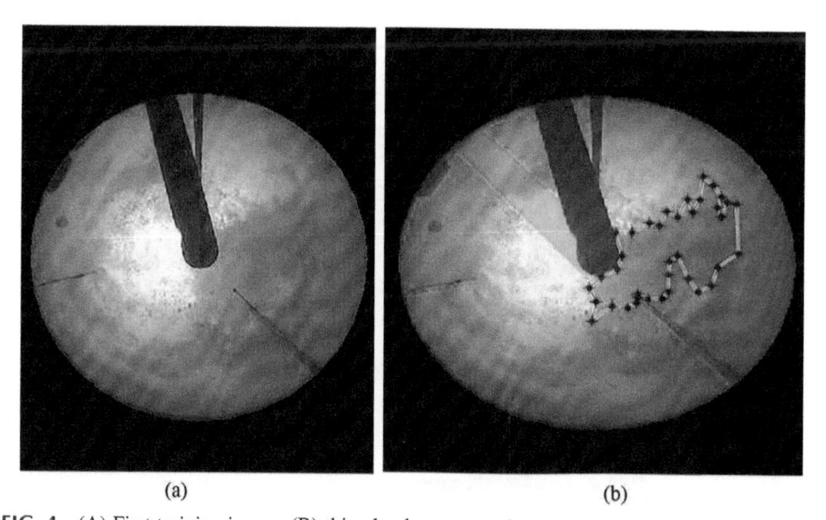

FIG. 4 (A) First training image, (B) thin cloud representation.

Fig. 4A shows the first training image and Fig. 4B shows an example of the cloud pixels region that is manually drawn to represent the thin clouds. Fig. 5A on the other hand shows the second training data and Fig. 5B shows an example of the thick cloud pixels region. The thin cloud and thick cloud regions are drawn out parts-by-parts just like the images shown. All the parts are summed up in the end to represent the total thin cloud and thick cloud coverage of particular sky image.

After collecting all the ΔRBR_{thin} and ΔRBR_{thick} values from both images, the histogram showing the distribution of these values are plotted as shown in Figs. 6 and 7.

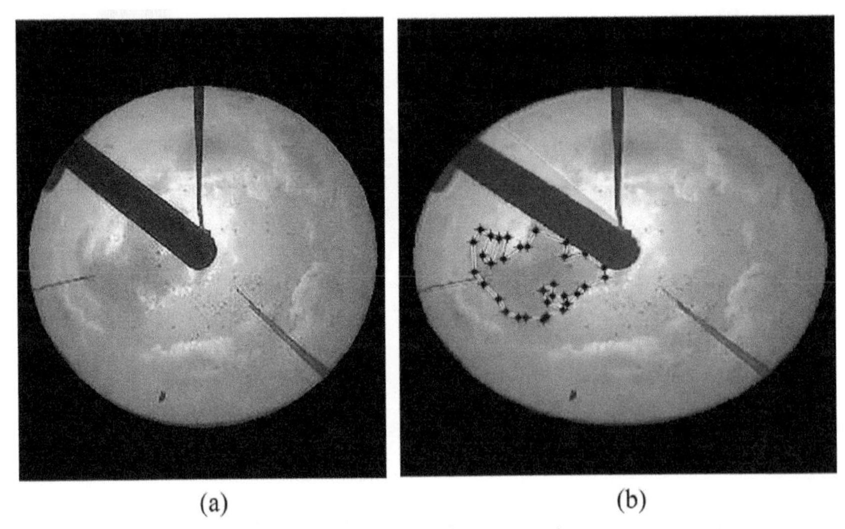

FIG. 5 (A) Second training image, (B) thick cloud representation.

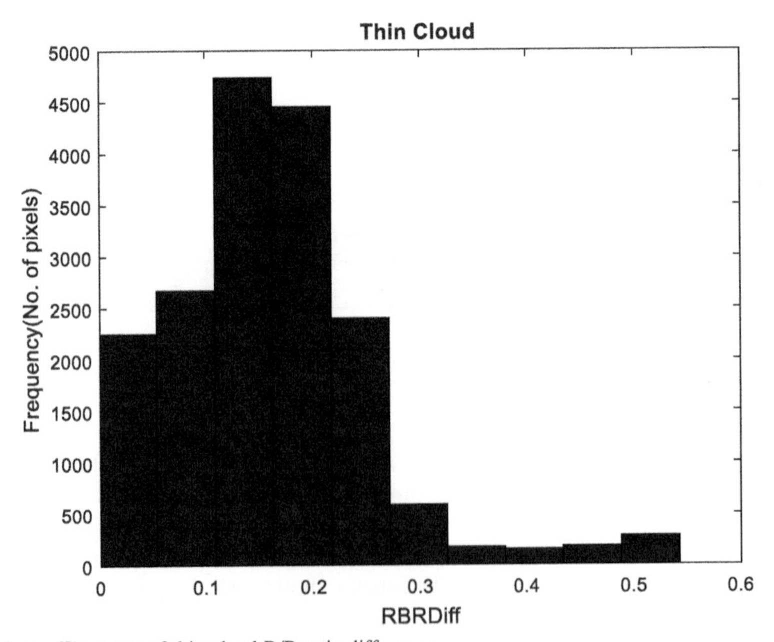

FIG. 6 Histogram of thin cloud R/B ratio difference.

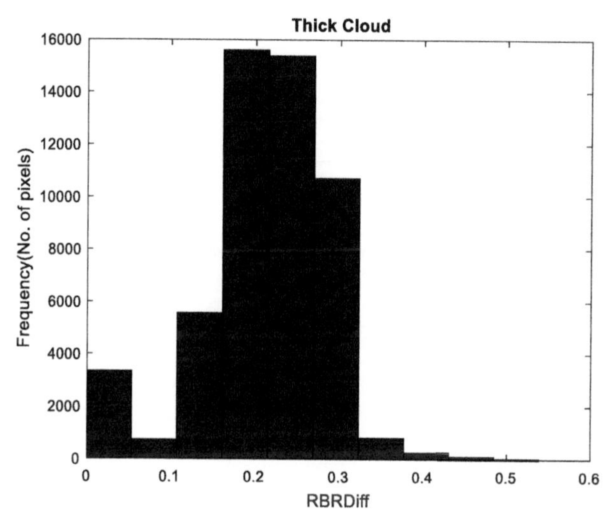

FIG. 7 Histogram of thick cloud R/B ratio difference.

From Fig. 6, it can be noticed that ΔRBR_{thin} falls mostly in between 0.1 and 0.2. By taking the mean of all training data, the threshold value for the thin cloud, $Threshold_{thin}$, is chosen to be 0.1575.

For ΔRBR_{thick}, Fig. 7 shows that most results falls in the range between 0.15 and 0.3. The threshold value for the thick cloud, $Threshold_{thick}$, is therefore 0.2115 after calculating the mean of all the training data items.

2.2.3 Cloud pixels classification

The pixels in the sky image will be classified into "Clear Sky," "Thin Cloud," or "ThickCloud" depending on the R/B ratio of that specific pixel and compared with the threshold value. After extracting the R/B ratio from a sky image, its R/B ratio will be compared with the R/B ratio of CSL and the difference will be computed. If the ΔRBR of that specific pixel exceeds $Threshold_{Thick}$, then this pixel will be classified as "Thick Cloud." If the ΔRBR falls between $Threshold_{Thick}$ and $Threshold_{Thin}$, then this pixel is said to be containing "Thin Cloud." If the ΔRBR is less than $Threshold_{Thin}$, then this pixel is categorized under "$Clear_{Sky}$." Fig. 8 shows the flow chart for classifying the image pixels into one of the three categories as mentioned.

To display the cloud pixels classification results, the pixels containing "Clear Sky" will be displayed as blue color, "Thin Cloud" will be displayed as red in color, while "Thick Cloud" is colored in purple. The pixels in green color are the pixels excluded from analysis as mentioned before. An example of the cloud determination result is shown in Fig. 9.

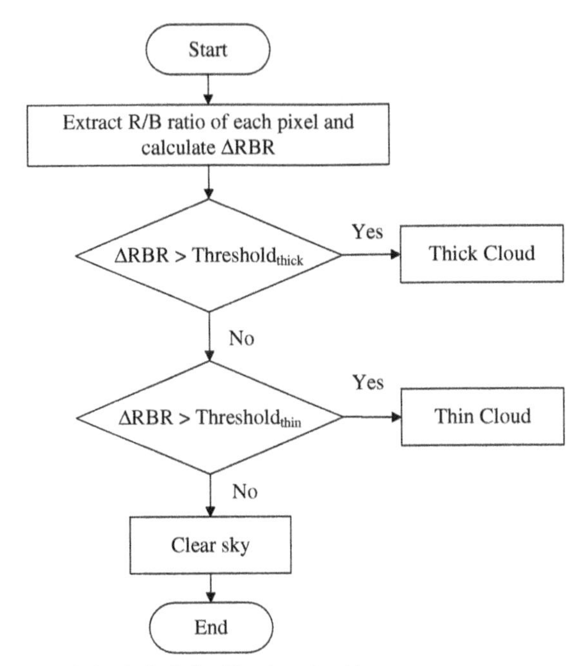

FIG. 8 Flow chart of cloud pixel classification algorithm.

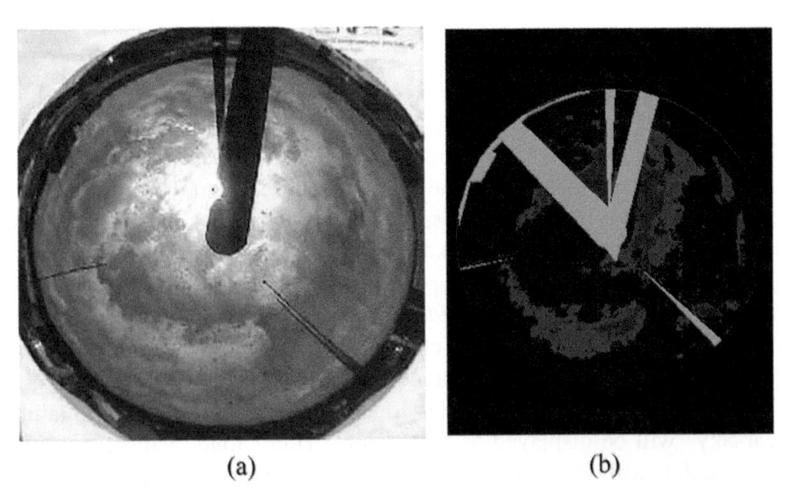

FIG. 9 Cloud decision image based on cloud pixels classification. (A) Original image, (B) cloud decision image.

2.3 Cloud movement determination

Two consecutive sky images with 1-min interval are fed into the optical flow algorithm to estimate the cloud movement. These two images are compared and the cloud movement has been estimated.

2.3.1 Application of optical flow algorithm

To display the cloud movement, the total pixels in each sky image is divided into small "boxes" of 6×6 pixels for analysis. The original 288×288 pixels are broken down into 48×48 small "boxes" to represent the movement. The optical flow algorithm [12] will display the cloud movement by drawing red arrows in the direction of movement from the one of the "boxes" in the previous image to the corresponding "boxes" in the current image. The magnitude of the red arrows means the velocity of the movement. The higher the cloud movement velocity of a certain pixel, the longer is the arrow drawing out from that pixel. Fig. 10 shows the results of cloud movement.

2.3.2 Compute representative velocity vector

After getting the velocity vectors from the optical flow algorithm, the average speed and direction of the cloud movement has to be determined, in order to estimate the cloud location of the next time step. A representative velocity vector will be chosen to represent the overall cloud movement with the

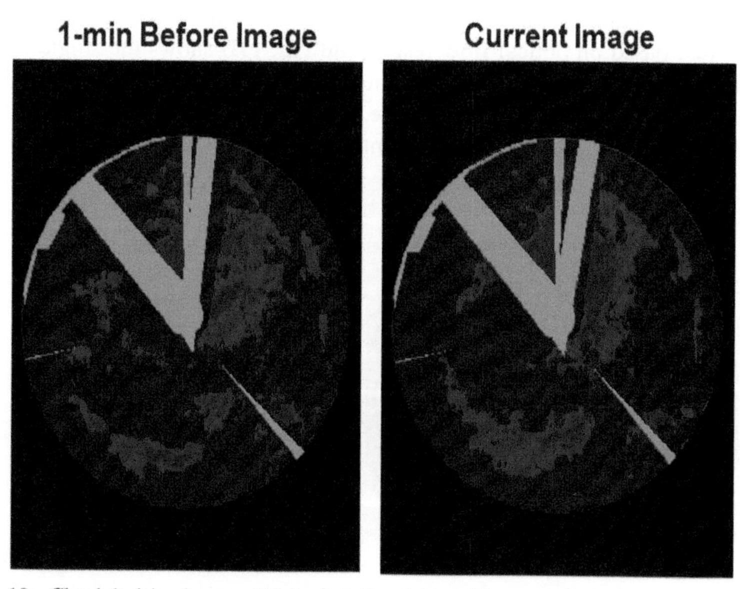

FIG. 10 Cloud decision images. (A) 1-min before image, (B) current image.

assumption that all the clouds will follow the pattern of the representative vector and move in a single direction with the average speed. Small velocity vectors with average speed less than "1 pixel" will be removed as these movements are treated as "noise" which will affect the accuracy of the result. The optical flow algorithm will produce movement vectors in terms of vertical movement (v_i) and horizontal movement (u_i). Therefore, by checking v_i and u_i values, we can determine the average movement of each velocity vector. If both v_i and u_i values are less than one, it means the average speed is less than 1 and will be categorized as "noise."

After removing the "noise" vectors, the rest of the vectors are averaged out to get the mean speed and direction. The speed and direction of the velocity vectors will be calculated by using the following equations:

$$V = \sqrt{u_i^2 + v_i^2} \tag{1}$$

$$\theta = \tan^{-1}\left(\frac{v_i}{u_i}\right) \tag{2}$$

The speed and direction of all the "nonnoise" vectors will be averaged to represent the speed and direction of the representative velocity vector. All the clouds in the sky image are assumed to move at the direction with the mean speed per minute.

Fig. 11 shows an example of the representative velocity vector computed based on the optical flow results. Fig. 11A shows all the velocity vectors that are plotted based on optical flow results and Fig. 11B shows the single representative velocity vectors taking the mean of all the vectors with an average speed more than 1 as described before. The flow chart in Fig. 12 describes the steps taken to compute the representative velocity vector.

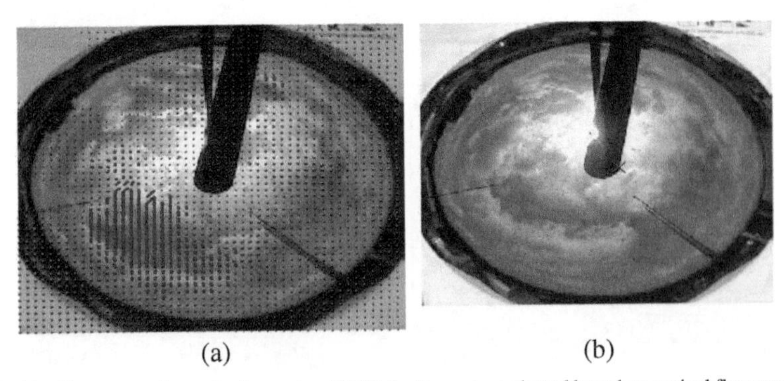

FIG. 11 Representative velocity vector. (A) Velocity vectors plotted based on optical flow results. (B) Single representative velocity vectors.

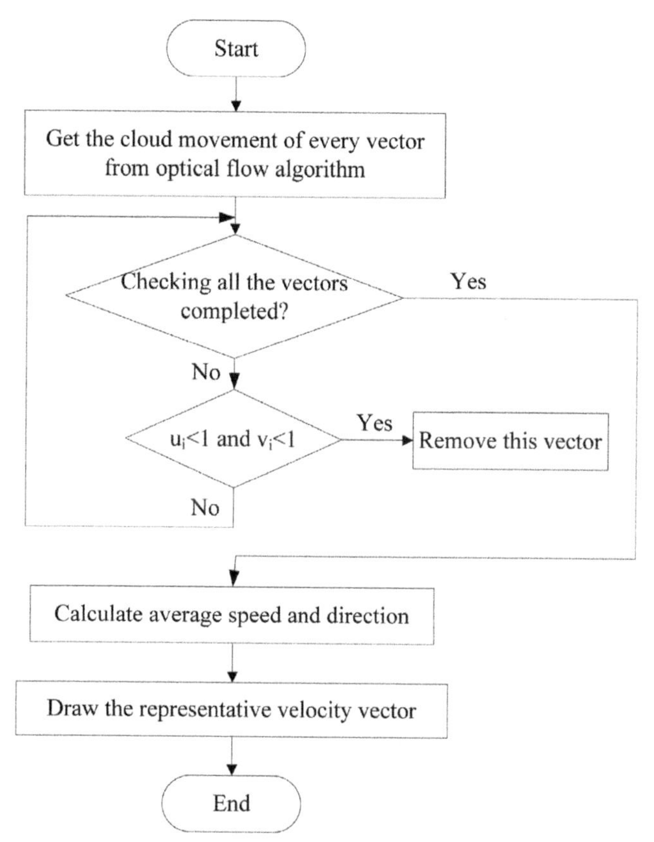

FIG. 12 Flow chart of computing representative velocity vector.

2.4 Forecast solar irradiance

A clear sky GHI, that is, GHIcsl, value had to be calculated according to the geographical location, time and date of the year. The GHIcsl value will then be multiplied by a clear sky index, Kt, based on the cloud condition. The value of Kt will be determined based on whether the pixel that covers the sun is consisted of a thin cloud, thick cloud, or clear sky. After computing the Kt values for all the three categories, the solar irradiance can be forecasted based on the movement of the cloud by determining how the cloud condition will be, at the center of the sun by the time of prediction.

2.4.1 Clear sky model

To find out the Kt value for all types of cloud, a reference clear sky model had to be constructed. In this chapter, the Ineichen and Perez clear sky model [4,13] has been used. The GHI_{csl} has been calculated from time to time based on the following equation:

$$GHI_{csl} = a_1 I_0 \cos\theta_z e^{[-a_2 a_m (f_{h1} + f_{h2}(T_L - 1))]} e^{0.01 a_m^{1.8}} \quad (3)$$

$$a_1 = 5.09 \times 10^{-5} \times Altitude + 0.868 \quad (4)$$

$$a_2 = 3.92 \times 10^{-5} \times Altitude + 0.0387 \quad (5)$$

$$f_{h1} = e^{-(Altitude/8000)} \quad (6)$$

$$f_{h2} = e^{-(Altitude/1250)} \quad (7)$$

where θ_z: Solar Zenith angle and I_0: extra-terrestrial normal incident irradiance.

On the other hand, optical air mass [14] will be calculated based on the formula given as follows:

a_m: Optical airmass

$$a_m = \frac{1}{\cos\theta_z + 0.50572(96.07995 - \theta_z)^{-1.6364}} \quad (8)$$

T_L: Linke turbidity co-efficient

$$I_0 = I_{sc}\left[1 + 0.033\cos\left(\frac{360n}{365}\right)\right] \quad (9)$$

n = day number.

(e.g., Jan 1 = 1, Dec. 31 = 365)

I_{sc}: 1.367 kW/m^2 (Solar constant).

The Linke turbidity factor, T_L will be obtained from solar radiation data (SoDa) database by inputting the optical air mass value as well as latitude of the TSI [15]. Based on the formulas stated, the GHIcsl will be calculated accordingly during each time step. This value will serve as the reference value for the solar irradiance depending on the cloud type.

2.4.2 Computation of clear sky index, Kt

The clear sky index, Kt, for different cloud conditions has been computed based on the historical irradiance value. To construct a lookup table for Kt value, a few days images from August and September 2015 are chosen to determine the Kt value.

Since the algorithm applied in this chapter is to determine the presence of cloud in between the sun position and the ground, the Kt value has also been computed in a similar pattern. However, since the center of the sun is always covered by the shadow band of TSI, it is unable to determine the cloud condition of that pixel at the instance. Therefore, the average cloud movement will be estimated and extrapolated for 5 min. The position of the clouds 3 min ago will determine whether the sun will be covered by "Thin Cloud," "Thick Cloud,"

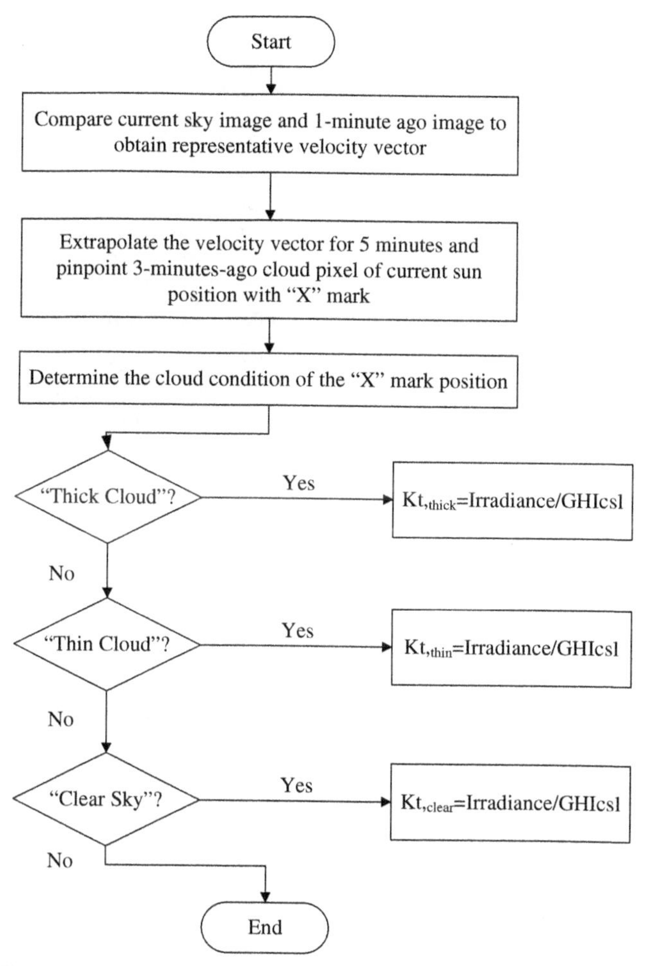

FIG. 13 Flow chart for computing Kt.

or "Clear Sky" at that instance. Fig. 13 shows the flow chart for computing the Kt lookup table based on the historical irradiance data.

An example of the extrapolated cloud movement and the corresponding pixel 3 min ago has been marked with a "X" spot as shown in Fig. 14. Based on the 3-min-ago sky image on the right, the cloud condition of "X"-marked pixel will be determined to indicate whether the cloud will fall on the sun position at the current time. In this example, the "X" mark falls on a thin cloud pixel; therefore, the sun is predicted to be covered by the thin cloud at the current time and the historical irradiance value will be used to compute $Kt_{,thin}$.

The same procedures are repeated for several training data to collect a lookup table of Kt values for different class of cloud condition. Fig. 15 shows

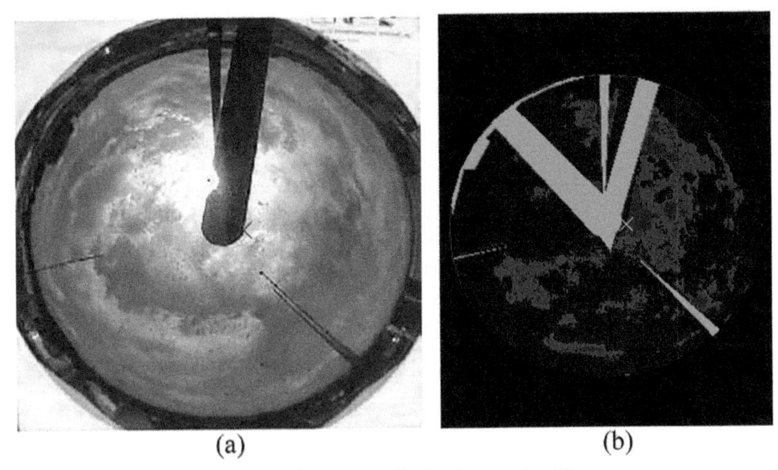

FIG. 14 (A) Extrapolation of velocity vector, (B) 3-min-ago cloud image.

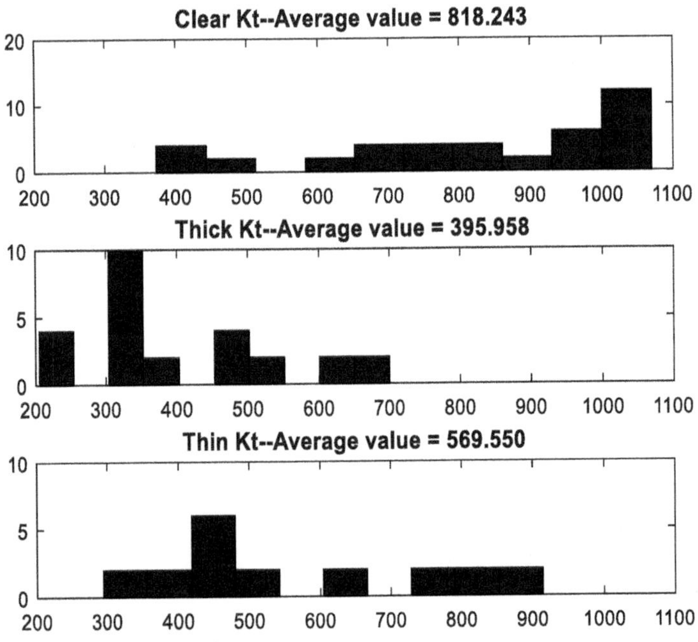

FIG. 15 Computation of Kt values.

the histogram of Kt values for "Clear Sky," "Thick Cloud," and "Thin Cloud" cases. The Kt values for each case will be averaged out to get a mean value for $Kt_{,thick}$, $Kt_{,thin}$, and $Kt_{,clear}$. Based on the results, the value of $Kt_{,thick}$ will be 395.958, $Kt_{,thin}$ is 569.550, and $Kt_{,clear}$ is 818.243.

2.4.3 *Estimate solar irradiance for forecast horizon of 2 min*

Similar algorithms mentioned in Section 2.4.2 have been used for forecasting future solar irradiance. Instead of determining the cloud condition from 5-min-ago sky image, the 3-min-ago image will be used for estimating the solar irradiance of the next 2 min.

After determining the cloud condition, the solar irradiance is calculated by multiplying the clear sky index, Kt, with the GHIcsl as mentioned before. Fig. 16 shows the flow chart of the solar irradiance prediction process.

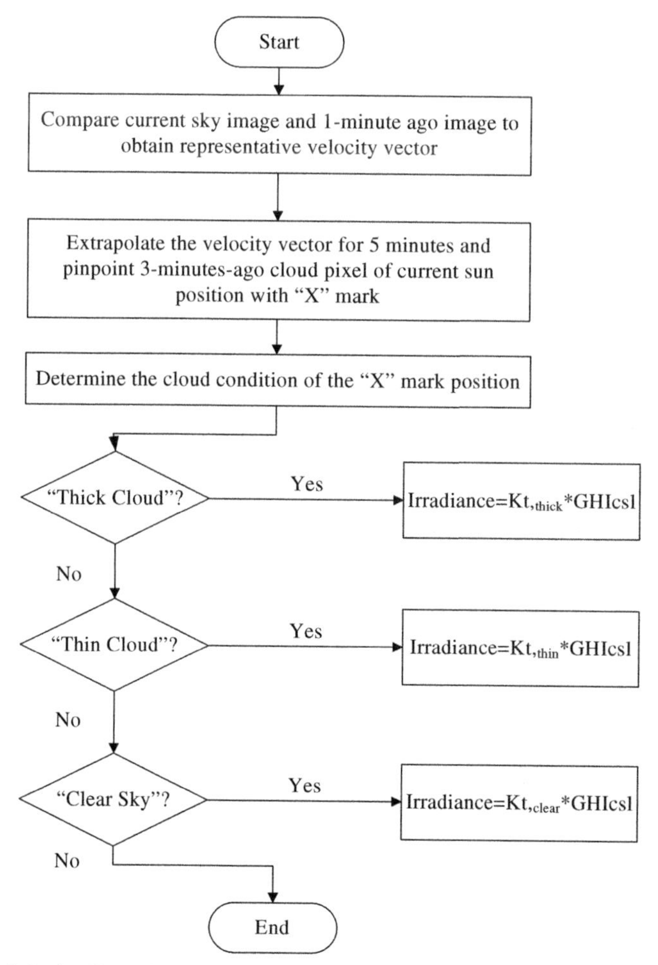

FIG. 16 Solar irradiance for forecast horizon of 2 min.

FIG. 17 Cloud image at 10.53 am.

2.5 Solar irradiance forecast

The "X" mark point will be pin-pointed in the 10:53 am sky image. Based on the cloud determination result at 10.53 am as shown in Fig. 17, it can be seen that the "X" mark falls in clear sky pixel. Therefore, it is predicted that by 10.58 am, the clear sky will move to the position of the sun. Therefore, the GHI_{csl} calculated will be multiplied by $\text{Kt}_{,\text{clear}}$. In this case, the solar irradiance predicted will be calculated as follows:

$$Irradiance_{predict} = GHI_{csl} \times \text{Kt}_{\text{clear}} = 0.9259 \times 818.243 = 757\ \text{W/m}^2$$

Based on the historical result, the solar irradiance sensed by solar pyranometer is 744 W/m^2. Therefore, the prediction error is calculated as follows:

$$Error = \frac{Irradiance_{predict} - \text{Irradiance}_{\text{real}}}{\text{Irradiance}_{\text{real}}} \times 100\% \tag{10}$$

$$Error = \frac{757 - 744}{744} \times 100\% = 1.74\% \tag{11}$$

As calculated, the prediction error is 1.74%.

To test the overall performance of the forecasting algorithm, the sky images are applied to forecast the solar irradiance for a total duration of 1 h on 2nd September 2015. Based on Fig. 18, it can be seen that the predicted solar irradiance is quite close to the real irradiance value recorded by the pyranometer with the mean error of 8.1546%. The reason of having wrong solar irradiance value as compared to real historical data even with the correct forecast result of the cloud condition is due to the Kt value used to calculate the solar irradiance.

Clearness index Kt plays a vital role in forecasting the value of solar irradiance for any forecast horizon. The Kt value has been updated using different signal processing and soft computing techniques to get a higher accuracy in

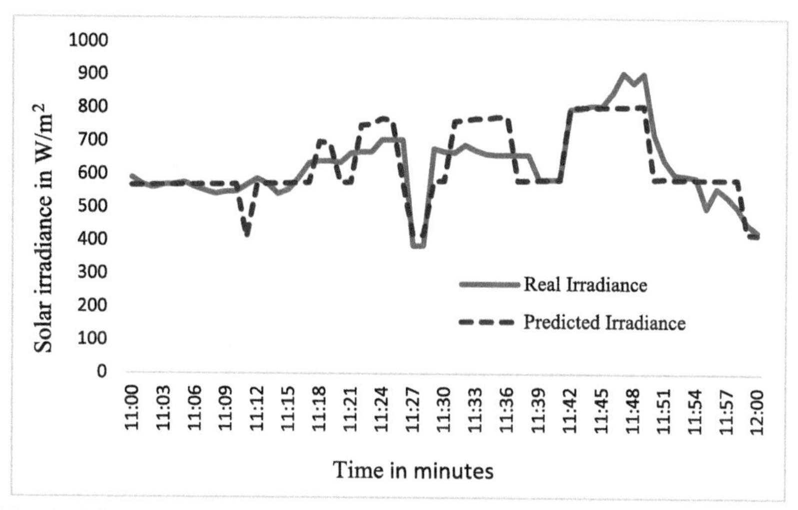

FIG. 18 Solar irradiance forecasting result for 1 h on 27th August 2015.

forecasting the solar irradiance. Different techniques of updating Kt values and forecasting of solar irradiance are discussed in the following sections.

3. Updating clearness index Kt using RLS algorithm

For getting accuracy in forecasting, Kt has been updated using Recursive Least Square (RLS) Algorithm. Solar irradiance forecasting results using RLS-updated Kt values and cloud condition Kt values (Forecast Horizons of 2 min) for a period of 1 h using sky images obtained from TSI has been compared.

The regressor form of the solar irradiance model can be represented as

$$f = H\theta \tag{12}$$

where f: Solar Irradiance, H: GHI_{Clear}, θ: Kt value of different cloud conditions (to be updated).

Estimate for the required parameter using RLS [8] algorithm becomes

$$\widehat{\theta}(k) = \widehat{\theta}(k-1) + K(k)\varepsilon(k) \tag{13}$$

Error in measurement can be found as

$$\varepsilon(k) = f(k) - H(k)^T\widehat{\theta}(k-1) \tag{14}$$

Gain K related with the covariance of the parameter vector is

$$K(k) = P(k-1)H(k)\left[\eta I + H(k)^T P(k-1)H(k)\right]^{-1} \tag{15}$$

$(0 < \eta < 1)$: Forgetting factor.

The updated covariance of the parameter vector using the matrix inversion lemma can be written as

$$P(k) = \left[I - K(k)H(k)^T\right]P(k-1)/\eta \tag{16}$$

The choice of initial covariance matrix is large. It is taken as $P = \alpha I$, where α is a large number and I is a square identity matrix.

The Kt values for different sky conditions obtained using RLS algorithm are given as follows:

Kt (Clear sky) = 820.8,
Kt (Thick cloud) = 430.8,
Kt (Thin cloud) = 576.8.

Using the above Kt values for different cloud conditions, the solar irradiance has been forecasted. Fig. 19 shows the comparison of solar irradiance forecasting (without and with RLS-updated Kt) with the real irradiance. It is found that the average prediction error without updating Kt becomes 7.562% and with RLS-updated Kt, it has been reduced to 6.565%.

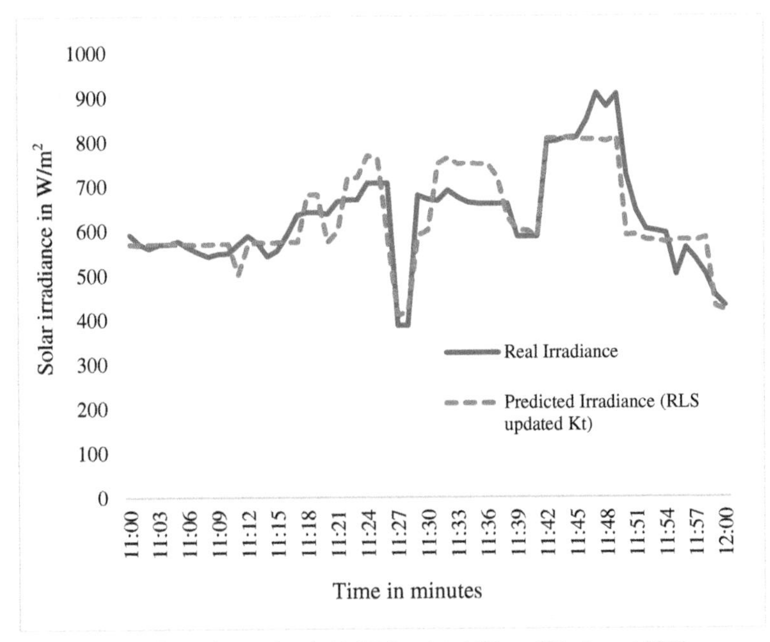

FIG. 19 Solar irradiance forecasting (with RLS updated Kt) on 27th August 2015.

4. Updating clearness index Kt using variable leaky LMS (VLLMS) algorithm

To achieve a higher accuracy in forecasting, a VLLMS algorithm has been used for updating the clearness index, Kt. VLLMS algorithm is a neural network based soft computing technique that adaptively updates the weight parameters to track the changes occurring in a system.

The regressor form of the solar irradiance model can be written as

$$f = H\theta \tag{17}$$

The estimate for the required parameter using VLLMS algorithm becomes

$$\theta(k+1) = (1 - 2\mu(k)\gamma(k))\theta(k) + 2\mu(k)e(k)f(\hat{k}) \tag{18}$$

Error in measurement

$$e(k) = f(k) - H(k)^T \theta(\hat{k} - 1) \tag{19}$$

$$\mu(k+1) = \lambda\mu(k) + \gamma(k)R(k)^2 \tag{20}$$

$$R_k = \beta R_{k-1} + (1-\beta)e_k e_{k-1} \tag{21}$$

β: an exponential weighting parameter $0<\beta<1$, $\lambda(0<\lambda<1)\beta$,
R_k represents the autocorrelation.
$\gamma>0$ control the convergence time.

The variable leakage factor

$$\gamma(k+1) = \gamma(k) - 2\mu(k)\rho e(k)y(\hat{k})X(k-1) \tag{22}$$

VLLMS parameters are taken as:
$\rho=0.9$, $\lambda=0.97$, $\beta=0.1$, $\gamma=0.01$, $R(1)=0$, $\mu(1)=0.08$.
Using the VLLMS algorithm, the Kt values obtained are given as below:

Kt (Clear sky) = 826
Kt (Thick Cloud) = 435
Kt (Thin Cloud) = 591

Taking the above Kt values, solar irradiance has been forecasted and compared with real irradiance. From Fig. 20, it has been found that the predicted irradiance using VLLMS-updated Kt value is closely approaching the real irradiance value and the mean absolute percentage error (MAPE) of this forecasting has been reduced to 3.537%.

Fig. 21 shows a comparative assessment of MAPE prediction, without updating Kt, using RLS-updated Kt, and VLLMS-updated Kt values. It has been found that MAPE, using VLLMS-updated Kt, is minimum, that is, 3.537%. So, it can be concluded that solar irradiance forecasting using VLLMS-updated Kt values outperforms those using RLS-updated Kt and cloud condition Kt values.

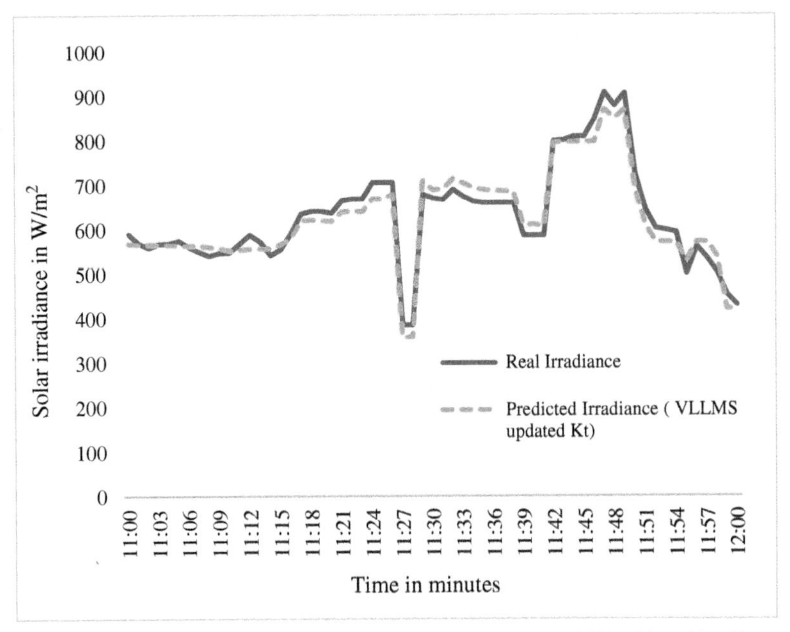

FIG. 20 Comparison of solar irradiance forecasting (VLLMS-updated Kt) with real irradiance on 27th August 2015.

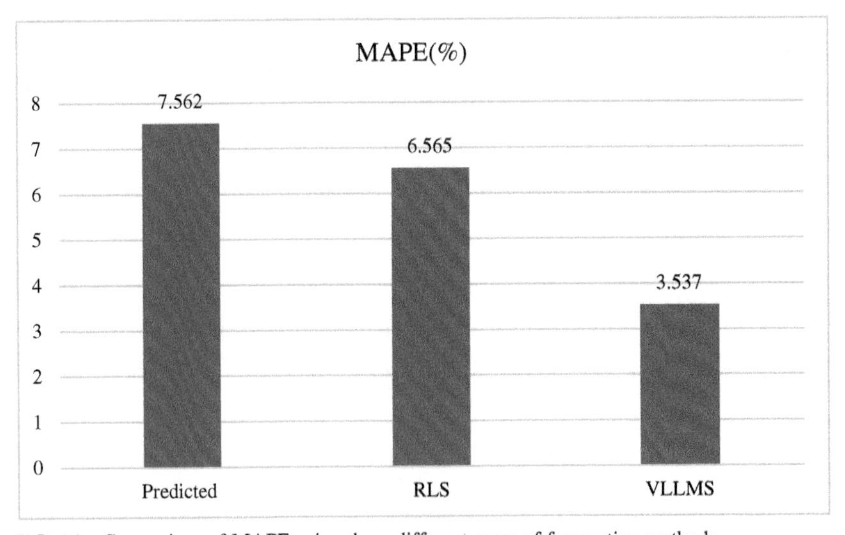

FIG. 21 Comparison of MAPE using three different cases of forecasting methods.

5. Conclusions

A solar irradiance forecasting process has been carried out by determining the cloud cover, analyzing the cloud motion, and finally calculating the solar irradiance, based on the predicted cloud condition at the position of the sun. It has been found that using VLLMS-updated Kt values, predicted irradiance is very much close to real irradiance with minimum MAPE and MSE in forecasting. Forecast Horizon can be changed depending on the need of users. Even though the forecast is based on the sky images captured by TSI, the same analysis can be extended based on the images captured by other equipment such as WAHRSIS (Wide Angle High-Resolution Sky Imaging System). Successful applications of short-term solar irradiance forecasting based on sky images are able to predict the change in solar irradiance and prepare the system for the ramp-down events. By switching on and off the backup energy storage beforehand, depending on the forecasting results, the solar intermittencies issue can be solved and impact on the system caused by delayed response of the systems due to weather changes can be avoided.

References

[1] H. Yang, B. Kurtz, D. Nguyen, B. Urquhart, C.W. Chow, M. Ghonima, J. Kleissl, Solar irradiance forecasting using a ground-based sky imager developed at UC San Diego, Sol. Energy 103 (2014) 502–524.

[2] F.M. Savoy, J.C. Lemaitre, S. Dev, Y.H. Lee, S. Winkler, Cloud base height estimation using high-resolution whole sky imagers, in: IEEE International Conference on Geoscience and Remote Sensing Symposium (IGARSS), 2015, pp. 1622–1625.

[3] C. Paoli, C. Voyant, M. Muselli, M.-L. Nivet, Forecasting of pre-processed daily solar radiation time series using neural networks, Sol. Energy 84 (2010) 2146–2160.

[4] R. Perez, P. Ineichen, K. Moore, M. Kmiecik, C. Chain, R. George, F. Vignola, A new operational model for satellite-derived irradiances: description and validation, Sol. Energy 73 (2002) 307–317.

[5] R. Perez, S. Kivalov, J. Schlemmer, K. Hemker Jr., D. Renné, T.E. Hoff, Validation of short and medium term operational solar radiation forecasts in the US, Sol. Energy 84 (2010) 2161–2172.

[6] R. Marquezand, C. Coimbra, Intra-hour DNI forecasting based on cloud tracking image analysis, Sol. Energy 91 (Oct. 2012) 327–336.

[7] C.W. Chow, B. Urquhart, M. Lave, A. Dominguez, J. Kleissl, J. Shields, B. Washom, Intra-hour forecasting with a total sky imager at the UC San Diego solar energy testbed, Sol. Energy 85 (2011) 2881–2893.

[8] P.K. Ray, B. Subudhi, BFO optimized RLS algorithm for power system harmonics estimation, Appl. Soft Comput., Elsevier 12 (8) (2012) 1965–1977.

[9] A. Bag, B. Subudhi, P.K. Ray, An adaptive variable leaky least mean square control scheme for grid integration of a PV system, IEEE Trans. Sustain. Energy, vol. 11, no. 3, pp. 1508–1515, July 2020.

[10] B. Subudhi, P.K. Ray, S. Ghosh, Variable leaky least mean-square algorithm-based power system frequency estimation, IET Sci. Measure. Technol., vol. 6, no. 4, pp. 288–297, July 2012.

[11] Automatic Total Sky Imager Model TSI-880. Yankee Environmental Systems, Inc., 2004.
[12] D. Sun, S. Roth, M.J. Black, Secrets of optical flow estimation and their principles, in: IEEE Conference on Computer Vision and Pattern Recognition (CVPR), 2010, pp. 2432–2439.
[13] P. Ineichen, R. Perez, A new air mass independent formulation for the Linke turbidity coefficient, Sol. Energy 73 (2002) 151–157.
[14] F. Kasten, A.T. Young, Revised optical air mass tables and approximation formula, Appl. Optics 28 (1989) 4735–4738.
[15] SoDa, Solar Energy Services for Professionals. Solar Radiation Data (SoDa), Available: http: // www.soda-is.com/eng/services/services_radiation_free_eng.php.

Chapter 4

Design and experimental validation of robust PID control for a power converter in a DC microgrid application

R. Jeyasenthil[a], Tarakanath Kobaku[b], Bidyadhar Subudhi[b], Subham Sahoo[c], and Tomislav Dragicevic[d]

[a]*Department of Electrical Engineering, NIT Warangal, Warangal, India,* [b]*School of Electrical Sciences, Indian Institute of Technology Goa, Farmagudi, Ponda, Goa, India,* [c]*Department of Energy Technology, Aalborg University, Copenhagen, Denmark,* [d]*Department of Electrical Engineering, Technical University of Denmark, Kongens Lyngby, Denmark*

1. Introduction

In the recent years, the investigations and interest on DC microgrids are increasing as they serve as interface between various renewable energy sources and dc loads, with high efficiency and high reliability [1]. The major component in a DC microgrid is the DC-DC power electronic converter. Among the converter topologies, the boost-type DC-DC converter plays key role in DC microgrid when a source supply is of low voltage, such as PV and when battery is interfaced with the load (Bi-directional DC-DC converter) [2,3]. The end performance of boost DC-DC power converter in DC microgrid application is based on the control scheme used in the closed-loop. The purpose of DC-DC boost converter is to give constant voltage across the load during the normal operation and also with the load variations and input voltage. Indeed, the controllers design for such power converters are based on the single plant dynamics despite the uncertainty (parametric/nonparametric) in the model due to the tolerance/temperature variations in switches, inductor and capacitor values, or linearization of the plant model. Hence, the controller employed for boost DC-DC converter needs to be insensitive against both system uncertainties and the disturbances (external). So, the closed-loop system needs to be insensitive towards rejecting the disturbance and also for system parameter variations.

Microgrid Cyberphysical Systems. https://doi.org/10.1016/B978-0-323-99910-6.00006-2

The conventional approach uses current-mode control (CMC) approach for controlling the load voltage of DC-DC converters. CMC is a two-loop control technique in which outer-loop controls the voltage and an inner-loop is for controlling the current. For any fixed frequency operation, the inner current loop goes into sub-harmonic instability mode for (steady state) duty ratio greater than 0.5, irrespective of the status of the outer voltage loop and the DC-DC converter topology [4]. The DC-DC converter load voltage can be controlled using single loop voltage mode control (VMC) approach, which does not pose a limitation on the duty ratio as in CMC. When VMC is adopted for controlling power converters, such as boost-type DC-DC converters, it exhibits poor dynamic response because of the right-half-plane (RHP) zero in its perturbed linear model. The poor transient response is mainly because of the closed-loop bandwidth limitation arising from RHP zero [4–7]. Achieving the performance near to the RHP zero boundary is an interesting control problem under VMC.

In the literature, several robust control methods exist, such as sliding mode control (SMC) and internal model control (IMC), which were used to regulate the voltage across the load terminal of boost-type converters [6,8]. The popularly employed SMC uses extra current and voltage sensors to maintain the desired load voltage of boost-type converters [6,9]. Usage of more sensors makes the SMC scheme to be not a cost-effective solution and moreover, fault in any additional sensor may either degrade the overall performance or causes the closed-loop unstable. SMC using the stable system center method has been presented in Ref. [10] for tracking the reference output voltage of NMP dynamical systems, such as boost- and buck-boost-type converters. However, this work was validated only through simulations. In addition, the SMC produces the high frequency chattering phenomena which is undesirable for power electronic converters. In Ref. [11], the H-infinity control scheme has been implemented as a reduced order controller on NMP boost-type converter. To reduce the long prediction horizons in model-predictive control was formulated [12] and implemented as a current-mode controller which requires the additional sensors for measuring inductor current and input voltage. IMC is a model based robust control scheme and was used to a voltage model controller for an NMP boost converter [8]. The IMC structure has a model in parallel to the plant and provides explicit information of disturbance estimate under no plant-model scenario. In Ref. [8], the IMC is applied as a 2DOF control scheme that has a disturbance filter and improves the overall performance while rejecting the external disturbances. However, the implementation of IMC requires the computation of model, controller, and disturbance filter, thereby significantly increases the computational burden. Also, the hybrid control [7] and feedforward control [13] uses additional voltage and current sensors for their implementation. All the above-mentioned challenges motivated the authors to employ a controller that is free of chattering, computationally easy, and requires only the sensing of the load voltage for feedback.

In this work, a well-known quantitative feedback theory (QFT)-based robust controller is synthesized due to its success in many practical applications [14]. The robust control methods such as the QFT and H_∞ control are a complement to each other but the difference is in the uncertainty of representation [15]. The inherent conservatism in H_∞ control algorithm leads to a higher order controller (which makes it difficult in experimental validation/requires reducing the controller order), whereas the QFT design produces the controller of low order which uses both open loop phase and gain in the loop-shaping. The feedback cost related to the controller bandwidth is directly controlled in the QFT design with the help of controller gain reduction at high frequency. This is an important practical perspective as it minimizes the sensor noise [16]. The attribute of this is due to the shaping of the nominal loop transmission function in QFT as opposed to the nominal sensitivity function in H_∞ control, which is more onerous to the sensor noise problem. This makes the QFT design different from the other robust control methods and more towards the practical approach.

QFT has been applied to power electronic dc-dc converters [17–21]. These reported works transform the NMP system into a minimum phase integrated with the all-pass filter and then designed the controller for an equivalent system. Such design was validated only on simulations and did not consider the output impedance and the audio susceptibility disturbance models in the controller design which could help in improving the disturbance rejection.

There are various control schemes that appear to perform satisfactorily in the regulation of load voltage of dc-dc converters. But the existing schemes lack the systematic approach for the design and the ability in handling internal disturbances, such as parametric variations, and external disturbances, such as load current variations, voltage at input, and computationally simple. This challenging issue needs to be addressed thoroughly. For various applications in industries, PID is popularly employed as it possess clear functionality, applicability, and computationally simple for realization in hardware [22]. Due to these additional features, PID is highly successful and making biggest impact on the industrial technology [23]. The tuning of P, I, D parameters for achieving satisfactory performance is a difficult task. Due to this, almost 75% of the PID controllers in total were poorly tuned in various industries as pointed in Ref. [24]. The renowned PID tuning rules such as Ziegler and Nichols method, CHR method, and Cohen-Coon methods assume that plant is a linear time invariant (LTI) model and does not consider the challenges arising from neglecting the external disturbances and the uncertainty. In general, the PID controller is employed to maintain the constant load voltage in dc-dc converter applications [8,25–28]. In these important applications, the tuning of P, I, D parameters is a tough task to obtain better steady state and transient responses on the occurrence of sudden variations in the source voltage, the filtering components (L, C), and load current. To enhance the closed-loop performance of a NMP boost dc-dc converter, the P-I-D parameters were tuned using phase margin (PM), obtained from extra phase sensitive multi-meter device [27]. In case of conventional PID controller, the design procedure does not take the RHP zero into account and ad-hoc tuning is adopted for controlling the load voltage of DC-DC converter with NMP behavior. Further, the conventional PID design does not account for the parameter uncertainty and external disturbances of the boost-type converter using only load voltage as feedback.

For external disturbances, the performance of DC-DC converters while rejecting the disturbance can be enhanced with the dynamics of output disturbance such as audio susceptibility and output impedance in the controller design as carried in Ref. [9]. The improved disturbance rejection performance can also be accomplished with a computationally simple PID control structure using the load voltage as the only feedback information. In this direction, the QFT control design method systematically takes into account the disturbance models and uncertainty information and incorporates into PID. The challenge becomes profound when such a PID is used in voltage control mode, as it is a single loop controller unlike CMC that uses both inductor current and load voltage information. The proposed robust PID controller in VMC framework is validated in simulations and further on hardware prototype of boost-type DC-DC converter with dSPACE1103 controller board. It is noticed that the proposed QFT-based PID controller which is computationally simple offers satisfactory dynamic response in both the simulations and on the hardware experimental test-bed. The key contribution of the proposed work is to design and experimentally validate the QFT-based robust controller for the NMP boost converter with the disturbance dynamics which is not reported in the literature. The proposed methodology is limited by the manual loop-shaping involved for controller design which may be tedious, time consuming, and need experts to carry it out because of its graphical nature. The proposed QFT controller design is successfully carried out to overcome the nonminimum phase characteristics of the DC-DC boost-type converter.

2. QFT design of external disturbance minimization problem

Among many robust control design methods, the QFT is a renowned two degree of freedom (2-DOF) method due to its suitability for the practical systems and its design is based on the frequency domain [14,15]. The main steps of QFT are template generation, bound generation, design of controller, and pre-filter design. We will discuss each step in detail in Section 3. The key step of QFT design is to synthesize a controller using the graphical loop-shaping method and minimizing the high frequency controller gain (cost of feedback). Basically, the QFT bounds capture the conversion of the robust stability and performance specifications into a bound on the open loop transmission function. The QFT produces the low order controller structure using the nominal loop-shaping method in the Nichols chart (NC) which captures both the phase and gain information. Fairly complex uncertain systems such as unstable, largely uncertain, and nonminimum phase system can also be treated in QFT.

The plant is considered as LTI model and is represented by $P(s) \in \{\mathcal{P}(s,\alpha) : \alpha \in \beta\}$. To accommodate the variation in the LTI model parameter values over a box β,it is chosen as follows:

$$\beta = \left\{\alpha \in R^l : \alpha_i \in \left[\underline{\alpha_i}, \overline{\alpha_i}\right], \underline{\alpha_i} \le \overline{\alpha_i}, i = 1., \ldots, l\right\}$$

The loop transmission function $L(s,\ \alpha)$ is represented by $G(s)P(s,\ \alpha)$. The goal is to synthesize the feedback controller $G(s)$ such that the design meets the disturbance rejection and stability specifications for the entire uncertain plant set $\mathcal{P}$. To regulate the output voltage of boost-type DC-DC converter using the QFT method, following control-oriented specifications are considered:

1. Robust stability margin:

$$\left|\frac{P(j\omega)G(j\omega)}{1 + P(j\omega)G(j\omega)}\right| \le W_s \tag{1}$$

2. Robust external disturbance minimization:

$$\left|\frac{1}{1 + P(j\omega)G(j\omega)}\right| \le W_d(\omega); \omega \in \Omega_{pf} \tag{2}$$

The control specification given by Eq (1) indicates relation between the M-circle magnitude and the stability margin. Here, the specification for stability margin, W_s, is related to the desired gain and PM of an M-circle. The output disturbance rejection performance specification given in Eq (2) is denoted as W_d. The objective is to fulfill the desired specifications over an entire frequency range, Ω_{pf}, and the robust stability margin specification. The design procedure involved in QFT is given as follows:

For the given uncertain LTI plant $P(s) \in P$, and at each design frequency $\omega \in \Omega$, the phase and magnitude of the plant element in $P(s) \in P$ from the NC is calculated. This results in a region, at each frequency, in the NC known as the "template." Then, the performance and the stability specifications in Eqs. (1) and (2) are converted into the bounds in the NC (corresponding quadratic inequality [14]), at each design frequency, ω. Next, design feedback controller, $G(s)$, such that the nominal loop transfer function L_0, is shaped to satisfy the bounds, that is, on or above the disturbance rejection bound and lies outside the stability margin bound. Finally, validate the design in frequency domain and also in time domain.

2.1. QFT design for NMP system

Two approaches were available in QFT for handling the NMP uncertain system to facilitate the controller design [14–16]. One approach is to perform the shaping of nominal loop transmission function of NMP system directly. Another approach based on considering the bound generated using the minimum system component of the NMP system and scaling/shifting the bound (minimum phase) by the all-pass phase angle so as to carry out the loop shaping. For instance, linear model of an NMP boost-type power converter is decomposed as

$$P(s) = \frac{N(s)\widehat{N}(s)}{D(s)} \frac{\widehat{N}(-s)}{\widehat{N}(s)} = P_M(s)A(s)$$

where, $\widehat{N}(-s)$ denotes the parts with RHP zero and P_M and $A(s)$ are the minimum phase and all-pass factor of the system, respectively. The all-pass filter has the properties of $|A(j\omega)|=1$ with the arg$(A(j\omega))$, $\omega \in [0,\infty]$. In this work, perform the loop-shaping directly for NMP system is adopted with the information of dynamics of disturbance. Inclusion of the dynamics of disturbance in the synthesis stage helps the proposed method differ with the reported methods in Refs. [19–21]. In addition to the disturbance dynamics in the design, the QFT controller is designed to be more robust than without it [19–21].

The external disturbance minimization problem (2) with disturbance dynamics is:

$$\left| \frac{\beta(j\omega)}{1 + P(j\omega)G(j\omega)} \right| \leq \omega_d(\omega) \tag{3}$$

Here, the disturbance dynamics is denoted by β.

3. Application of the proposed QFT design to NMP boost-type DC-DC power electronic converter

This section uses the proposed QFT design methodology, discussed in Section 2, and presents the step-by-step procedure to synthesize a robust PID for a boost-type power electronic DC-DC converter. The circuit of boost-type DC-DC power electronic converter with a resistive load is shown in Fig. 1.

The DC-DC boost converter (CCM mode) values are: $R_L = 0.3\,\Omega$, $L = 3.1\,\text{mH}$, and $R_c = 0.08\,\Omega$, $C = 1930\,\mu\text{F}$. The nominal resistance value of load is (R_{nominal}) $90\,\Omega$ and the switching frequency of 25 kHz. The source and load voltages are $V_i = 10\,\text{V}$ and $V_o = 15\,\text{V}$, respectively. The linear model of a boost-type DC-DC converter relates the duty ratio control input to load voltage, obtained using the SSA method [4] and it is given as follows:

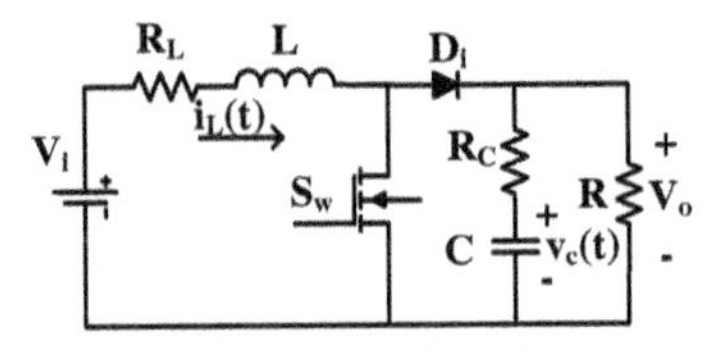

FIG. 1 Circuit diagram of the power stage of a DC-DC boost-type power electronic converter.

$$P_n(s) = \frac{22.0617\left(1.5440 \times 10^{-4}s + 1\right)\left(-7.8287 \times 10^{-5}s + 1\right)}{1.3345 \times 10^{-5}s^2 + 1.8847 \times 10^{-3}s + 1} \tag{4}$$

The nominal boost converter transfer function (4) contains the RHP zero at 1.2774×10^4 rad/s. By considering an uncertainty of 10% for each parameter, the linear uncertain model of the boost-type power converter becomes as,

$$P(s) = \frac{g(z_1 s + 1)(z_2 s + 1)}{(y_1 s^2 + y_2 s + 1)} \tag{5}$$

where, $g \in [19.85, 24.27]$, $z_1 \in [1.3896, 1.6984]e-4$, $z_2 \in [-7.04583, -8.61157]e-5$, $y_1 \in [1.20105, 1.46795]e-5$, and $y_2 \in [1.67, 2.073]e-3$.

The frequency response plot of the converter system (for 243 randomly generated plant models from the uncertain plant set (5)) shows a resonant characteristic around 274 rad/s as shown in Fig. 2, which required to be considered in the design frequency set. The magnitude response rolls-off with a slope of −40 dB/dec.

The feedback control structure for regulating the output voltage of DC-DC power electronic converter with output disturbance dynamics such as variations in input voltage ($v_i(s)$) and load current ($i_L(s)$) are shown in Fig. 3. Here, $G(s)$ represents the PID controller and $d^*(s)$ represents the plant input (duty ratio).

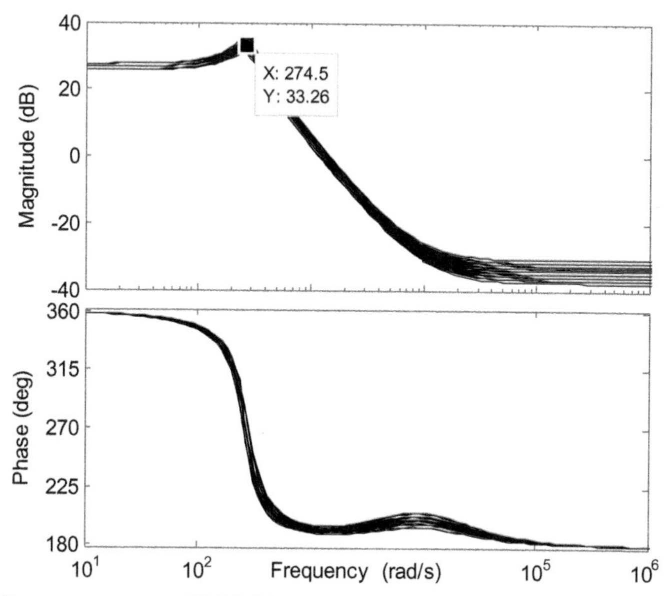

FIG. 2 Frequency response of DC-DC boost-type power electronic converter with chosen 10% uncertainty.

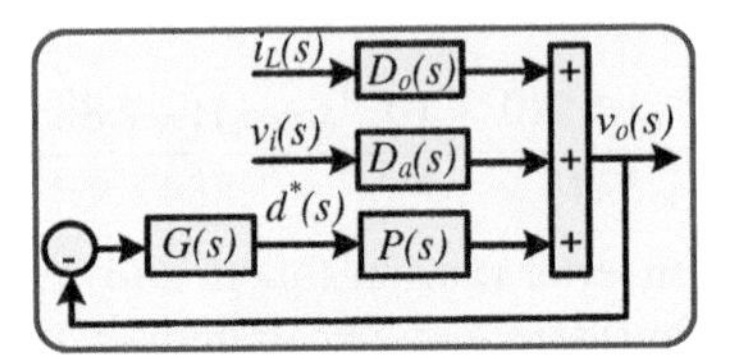

FIG. 3 Feedback control structure of DC-DC boost-type power electronic converter with output disturbance dynamics.

To design the PID controller with the principal of QFT, the following specifications are considered:

3.1. Desired specifications

3.1.1 *Robust external disturbance rejection problem*

The disturbances (external) affect the load voltage of boost converter are the variations in the input voltage and output load current. These are discussed later from the view point of QFT and DC-DC boost power converter.

(i) Input voltage disturbance minimization problem: The source voltage changes with respect to the load voltage are given by the following linear perturbed model [28]:

$$D_a(s) = \frac{v_O(s)}{v_i(s)} = \frac{1.4857(0.0001544s + 1)}{(1.3345 \times 10^{-5}s^2 + 0.0018847s + 1)}$$

The desired requirement is (when a unit step disturbance is applied, $|V_o| < 0.2\,\text{V}$ at time $>20\,\text{ms}$):

$$\left|\frac{D_a(j\omega)}{1 + L(j\omega)}\right| \le \omega_d(\omega) = \left|\frac{s}{s + 75}\right|_{s=j\omega} \tag{6}$$

(ii) Output load current disturbance rejection problem: The load current perturbation (linear) with respect to the load voltage [28] near the operating point is

$$D_o(s) = \frac{v_O(s)}{-i_L(s)} = \frac{-0.8567(0.0001544s + 1)(0.0080639s + 1)}{(1.3345 \times 10^{-5}s^2 + 0.0018847s + 1)}$$

For simplicity reason, the desired specification mentioned in Eq. (6) for the input voltage disturbance rejection problem is considered.

3.1.2 Robust stability margin

The stability margin is chosen as (PM $\geq$ 60 degree, Gain margin $\geq$ 5 dB)

$$\left|\frac{L(j\omega)}{1+L(j\omega)}\right| \leq 1.2 \tag{7}$$

3.2. Design frequency selection

QFT design starts with the selection of design frequency set (Ω). This set should include a maximum number of frequency points which cover both the low and high frequency (HF) specifications. The disturbance rejection and tracking problem correspond to the low frequency specification whereas the sensor noise minimization problem belongs to the HF specification. The frequency responses of the system (ref. Fig. 2) serve as a guide to choose the Ω set, which leads to the point wise design. Based on this, the selected frequency set for QFT design is as follows:

$$\Omega = 2\pi\,[1, 2.5, 7.5, 10, 20, 30, 50, 100, 200, 274, 350, 500, 1000, 2000, 5000, 12500]\ \text{Hz}$$

As we can observe that the frequency set is formed with the half of the switching frequency (12.5 kHz) as maximum frequency.

3.2.1 Step 1: Template generation

The template represents the region/area in the NC formed by the magnitude and phase of the system within the uncertainty range at each design frequency, $\omega \in \Omega$. Fig. 4 shows the templates at each chosen design frequency and as the frequency varies, the region covered in the NC varies (in terms of phase and gain). This representation is useful to understand the uncertainty present in the system at each frequency. At $\omega = 274$, the template region is widened (i.e., the angular and magnitude width) and becomes large as compared to the other frequencies due to the resonance. Next step is the generation of the performance and disturbance rejection bounds.

3.2.2 Step 2: Generation of the QFT performance and stability margin bounds

In this step, the desired closed loop design specifications defined in the inequalities (6-7) are converted into the open/closed curves on NC (also known as bounds) using the quadratic inequalities [14]. The quadratic inequalities are derived for each of the desired specifications by using the polar coordinates. The bounds usually put a restriction on the nominal L_0 at each design frequency $\omega \in \Omega$. The disturbance rejection specifications are transformed as open bounds on the NC. Figs. 5 to 6 captures the QFT open bounds for the external

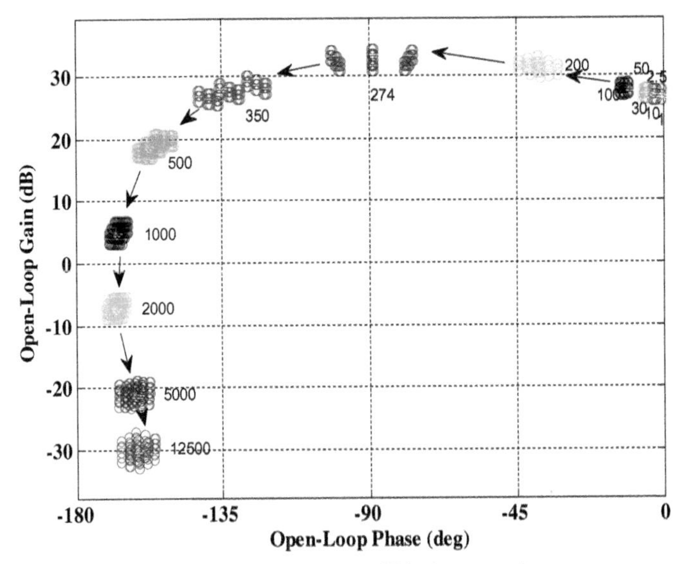

FIG. 4 Templates of the DC-DC boost converter within the uncertainty.

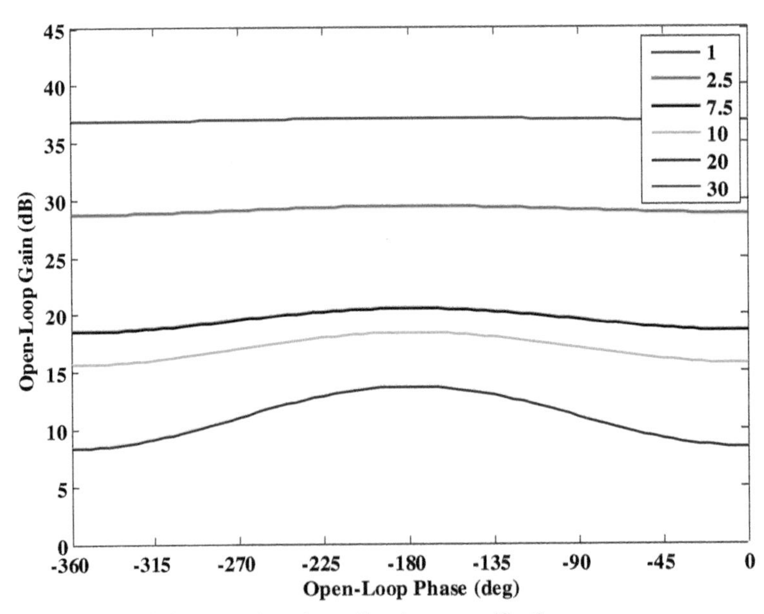

FIG. 5 QFT Bounds for output impedance disturbance specifications.

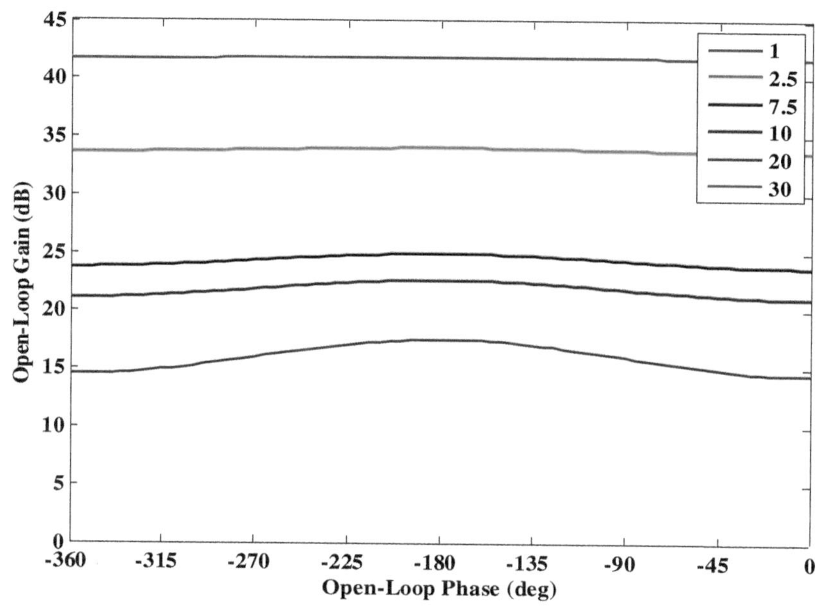

FIG. 6 QFT bounds for audio susceptibility disturbance specifications.

disturbance minimization requirement. The closed bounds (also known as high frequency barrel) are generated for the robust stability margin specification, illustrated in Fig. 7.

Once the bounds are generated at each ω, the worst-case bound (composite) is formulated by combining each of the individual bounds at each frequency. At the end of this step, the most dominating bounds at each ω are generated, and it makes sure that satisfying the worst-case bounds for the nominal system also works for any of the plant elements from the uncertain set. The bounds are drawn on NC with the MATLAB-QFT Toolbox [28]. Fig. 8 shows the worst case bound (composite) for robust disturbance and stability margin specification.

3.2.3 Step 3: Loop-shaping technique: Controller design

In QFT, the graphical method known as *loop-shaping* technique is used to synthesize the controller either manually or automatically. This method appends the gain, poles, and/or zeros (either complex and/or real) terms to the nominal plant so that L respects the composite QFT bounds (Step 2). The satisfaction of the QFT bounds implies that the nominal L is shaped so that it is placed above or exactly on the flat (open) bounds especially at low (performance) frequency as well as outside the closed barrel (stability margin) especially at high frequency.

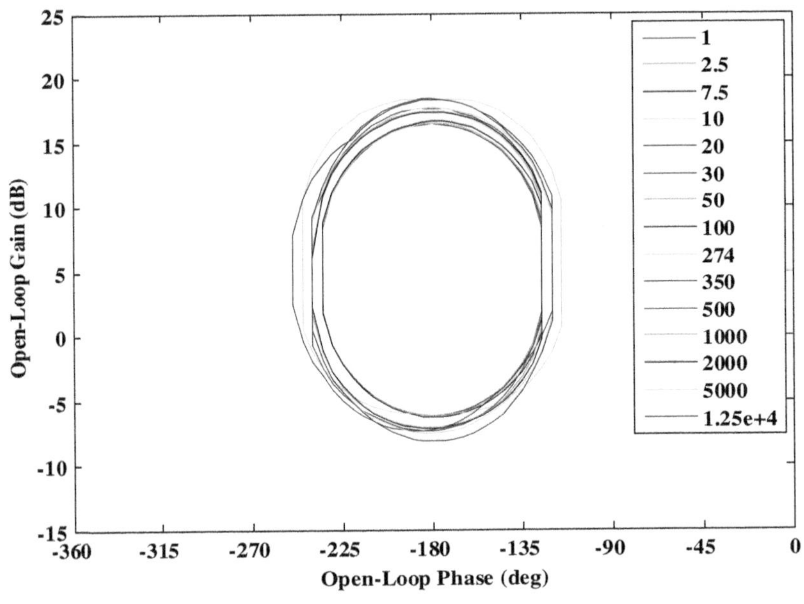

FIG. 7 Robust stability margin bounds.

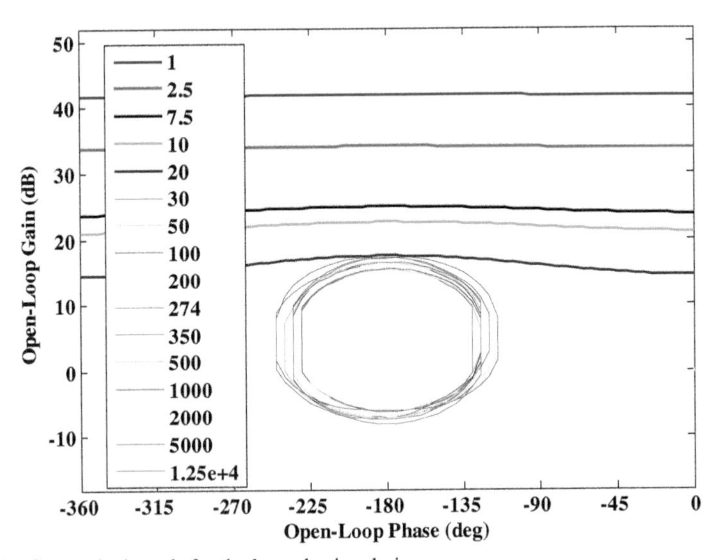

FIG. 8 Composite bounds for the loop-shaping design.

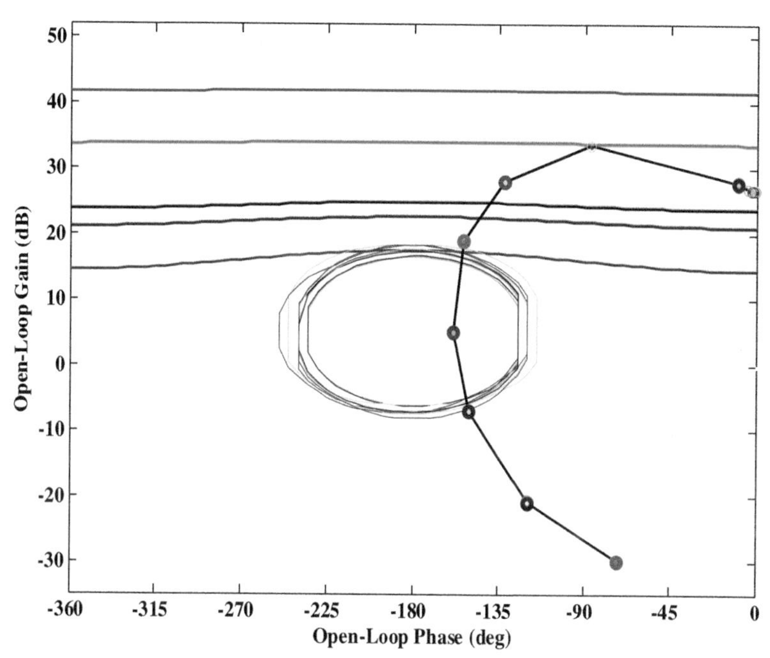

FIG. 9 Without loop shaping of the DC-DC boost converter.

Fig. 9 shows the nominal plant magnitude and phase prior to loop shaping stage. It is clear that the specifications at different frequencies are not met alone with the nominal plant and requires the controller to satisfy the requirement. The manual loop-shaping process results in the feedback controller with low order and proper (strictly) transfer function which in turn leads to the low complexity controller, from the practical implementation point of view. On the other hand, another renowned H_∞ robust control always results in the controller with higher order and often relies on the model order reduction methods to get the lower order controller. The nominal L shaping of the NMP boost converter is shown in Fig. 10. Here, loop-shaping is performed on the NMP system and the designed PID controller is

$$G(s) = \frac{6.14\left(\frac{s}{117.9} + 1\right)\left(\frac{s}{150} + 1\right)}{s\left(\frac{s}{2000} + 1\right)} \tag{8}$$

3.2.4 *Step 4: Design validation*

The final step in QFT design is to validate the designed controller (8) in the frequency domain as well as in the time domain. Fig. 11 shows that the maximum

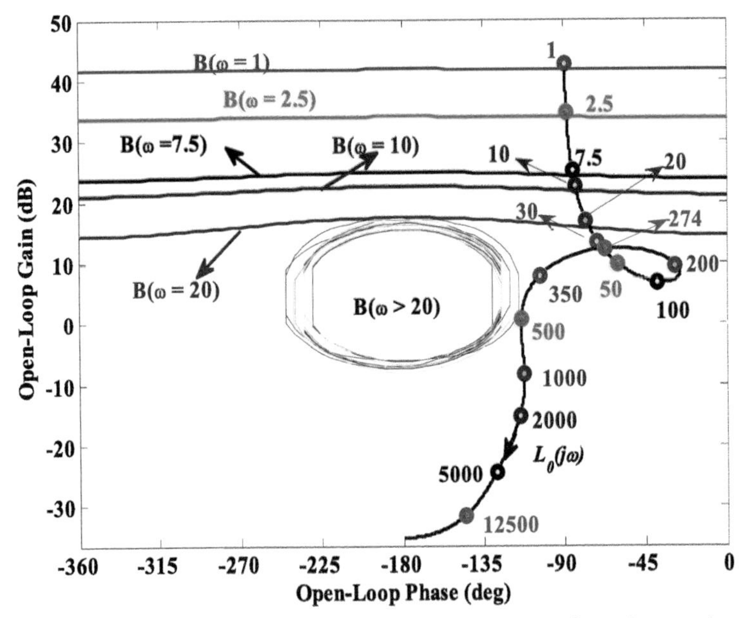

FIG. 10 Loop shaping of the DC-DC boost converter ($B(\omega)$ represents the performance bounds at design frequency ω).

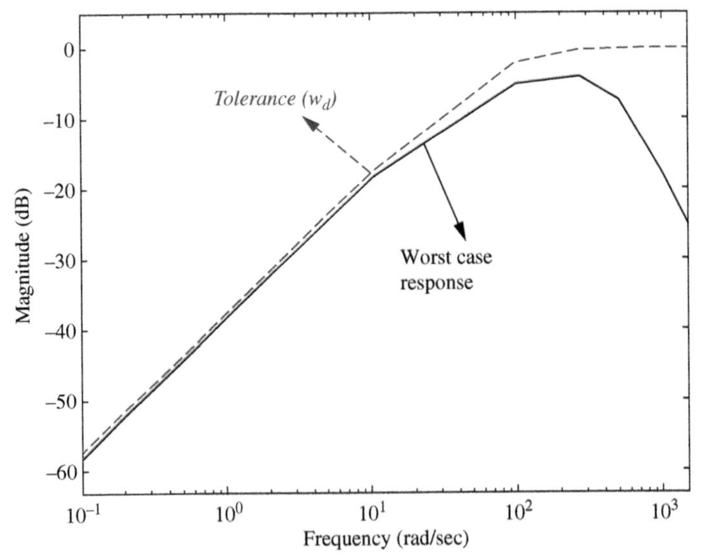

FIG. 11 Frequency domain validation (worst case) of the audio-susceptibility disturbance specification. The plots show the disturbance tolerances, ω_d (*dashed red line*) and the maximum magnitude of the closed-loop disturbance transfer function (*solid blue line*) over all the 243 plants.

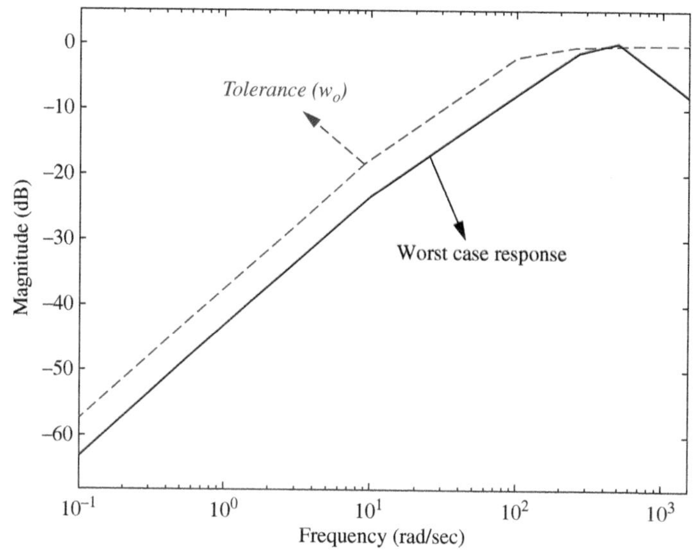

FIG. 12 Worst case validation of the output impedance disturbance specification. Here, the dashed red line denotes the tolerance (ω_o) and the solid blue line denotes the maximum magnitude of the closed-loop disturbance transfer function over all the 243 plants.

magnitude response of the closed-loop with the disturbance (audio susceptibility) at the output of the converter system lies within the tolerance limits ω_d given by Eq. (6).

Similarly, the response (maximum magnitude) of the closed-loop system with output impedance disturbance also satisfies the tolerance limits as shown in Fig. 12. The figure confirms that the designed PID controller, $G(s)$, indeed satisfies the desired specifications (6). Hereafter, the designed controller G is denoted as PID_{QFT}. The time domain validation is discussed in detail in the forthcoming section.

The complete QFT design process is given in Fig. 13. To start with, the problem formulation stage involves the DC-DC boost converter with uncertainty in its parameters, desired specifications, and discrete design frequency selection. Next, the plant templates/value sets are produced by calculating the phase and the magnitude of each plant element from the uncertainty set (plot it in the NC). Further in the bound generation step, translate the robust performance and robust stability specifications to obtain the bounds (closed/open curves), which keeps restriction on the feedback controller in the NC. Now, a robust controller $G(s)$ is designed using the manual loop shaping approach by adding the Gain, zeros, poles, and other elements (such as complex poles/zeros, integrator/derivate, etc.) to the nominal plant. The nominal open loop transmission function is shaped such that it sits on or above the open bounds (robust performance) and outside the barrel (stability bounds) generated from the previous step. Further, a

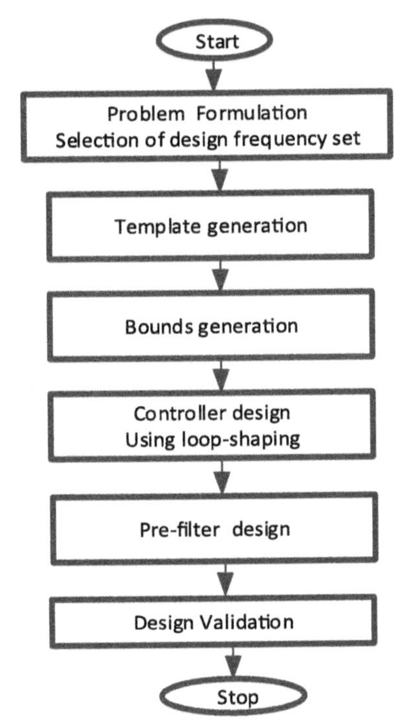

FIG. 13 Flow chart of general QFT design procedure.

prefilter $F(s)$ is designed such that the closed loop response of the uncertain system lies between the lower and upper tracking limits, respectively. The prefilter design step is required for the tracking specification and can be omitted for problems involving only the disturbance rejection specification such as the boost converter. Finally, validate the design in frequency domain as well as in time domain.

4. Simulation studies

Before proceeding with the experimental validation, the simulation studies are performed to validate the performance of the designed controllers on a boost-type DC-DC converter. This simulation studies section presents the robustness verification of the designed controllers for step variations in the input voltage and the load current. For a fair comparison with the designed QFT-based robust PID, a conventional PID controller is designed using the linear perturbed model of the boost converter [4]. The linear model around the operating point has a PM of 12 degree at a gain cross-over frequency of 1.33 krad/s and has an RHP zero. The PID controller reported in Ref. [12] is chosen for comparison and it offers a

PM of about 60 degree and a loop gain crossover frequency of 600 rad/s. The respective PID is given as follows:

$$K_p = 78.4 \times 10^{-3}, K_i = 3.34, K_d = 0.245 \times 10^{-3} \text{and } T_f = 0.811 \times 10^{-3} \quad (9)$$

For performance comparison in closed-loop operation, the simulations are carried using the linear and nonlinear models developed in MATLAB/Simulink. Following two scenarios are considered:

4.1. Linear simulation studies

These are carried using the linear perturbed model of the DC-DC boost-type power electronic converter obtained from the state-space averaging method [4].

Under the linear simulation studies, a regulatory case study related to a step change of 3 V in source voltage given at $t=0.02$ s is considered. The respective load voltage response along with the plant input is shown in Fig. 14. In the transient phase, the designed QFT-PID controller gives more controller energy as input to the plant (d^*) than the PID controller (conventional) and can be seen in Fig. 14B. As a result, the load voltage with the proposed PID controller design responds quickly and reaches back to the operating voltage of 15 V with insignificant undershoot, shown in Fig. 14A.

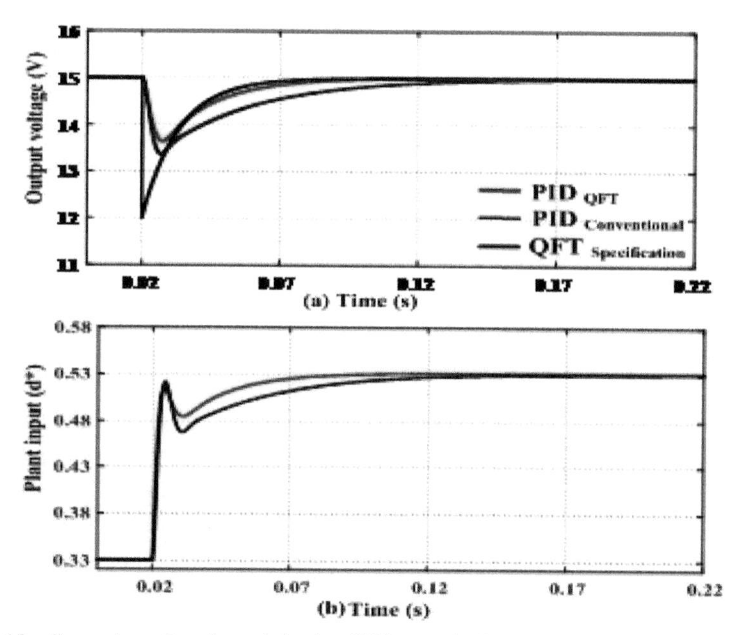

FIG. 14 Comparison of regulatory behavior of PID control scheme in linear simulations for a step variation in source voltage from 10 to 7 V: (A) Load voltage and (B) Control input (d^*).

Based on this regulatory scenario case study, it can be concluded that the proposed QFT-based robust PID controller provides large input to the plant to reject the external variations in the source voltage and allows a quick recovery back to the nominal operating point.

4.2. Nonlinear simulation studies

To investigate the robustness of all the designed controllers, the plant-model mismatch is created by developing the nonlinear circuit model of boost DC-DC converter in the MATLAB/Simulink using SIMSCAPE POWER SYSTEMS toolbox. The simulations performed using the nonlinear circuit model are named as Nonlinear simulation studies. Here, the tuning parameters of the conventionally tuned PID and the proposed QFT-based tuned PID are kept identical to the linear simulations. Following case studies are considered to verify the strength of the proposed QFT-based robust PID control algorithm.

4.2.1 Step variations (up/down) in the source voltage

The regulatory behavior for the step changes in the input voltage from 10 to 7 V and from 10 to 13 V is considered. The respective load voltage responses are shown in Figs. 14 and 15. From these figures, it can be observed that, in comparison to the conventionally tuned PID, the proposed robust PID controller has an upper hand in reaching back quickly to the nominal load voltage of 15 V with minimal deviation in the presence of plant-model mismatch.

4.2.2 Step variations in the load current

Further, a step change in the load resistance from 90 to 45 Ω is investigated on the hardware step-up to verify the strength of proposed QFT-PID controller. In this case, the obtained load voltage response for the transient and steady state

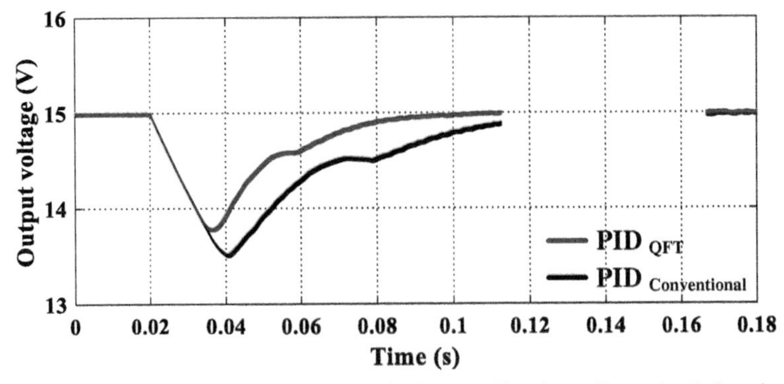

FIG. 15 Comparison of regulatory behavior of PID controllers in nonlinear simulations for step variation in source voltage from 10 to 7 V.

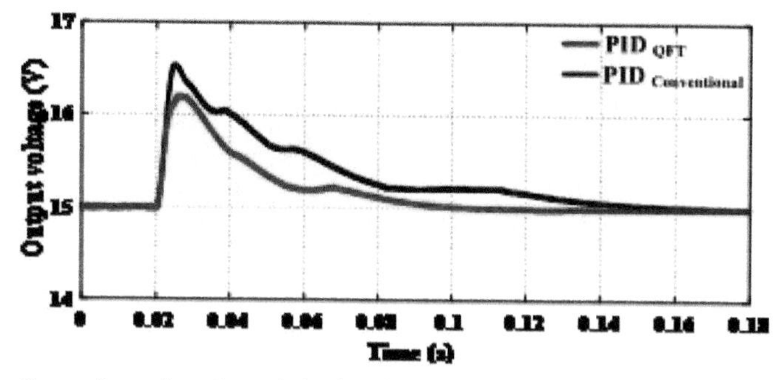

FIG. 16 Comparison of regulatory behavior of PID controllers in nonlinear simulations for step variation in source voltage from 10 to 13 V.

operation is shown in depicted in Fig. 16. In comparison to the conventional PID, the load voltage response recovers quickly for the sudden variation in load and reaches to the nominal operating voltage of 18 V with the proposed robust PID controller. This behavior is clearly depicted in Fig. 16.

It is interesting to observe that the closed-loop responses obtained in nonlinear simulations resembles closely to the linear simulation studies. Based on the regulatory case studies in linear and nonlinear simulation studies, it can be concluded that a PID controller tuned using QFT by including the available disturbance models information provides a quick and corrective duty ratio as an input to the boost-type converter for rejecting the step variations in input voltage, load current which are the external disturbances, and also robust against the mismatch between the plant and the model (Fig. 17).

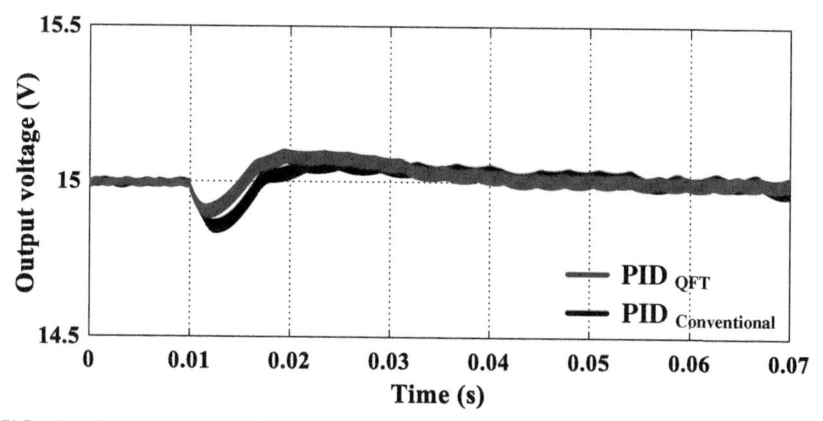

FIG. 17 Comparison of regulatory behavior of PID controllers in nonlinear simulations for step change in load resistance from 90 to 45 Ω.

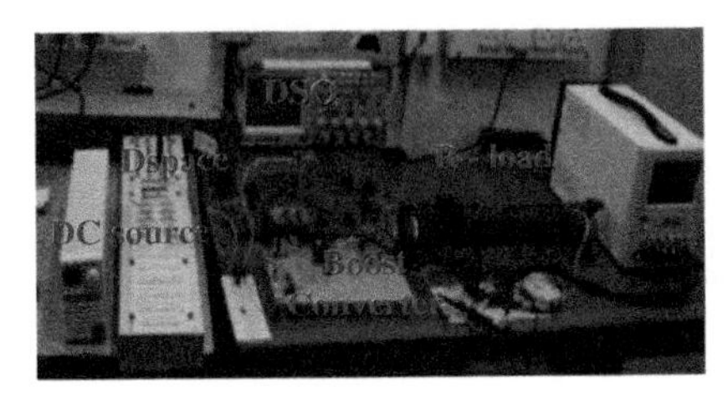

FIG. 18 Laboratory prototype of experimental setup.

5. Experimental validation

A prototype (laboratory) of boost converter is built to verify the designed QFT-based PID controller and the observations made from the linear simulation studies in Section 4. Fig. 18 shows the experimental setup. The experiments were conducted on the laboratory setup to verify the performance of the designed controllers for regulating the load voltage as well as the robustness. To implement the controller in real time, we used the dSPACE1103 controller board. The prototype DC-DC converter setup comprises a diode (MUR 860), gate driver circuit (TLP 250), and controllable MOSFET switch (IRF 640). Further, to verify the feasibility of QFT PID controller in real time, the parameters of PID are kept same as that of the simulations (ref. Section 4). The dynamic performances of the designed controllers in the closed-loop operation are tested for the following regulatory behaviors:

- **(i)** Step variations in the source voltage
  - **(a)** From 10 to 7 V with the 0.166 A nominal load current
  - **(b)** From 10 to 13 V for nominal 0.166 A load current
  - **(c)** 10 to 7 V with the load current of 0.2 A
- **(ii)** Step variations in the source voltage at various levels of reference output voltage.
  - **(d)** 7 from 10 V at 18 V output voltage
  - **(e)** 7 from 10 V at 11 V output voltage

5.1. Scenarios (a-b)

The experimental evaluations of the aforementioned case studies are graphically shown in Figs. 19 and 20. The source voltage is changed (step) to 7 from 10 V (scenario a) and to 13 from 10 V (scenario b) and are applied with 0.166 A nominal load current. The corresponding load voltage responses along with the duty ratio (plant input) variations are depicted in Figs. 19 and 20, respectively. Under these scenarios, the designed PID controller brings quickly the load voltage to the desired operating point with an improvement in the load voltage deviations as against the conventional PID controller (Eq. 9).

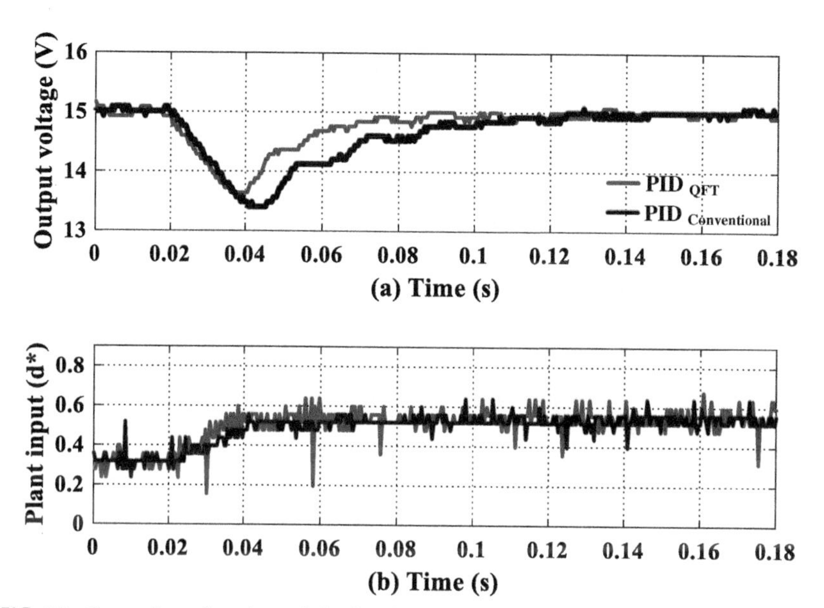

FIG. 19 Comparison of regulatory behavior of PID controllers in experiments for step variation in source voltage from 10 to 7 V at nominal load: (A) Load voltage; (B) Control input (d^*).

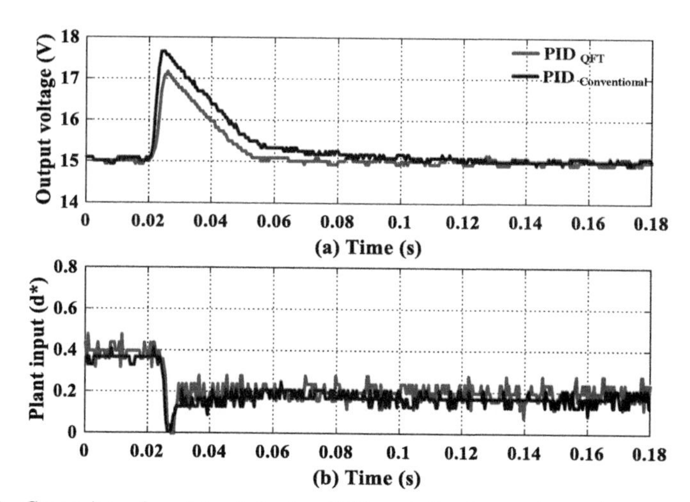

FIG. 20 Comparison of regulatory behavior of PID controllers in experiments for step variation in source voltage from 10 to 13 V at nominal load: (A) Load voltage; (B) Control input (d^*).

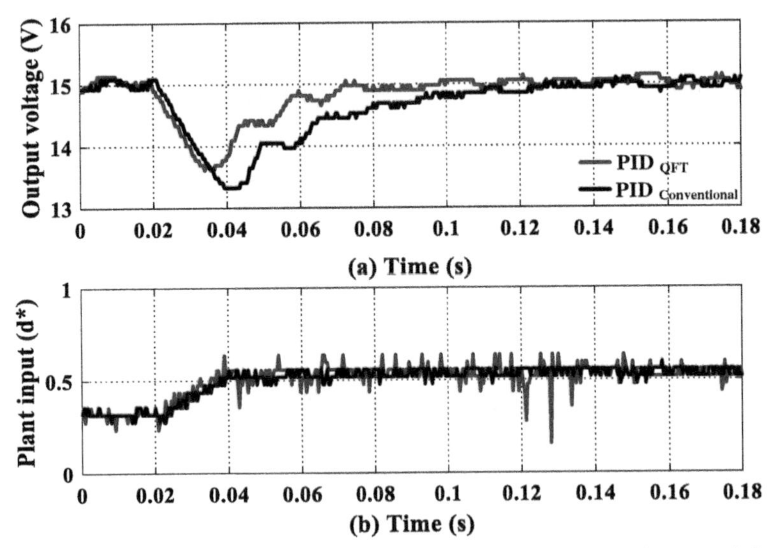

FIG. 21 Comparison of regulatory behavior of PID controllers in experiments for step variation in source voltage from 10 to 7 V at a load current of 0.2 A: (A) Load voltage; (B) Control input (d^*).

5.2. Scenario (c)

To illustrate the fast dynamic response and robustness of the proposed PID controller, various experiments were performed on the hardware setup for various combinations of changes in load current and input voltage such that the operating point moves to the extreme. The dynamic response of the controllers is shown in Fig. 21 for scenario (c). From Fig. 21, it is understood that the response of the closed loop with the PID controller (conventional) produces more deviation and requires more time to come back to the steady-state point as against the proposed PID design. The closed loop response with the proposed PID design disturbance rejection exhibits a major improvement in terms of the settling time with less deviation from the operating point. Thus, the closed-loop responses in the scenario (c) exhibit better robustness with the proposed PID design. Fig. 21B shows the plant inputs for this scenario under conventional PID controller and the proposed PID design.

5.3. Scenarios (d-e)

Further, the experiments were conducted on the boost-type converter to test the effectiveness of the proposed PID design for regulatory scenarios, particularly at various load voltage levels. For an operating point of 18 and 11 V, the input voltage (step change) is applied to 7 from 10 V. Figs. 22 and 23 show the subsequent closed-loop experimental responses. It is observed that, for both

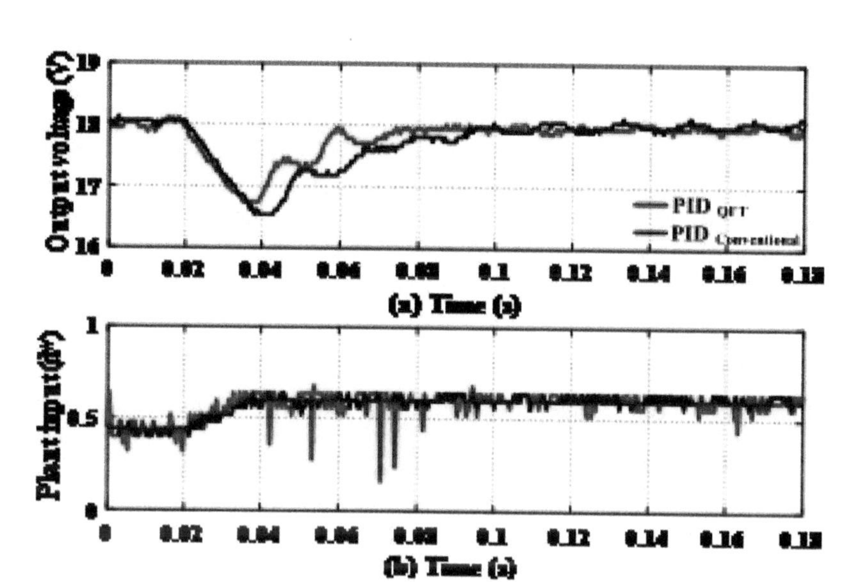

FIG. 22 Comparison of regulatory behavior of PID controllers in experiments for step variation in source voltage from 10 to 7 V at a set-point of 18 V: (A) Load voltage; (B) Control input (d^*).

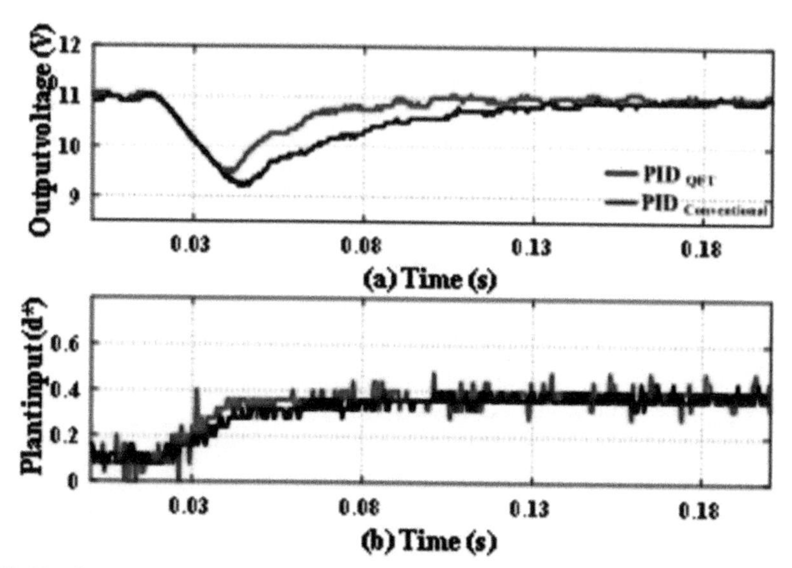

FIG. 23 Comparison of regulatory behavior of PID controllers in experiments for step variation in source voltage from 10 to 7 V at a set-point of 11 V: (A) Load voltage; (B) Control input (d^*).

scenarios (d) and (e), the load voltage moves the set point immediately with the proposed PID design over the conventional PID. Figs. 22B and 23B show the plant inputs, respectively, for these particular cases (d) and (e).

Based on the experimental validation, it is illustrated that the proposed robust PID controller shows a significant response improvement in the output voltage, particularly in terms of settling time and the deviations from the steady state output voltage. In the regulatory case studies considered, the proposed design exhibits the improvements in the closed-loop responses.

6. Conclusions

Achieving enhanced voltage regulation of power electronic interfacing DC-DC conversion is of paramount importance in a DC microgrid environment. Boosting of lower DC input voltage to higher DC output voltage is achieved using the DC-DC boost-type converters and this plays key role in DC microgrid application. To address the challenges posed by control of such step-up converters, a robust voltage mode PID controller is synthesized with the QFT principle for regulating the voltage at output of a nonminimum phase DC-DC boost-type converter is proposed in this work. The RHP zero in the boost converter dynamics limits the bandwidth of the closed loop system and the design of the controller becomes quite complex and challenging to get better response in closed-loop with an external disturbance and the parametric uncertainty. The design of the disturbance rejection control problem is addressed by accessing the disturbance dynamic model of boost converter while at the design stage itself. The PID controller proposed manifests clear improvement in the output voltage response. The proposed design method is verified for large extent in both simulations (linear and nonlinear) to validate the robustness and the disturbance rejection capability. It clearly shows that the proposed PID design outperforms the conventional one under various disturbance scenarios considered. The experiments are conducted to validate the proposed design method. The closed-loop responses obtained from the experiments confirms the improvement in all the disturbance scenarios, which agrees with the simulation results.

References

[1] X. She, A.Q. Huang, S. Lukic, M.E. Baran, On integration of solid-state transformer with zonal DC microgrid, IEEE Trans. Smart Grid 3 (2) (2012) 975–985.

[2] R. Bhosale, V. Agarwal, Fuzzy logic control of the ultracapacitor interface for enhanced transient response and voltage stability of a DC microgrid, IEEE Trans. Ind. Appl. 55 (1) (2019) 712–720.

[3] R. Bhosale, V. Agarwal, Enhanced transient response and voltage stability by controlling ultracapacitor power in DC micro-grid using fuzzy logic controller, in: 2016 IEEE International Conference on Power Electronics, Drives and Energy Systems (PEDES), 2016, pp. 1–6.

[4] R.W. Erickson, D. Maksimovic, Fundamental of Power Electronics, second ed., Kluwer, New York, 2001.

[5] K.J. Åström, Limitations on control system performance, Eur. J. Control. 6 (1) (2000) 2–20.
[6] S. Tan, Y.M. Lai, C.K. Tse, L. Martinez-Salamero, C. Wu, A fast-response sliding-mode controller for boost-type converters with a wide range of operating conditions, IEEE Trans. Ind. Electron. 54 (6) (2007) 3276–3286.
[7] C. Sreekumar, V. Agarwal, A hybrid control algorithm for voltage regulation in DC-DC boost converter, IEEE Trans. Ind. Electron. 55 (6) (2008) 2530–2538.
[8] T. Kobaku, S.C. Patwardhan, V. Agarwal, Experimental evaluation of internal model control scheme on a DC-DC boost converter exhibiting nonminimum phase behavior, IEEE Trans. Power Electron. 32 (11) (2017) 8880–8891.
[9] S. Tan, Y.M. Lai, C.K. Tse, General design issues of sliding-mode controllers in DC-DC converters, IEEE Trans. Ind. Electron. 55 (3) (2008) 1160–1174.
[10] Y. Shtessel, A. Zinober, I. Shkolnikov, Sliding mode control of boost and buck-boost power converters using method of stable center, Automatica 39 (6) (2003) 1061–1067.
[11] R. Naim, G. Weiss, S. Ben Yaakov, H∞ control applied to boost power converters, IEEE Trans. Power Electron. 12 (4) (1997) 677–683.
[12] L. Cheng, et al., Model predictive control for DC–DC boost converters with reduced-prediction horizon and constant switching frequency, IEEE Trans. Power Electron. 33 (10) (2018) 9064–9075.
[13] C. Yao, X. Ruan, X. Wang, Automatic mode-shifting control strategy with input voltage feed-forward for full-bridge-boost DC-DC converter suitable for wide input voltage range, IEEE Trans. Power Electron. 30 (3) (2015) 1668–1682.
[14] M. Garcia-Sanz, Robust Control Engineering: Practical QFT Solutions, CRC Press, Taylor and Francis, 2017.
[15] I.M. Horowitz, Quantitative Feedback Theory, QFT Publications, Colorado, USA, 1993.
[16] R.E. Nordgren, O.D.I. Nwokah, M.A. Franchek, New formulations for quantitative feedback theory, Int. J. Robust Nonlinear Control. 4 (1) (1994) 47–64.
[17] C. Olalla, R. Leyva, A.E. Aroudi, P. Garces, QFT robust control of current-mode converters: application to power conditioning regulators, Int. J. Electron. 96 (5) (2009) 503–520.
[18] M. Veerachary, A.R. Saxena, Design of robust digital stabilizing controller for fourth-order boost DC-DC converter: a quantitative feedback theory approach, IEEE Trans. Ind. Electron. 59 (2) (2012) 952–963.
[19] A.A. Towati, Dynamic Analysis and QFT-Based Robust Control Design of Switched-Mode Power converters, Thesis for Doctoral Degree, Helsinki University of Technology, 2008.
[20] A. Altowati, K. Zenger, T. Suntio, Analysis and design of QFT-based robust control of a boost power converter, in: 4th IET Conference on Power Electronics, Machines and Drives (PEMD), York, 2008, pp. 537–542.
[21] A.M. Basim, P.V.K. Kiran, R.J. Abraham, QFT based robust control for DC-DC boost converter, in: International Conference on Circuits, Controls and Communications (CCUBE), Bengaluru, 2013, pp. 1–6.
[22] K.H. Ang, G. Chong, Y. Li, PID control system analysis, design, and technology, IEEE Trans. Control Syst. Technol. 13 (4) (2005) 559–576.
[23] T. Samad, "A survey on industry impact and challenges thereof [technical activities]," IEEE Control. Syst. Mag., vol. 37, no. 1, pp. 17-18, Feb. 2017.
[24] R. Vilanova, A. Visioli, PID Control in the Third Millennium: Lessons Learned and New Approaches, Springer, New York, 2012.
[25] S. El Beid, S. Doubabi, DSP-based implementation of fuzzy output tracking control for a boost converter, IEEE Trans. Ind. Electron. 61 (1) (2014) 196–209.

[26] L. Guo, J.Y. Hung, R.M. Nelms, Evaluation of DSP-based PID and fuzzy controllers for dc–dc converters, IEEE Trans. Ind. Electron. 56 (6) (2009) 2237–2248.
[27] K.I. Hwu, Y.T. Yau, Performance enhancement of boost converter based on PID controller plus linear-to-nonlinear translator, IEEE Trans. Power Electron. 25 (5) (2010) 1351–1361.
[28] E.W. Zurita-Bustamante, J. Linares-Flores, E. Guzman-Ramirez, H. Sira-Ramirez, A comparison between the GPI and PID controllers for the stabilization of a DC–DC "Buck" converter: a field programmable gate array implementation, IEEE Trans. Ind. Electron. 58 (11) (2011) 5251–5262.

Further reading

T. Kobaku, R. Jeyasenthil, S. Sahoo, R. Ramchand, T. Dragicevic, Quantitative feedback design-based robust PID control of voltage mode controlled DC-DC boost converter, IEEE Trans. Circuits Syst. II: Express Briefs 68 (1) (2021) 286–290.
C. Borghesani, Y. Chait, O. Yaniv, The QFT Frequency Domain Control Design Toolbox-For Use with MATLAB, TERASOFT, San Diego, CA, USA, 2002.

Chapter 5

Control of PV and EV connected smart grid

Zunaib Ali[a], Komal Saleem[a], Ghanim Putrus[b], Mousa Marzband[b], and Sandra Dudley[a]

[a]*School of Engineering, London South Bank University, London, United Kingdom,* [b]*Department of Physics, Mathematics and Electrical Engineering, Northumbria University, Newcastle, United Kingdom*

1. Introduction

Environmental concerns and ambitious goals for sustainable energy generation are leading to massive deployment of renewable energy (RE) systems, such as photovoltaic (PV) and wind energy conversion systems, to fully or partially supply electricity demands. In addition, the vehicle industry is shifting toward the electrification of the transport system, introducing hybrid and fully electric cars. RE generation directly contributes to reducing carbon footprints by utilizing greener energy that runs with near zero carbon emissions. However, hybrid and fully electric vehicles present significant challenges in terms of sustainability, if they are charged by using power generated from conventional plants that consume fossil fuels and hence contribute to CO_2 footprint. This refers to scenarios where electric vehicles (EVs) are charged from the grid supply, especially during peak demand periods.

The increase in deployment of distributed RE systems provides an opportunity to charge EVs from green and sustainable energy that is locally generated. In addition, the massive deployment of EVs provides the opportunity to use their massive storage capacity to support the grid and deal with the intermittent nature of RE generation. This may be achieved by implementing smart control of EV charging and the use of Vehicle to Grid (V2G) concept, when EVs are used to supply power to the grid at times of peak demand. This can help prevent the costly grid system upgrades as well as providing technical benefits to the grid and financial benefit to participating EV owners.

This concept gives rise to the use of local microgrids, incorporating distributed generating sources, to actively support the main power grid by utilizing locally generated RE. This support requires effective and dynamic control of

Microgrid Cyberphysical Systems. https://doi.org/10.1016/B978-0-323-99910-6.00001-3

power flow as well as advanced techniques for estimating and forecasting both future power demands and intermittent renewable power generation. Smart control of EV battery charging can significantly help in reducing power exchange between the microgrid and the main grid, resulting in overall cost reduction for the grid as well as the microgrid (or a house) by effectively managing the local energy use and the power exchange with the utility grid.

1.1 PV market penetration

RE generation is increasingly seen as a critical technology to reducing CO_2 emissions. The benefits from using RE are recognized from various perspectives such as decreasing electricity prices, lowering carbon footprints, enhancing air quality (public health), increasing employment, and financial growth. The increasing installation and use of renewable energy systems (RES; PV in particular) for distributed generation is evident from the REN12 Global Status Report [1], where over the past few years, the penetration of renewables in the energy market has increased significantly, as shown in Figs. 1 to 4. The worldwide capacity of installed PV in 2020 is 760 GW with an annual increase of 18.2% [1]. The renewable market is heavily dominated by China contributing to almost half of overall world's 2020 installations (nearly equal to 117 GW). In terms of PV penetration, South Australia is leading the market and has the first large system in the world capable of fully eliminating the need of power from the main grid. Among various countries in the world, Rwanda contributes to the highest share of population using home solar system for daily electricity use [1]. Roof top residential and commercial PV systems are being deployed at an increasing rate around the world, particularly in urban areas in Fig. 5. In Germany, residential and commercial rooftops PV systems account for around 90% of the total PV installed capacity in Germany

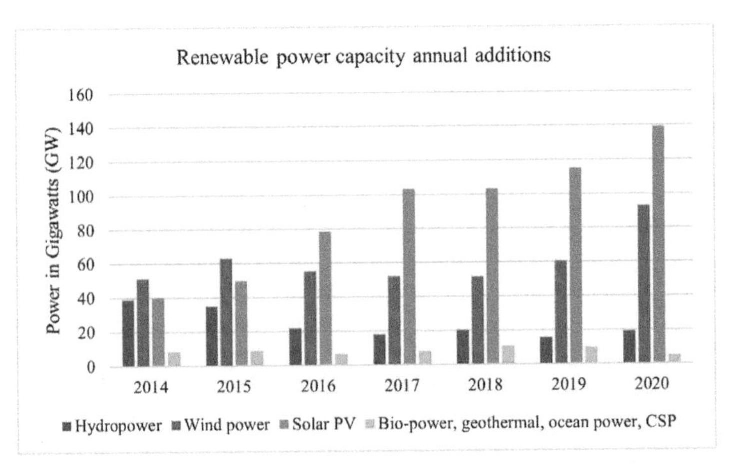

FIG. 1 Renewable energy annual addition categorized based on technology [1].

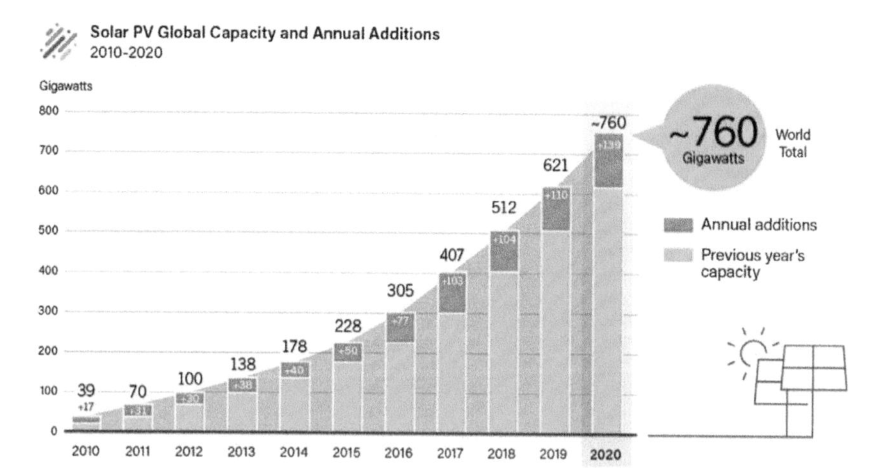

FIG. 2 Solar PV global capacity [1].

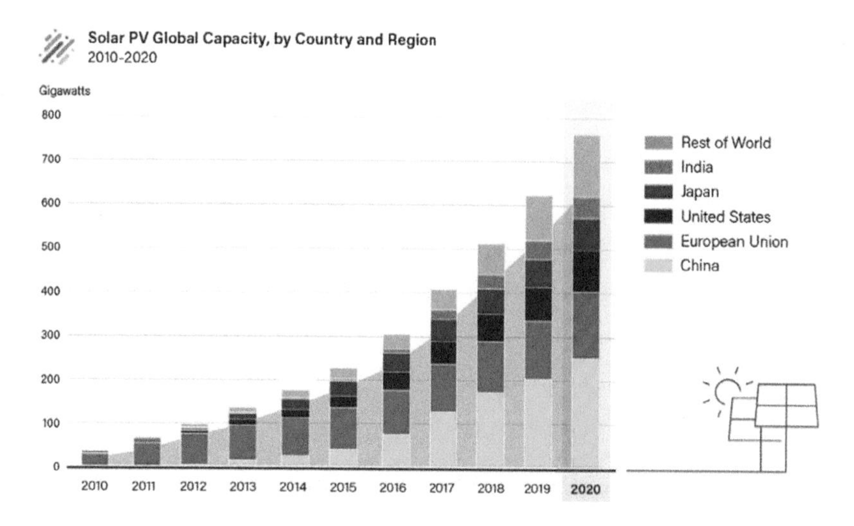

FIG. 3 Solar PV global capacity by region [1].

(~42GW), whilst in the UK, this is about 35% PV of the installed capacity (~13GW). The deployment of micro wind turbines in residential areas, especially in urban environment, has been very limited due to several technical and safety reasons. Since the scope of this work is on microgrids, particularly small urban microgrids, the focus is on PV renewable energy generation only.

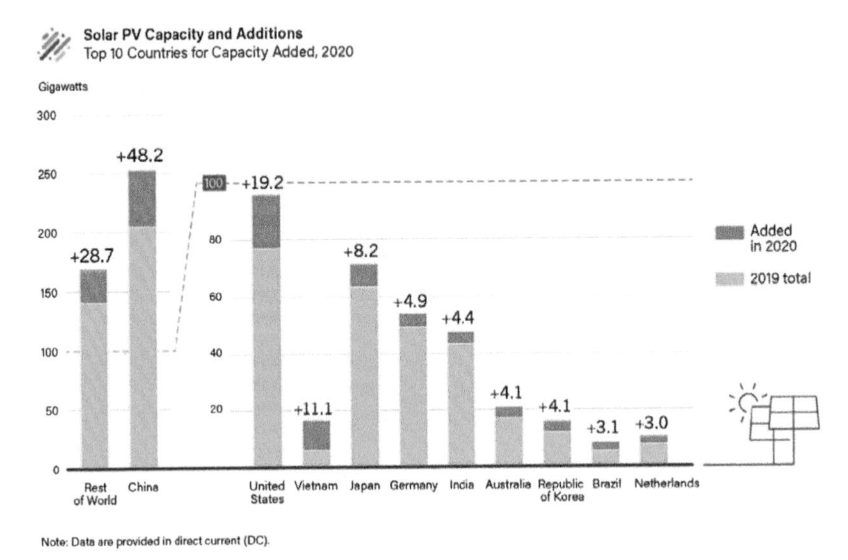

FIG. 4 Solar PV global capacity, top 10 countries [1].

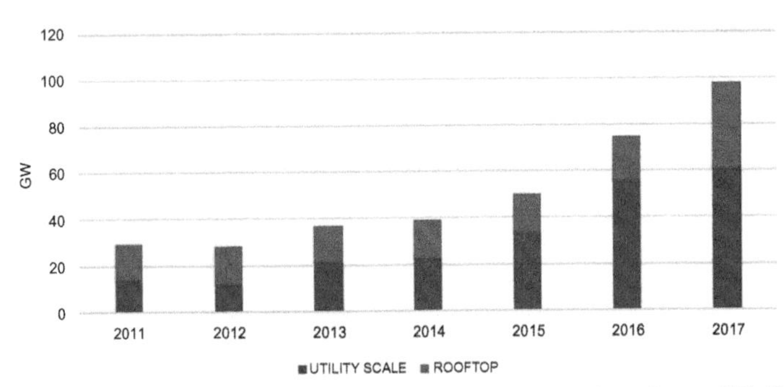

FIG. 5 Roof top PV systems. *(Source: Global Photovoltaic Markets, Report IEA PVPS T1-33:2018.)*

1.2 EV market growth

The massive interest in EVs can be noticed from the ambitious targets for phasing out conventional vehicles (that uses fossil fuel) as set by UK and other countries in Europe and the world. However, in many countries, there is still lack of a coherent approach for decarbonizing the transport sector, which will hinder their ability to fulfill the world's goals to combat climate changes. In 2020, the use of electric vehicles received considerable interest from policy makers together with biofuels that serve as a source of green energy for policy framework of road transport. Employing renewable energy to charge EVs presents a great potential for increased penetration of EVs as well as renewable energy sources (RES). The use of renewable energy to charge EVs not only result in green transport but also help to reduce the energy demand on the main grid and prevent or reduce costly grid upgrades. EV presents a very critical and effective use of renewables where the charging energy can be extracted from such sustainable sources, or it can be used as sink serving as storage unit presenting vehicle to grid services improving grid's flexibility, known as V2G [2,3]. The concept of V2G could encourage EV owners to use their EVs' batteries for charging at times when there is surplus of renewable energy and use them as storage to supply power to the grid at times of peak demand. In this way, EVs provide economic benefits to the grid as well as to the EV owner.

The transport sector is required to reduce its carbon emissions, and this is set in some countries by establishing requirements for fuel economy and standards for emission control. For example, in the Organization for Economic Co-operation and Development countries, the kilometers travelled by vehicles are increased by 0.73% annually between 2008 and 2017 and the carbon emissions have decreased at an annual rate of 0.64%, as shown in Fig. 6. As compared to vehicles with internal combustion engines, EVs have high energy efficiency and lower overall carbon emissions, with zero running emission (if charged from RES). The impact of EV on carbon reduction is significant, especially when charging from a high mix of energy from renewables, complementing further the use of EV as a greener and sustainable transport. As mentioned earlier, EVs may be used for grid support (referred to as V2G services),

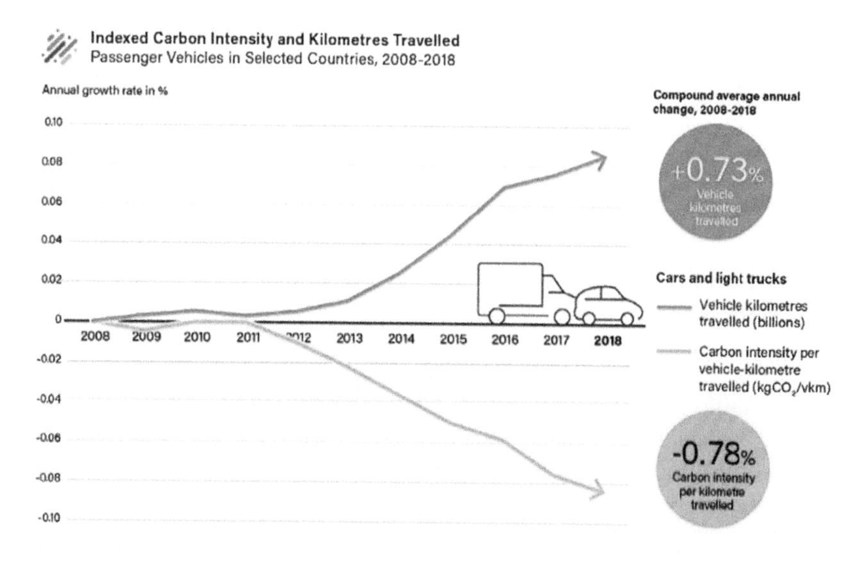

FIG. 6 Kilometers travelled vs. carbon indexed.

where the EV batteries are used as a load to absorb excess renewable energy and as storage to supply energy to the grid at times of peak demands.

The global sale of all type of electric cars has significantly increased over the past few years, despite the restrictions caused by COVID-19, whereas the overall car selling market received a downturn. The worldwide fleet of electric car in 2020 has increased by 43% in comparison to 2019, and EVs stock hit a maximum of 10 million with unprecedented number of new registrations [4]. The contribution of EVs to the overall car sales has increased to 4.6% in 2020, as compared to only 2.7% in 2019. The contribution of battery electric vehicles is nearly two-third of EV stock and one-third of new EV registrations in 2020. The republic of China marks the top position in EV market with 4.5 million electric cars [4].

The worldwide stock of EV cars by region and transport mode is depicted in Fig. 7. EV industry during the past year made noticeable announcement and contributions in terms of shifting to full production of electric vehicles in the near future, with lower battery costs and advances in battery technologies. The energy storage (battery) plays a vital role in the EV operation and a market growth of 191.1 W has been reported in Ref. [1] for various types of energy storage (important part of EV), including mechanical storage (nearly 174 GW), electro-chemical or electro-mechanical (14.2 GW), and thermal storage (2.9 GW). Share of Global Energy Storage Installed Capacity, by Technology, 2019 and 2020 is given in Table 1.

2. Impact of PV and EV on the electricity grid

The increasing penetration of PV and EV directly affects the operation and power quality of the grid. In Ref. [5], the impact of EV and PV has been summarized considering several key indicators and aspects, including grid stability, power quality, and electricity market (financial side).

2.1 Impact on power system stability

The stability of a network is defined as its capability to restore and reconfigure its normal stead-state operation after an event that involves a sudden disturbance on the system. Lack of efficient, high resolution, and robust monitoring and management system and improper control may result in blackouts or instability of the power system [6]. In terms of grid stability, three key indicators are voltage stability, frequency stability, and power oscillatory stability [5,7,8]. The nonlinear behavior and dynamic characteristics of EV charging can result in negative effects on the electricity grid. The amount of stress imposed on the grid depends mainly on the EV charging current, initial SOC of battery, permeation level, and location in the power system. A sudden demand for charging current resulting from the connection of several EVs at the same time can lead to unexpected increase in the power demand which may affect the grid stability [9]. This originates the need for monitoring the stability of DC and AC distribution

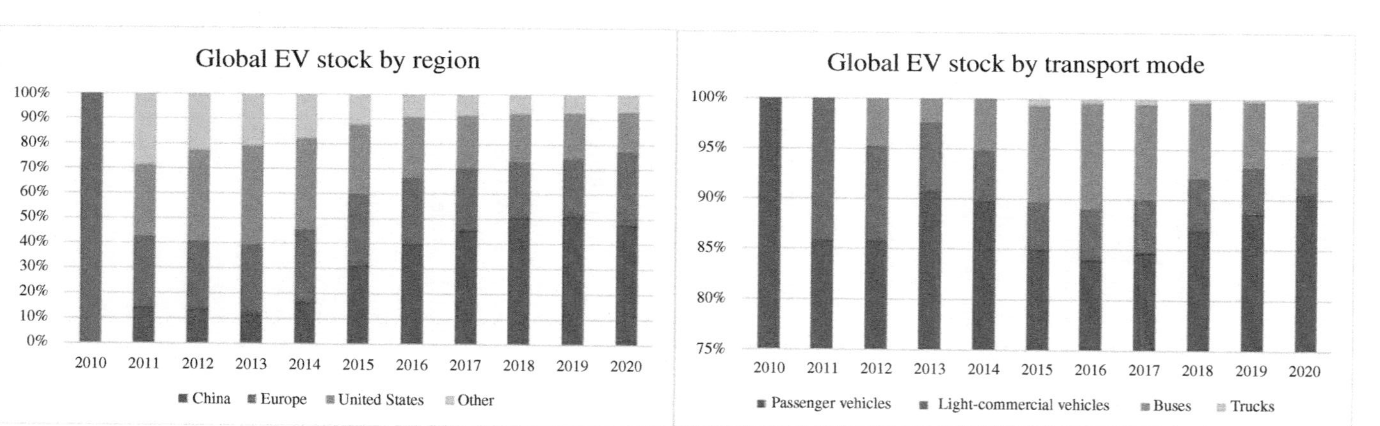

FIG. 7 The global electric vehicle stock by region and transport mode [IEA (2021), Global EV Outlook 2021, IEA, Paris https://www.iea.org/reports/global-ev-outlook-2021]. Europe includes EU27, Norway, Iceland, Switzerland, and United Kingdom. Other includes Australia, Brazil, Canada, Chile, India, Japan, Korea, Malaysia, Mexico, New Zealand, South Africa, and Thailand.

TABLE 1 **Share of Global Energy Storage Installed Capacity, by Technology, 2019 and 2020.**

Storage types	2019 (%)	2020 (%)
Pumped storage	92.60	90.30
Molten salt storage	1.7	1.8
Lithium-ion battery storage	4.6	6.9
Other electrochemical storage	0.6	0.6
Other energy storage	0.4	0.4

Source: REN 21, "Renewables 2021 Global Status Report," 2021. Accessed: 01/11/2021. [Online]. Available: https://www.ren21.net/reports/global-status-report/.

networks rather than only focusing on the transmission system. The EV integration analyzed by researchers in Ref. [10] shows that it presents significant deterioration to grid stability and increases system time to recover back to its normal state. It is therefore essential to use EV in a controlled manner along with the involvement of stationary storage for supply/demand balance and to support the grid in the event of grid faults. To this end, proper modeling, forecasting, and control of EV may help in eliminating the negative impact of EV on the grid [10,11].

Likewise, increased deployment of PV systems may significantly affect the stability of power system, caused by sudden changes in solar irradiance and reduced system inertia. The high PV penetration mainly affects voltage, frequency, transient, and small signal stability depending on location, penetration level, and system topology [7,8,12,13]. The PV system presents voltage fluctuations and may cause voltage rise specifically when connected far away from the distribution transformer with lightly loaded feeder. The voltage variations are dependent on the level of deployment of PV and on weather conditions. Furthermore, the replacement of conventional generators with PV systems results in lower system inertia which may lead to oscillatory response and frequency instability after a disturbance [14]. To compensate for these issues, appropriate control of PV systems and suitable system architecture involving energy storage and reactive power compensation are required [15,16].

2.2 Impact on power quality

The integration of PV and EV has significant impact on the power quality of AC grid due to their inherent characteristics and type of connection to the grid (via power electronic converters). This poses significant challenges to the power network in terms of reactive power control, harmonic distortions, voltage fluctuations, voltage imbalance, DC offset, power losses, consequent transformer and cable overloading, etc. For EVs, this impact is also related to the penetration level of EVs as well as charging profiles, increasing charging stations, and unregulated charging regimes. The main factors associated with the integration of PV as per IEEE 929-2000 are fluctuations in power as well as voltage rise and harmonic distortion [17]. The overall impact on the grid is voltage sag, swell, unbalance, flicker, and violation of statutory limits. The fluctuating PV power based on varying and unpredictable solar irradiance and temperature may results in grid frequency variations especially with high PV penetration. This requires robust control techniques and diversified network topologies so as to minimize these issues by either delivering high quality-controlled power or by providing ancillary services to the network (such as VAr compensation, smart and optimized EV charging, reconfigured phase approaches, etc.).

2.3 Impact on electricity market

EVs and PVs emerged as key players in deciding the operation and control of future energy market. The generation from PV systems is highly unpredictable and this can affect the price of energy. Thus, combining PV with storage batteries (stationary and in EVs) together with forecasting algorithms help in better scheduling of energy and control of energy prices. EVs may impose additional demand on the power system and energy requirements, and therefore may negatively affect the electricity prices if not properly controlled. An analysis presented in Ref. [18] shows 17% increase in the electricity prices because of uncontrolled charging of EVs. However, EVs may be controlled to support the grid by devising efficient smart charging strategies, providing the energy during the hours of peak demand, and charging when there is excess energy in the system (e.g., from PV). Therefore, off-peak charging is more beneficial and when using RES as a source would greatly help in reducing carbon emissions. The impact of EVs on the optimal load profiles of power system based on stochastic modeling is presented in Ref. [19]. It is shown that provision of ancillary services and optimal EV scheduling offer system support as well as financial benefits to the grid and EV owners. The effective utilization of RES, EVs, storage batteries, and other energy sources in a system helps in reducing the overall operational cost of a microgrid and provides economic benefits.

Thus, advanced monitoring and control techniques are required for RES, storage system, and grid, providing distributed problem solving, fault-tolerance, and system power compatibility between different resources in a microgrid, benefiting the overall power network.

3. Smart microgrids, classifications, and interconnection of sources

Microgrids present a suitable place and infrastructure for the interaction and operation of various distributed energy sources and loads in a smart and an intelligent manner.

3.1 Classification of microgrids

Several microgrid classifications exist [20–22], and some may be classified as follows:

3.1.1 University campus or institutional microgrids (UCI-μ grids)

The university campus microrgrids present an ideal Research and Development environment for the increased utilization and deployment of RES, resulting in a more sustainable environment [23,24]. These microgrids focus on combining local generations with several loads in a tight geographical proximity usually self-governed or managed by a single entity. The university campus (or traditional institution) presents the most favorable conditions and fulfill the technical constraints for the realization of a microgrid. The various buildings in a university campus contain local generations and loads that are closely located, connected to each other via the same campus network, and managed and operated by a single university authority. An additional advantage is the single-point of connection to the main power grid via the point of common coupling (PCC), which allows a more straightforward decision-making and smooth transition between grid-connected and standalone operation of the campus network [20,22,24]. Furthermore, the operation of various generation units and loads within a microgrid are monitored, analyzed, and managed by a single control unit. These microgrids have significant flexibility on their side of meters, so the nature of restrictions on their operation by the power system operator is relatively low [25]. All these advantages of a university campus enable a more convenient and simpler implementation of university-based microgrid.

To summarize, a university microgrid presents three main characteristics, that are, (i) presence of all the loads and generation units within a proximity, (ii) management and control by a sole university service, and (iii) connection to the same university network and with a single point of connection to the grid (PCC). A number of microgrid demonstrations projects have be employed by several universities in order to analyze and demonstrate new ideas, making microgrid more intelligent and smarter. These projects constitute several underlying objectives, such as, (i) improve the microgrid efficiency, reliability, and overall operation; (ii) enable increased integration and utilization of RES together with storage technologies (including EVs); (iii) improve on-campus energy efficiency by involving students, faculty members, and staff; and (iv) facilitate infrastructures for research, education, and training related to microgrid.

3.1.2 Islanded or remote microgrid (IR-μ girds)

The remote and islanded regions present several challenges and complex problems while fulfilling energy demand, and this is due to their underlying ecological factors and geographical locations [26]. Such remote areas (including small islands, rural areas, and small villages) cannot be supplied and reached by the main grid, because the constructional investments and huge operational cost restricts the expansion of grid. An example of such areas includes rural African villages, where people use wood, coal, kerosene oils, and paraffin for cooking and lightning purposes, and thus, IR microgrids provide a solution for such places. The generation in these microgrids can be supported using diesel generators, which however, adversely affects the environment, and also the diesel supply to the island is a difficult task. On the other hand, using natural and green energy from renewables can be advantageous to help satisfy the demand. Consequently, such microgrids usually involve combinations of conventional diesel generators together with RES. In general, IR microgrids are not grid-connected (thus, always operated in islanded mode) and are considered as the largest of all microgrid types. There are several remote microgrids worldwide. One of the examples is Indonesia which is rich with such remote microgrids, as it has more than 20,000 islands and powering them through a central grid is practically impossible. Another example is a microgrid developed under the H2020 framework, funded by European Commission in a Greek island named TILOS. The Tilos project supported the development of a microgrid with an 800-kW wind power system, a PV system rated at 160 kW, and two containers of battery storage system, each one rated at 400 kW, 1.44 MWh. The Tilos microgrid is electrically connected to the neighboring Kos and Kalimnos islands and its own population is only 500 people.

3.1.3 Commercial or industrial microgrids (CI-μ grids)

Commercial and industrial microgrids are, in principle, similar to the university microgrids but they are mainly developed for meeting the requirements of customers and clients. These microgrids are scalable in terms of increasing or decreasing the required net capacity and the usable capacity is usually dependent on the varying demand.

3.2 Microgrid configurations and infrastructure

Various microgrid configuration and infrastructures that are popular include AC, DC, and hybrid microgrids. The operation of these microgrids is dependent on the type of interconnection, distributed and conventional sources, power electronic converters, and loads used in the system.

3.2.1 AC microgrid

AC microgrids represent a small conventional power system where all the sources are connected to each other through an AC link and connected to the utility grid via transformer. The outputs of DC sources are converted to AC via DC-AC power-electronic converters and, usually, each source has its own converter. The direction of power flow for DC sources depends on their type, such as, PV systems, which allow unidirectional power flow, and storage battery, which has the flexibility of absorbing excess power from the system and delivering it later when most needed. The AC microgrids are difficult to control as it involves stability issues and hence require complex controllers. AC microgrids are probably the most common form of existing microgrids, since they are easy-to-integrate with existing power systems and loads. Examples of AC microgrids are illustrated in Fig. 8.

3.2.2 DC microgrid

The DC microgrid provides a flexibility of using DC-based distributed sources (PV, storage, etc.) and electronic devices and loads such as electric vehicles, laptops, chargers, and so on. The energy from DC sources, often after processing by DC-DC converter, is utilized directly by DC loads. AC generators, such as wind energy conversion systems and so forth, are converted to DC via power electronic converters. DC microgrids are connected to main electrical grid at PCC via main DC-AC inverter and share energy with the utility. The control of DC-DC and AC-DC converters is enabled by simpler techniques, whereas DC-AC conversion requires complex and advanced algorithms. The implementation and control of DC-microgrids in general is simpler and requires less power conversion stages. However, practically, it is not easy to integrate it with existing AC network, which requires restructuration of network and leads to high investment. Examples of DC microgrids are illustrated in Fig. 8.

3.2.3 Hybrid microgrid

A Hybrid microgrid is a combination of several sub-DC and AC microgrids, as shown in Fig. 10. The AC microgrid is directly connected to the main utility grid, whereas the connection of DC microgrid is enabled though a single DC-AC inverter which is fed by a common DC bus to which all the DC sources/loads are connected. Hybrid microgrids offer more flexibility in terms of satisfying modern electronic devices as well as the conventional AC network configurations.

4. Control architecture for efficient operation of microgrids

Power grid faces challenges such as increased electricity demand, power quality issues related to integration of nonlinear equipment, intermittent nature of renewables, and power system faults. The effective utilization of RES and EV requires intelligent, adaptive, and diversified control algorithms to measure, monitor, and control energy exchange, provide high quality power, maximum grid support, increased energy autonomy, low operational cost, and reduced carbon emissions. The PV generation profile and EV user behavior (charging profile) directly affects the operation and energy management in a microgrid. Thus, to achieve smart energy management and control of grid-connected systems,

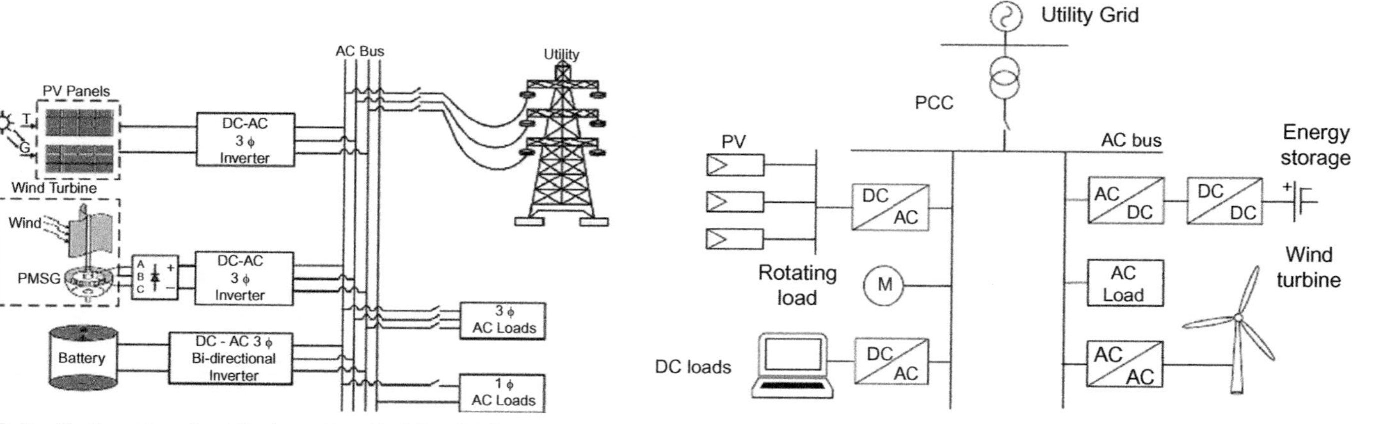

FIG. 8 Configurations for AC microgrids with AC and DC loads [27,28].

there is a pivotal need for high-resolution and reliable data sources that merge digital solutions and artificial intelligence-based control technologies. This will help the microgrids and distribution network operators' to effectively manage and control real-time power system operation (increasing self-sufficiency from renewables) as well as other stakeholders better plan capital investments (defer costly system upgrades). Thus, digital control technologies such as converter control, smart energy management techniques, energy and load forecasting, and parameter estimation algorithms [29] based on realistic available power system data are required for making the grid smart and capable of integrating significant penetration of renewables effectively; thus reducing net energy demand, carbon footprints, and costly infrastructure upgrades.

This section describes various control techniques aimed at the grid-connected operation of microgrid, including the controllers, synchronization algorithms, solar forecasting, and energy management algorithms. In addition, EV battery modeling techniques, EV charging techniques, and various EV charging/driving profiles are also discussed which play a very important role for effective operation of EV chargers in order to reduce the impact of EVs on to the microgrid.

4.1 Grid connected system and control of the DC side

The DC-side controller mainly involves controlling the buck-boost DC-DC converter for maintaining the DC-link voltage, as shown in Figs. 8 to 10 [30–32]. The operation of controllers is dependent on the type of source connected at the DC-link (such as PV, storage system, EV, wind, etc.). The PV connected systems are mainly controlled to extract the maximum power from the system referred to as MPPT algorithms [33–36]. This chapter, however, focuses on the AC-side control of PV/EV grid integration.

4.2 Grid connected system and control of AC side

The control of grid-connected PV (or EV) is enabled through a fully controlled power electronic converter which allows bidirectional power flow (in case a battery storage is connected). A typical block diagram of a grid connected system is presented in Fig. 11 [15,16,37]. The energy from the solar and/or EV is first conditioned though the DC-DC converter and is then fed to the DC bus of the inverter. The DC bus is connected to the AC-side via power electronic inverter (referred to as a Grid Side Converter (GSC)). The control of GSC is of pivotal importance as it is responsible for the synchronization and overall operation of the energy system as regulated by the grid codes. For PV, the flow of power is unidirectional, whereas EV supports bidirectional power flow. The GSC controller constitutes three key components; they are the PQ controller, the current controller, and the synchronization unit. The *PQ* controller specifies the reference signals for the current controller ($\mathbf{i}_{ref}$) considering the active/reactive

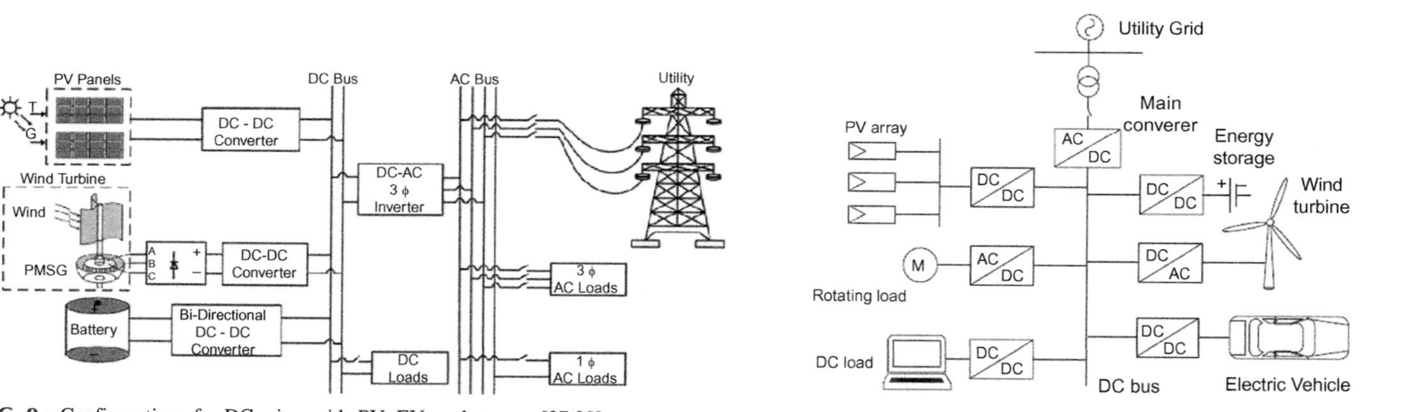

FIG. 9 Configurations for DC microgrids PV, EV, and storage [27,28].

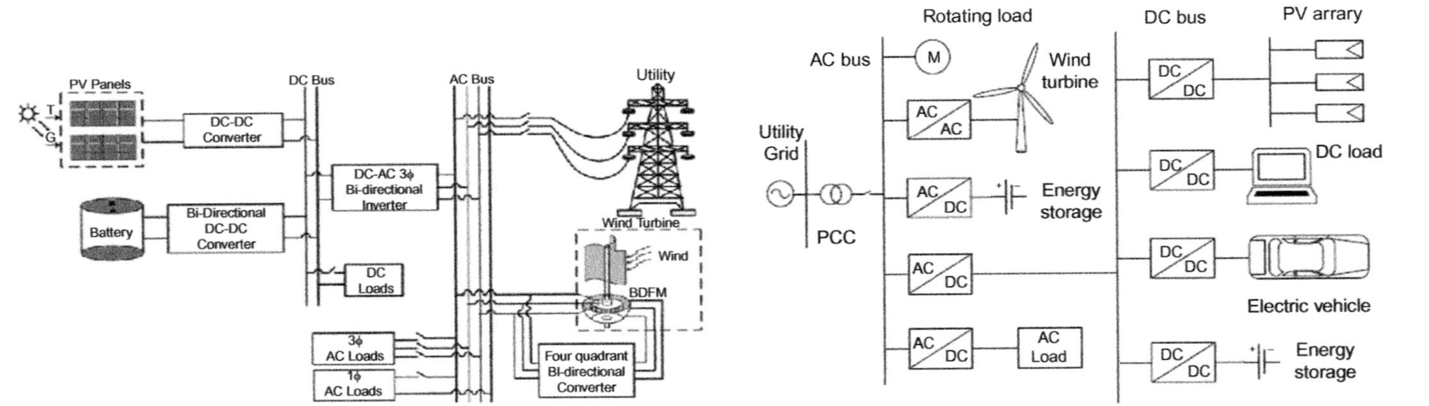

FIG. 10 Configuration for hybrid DC-AC microgrids with PV, EV, wind, and storage [27,28].

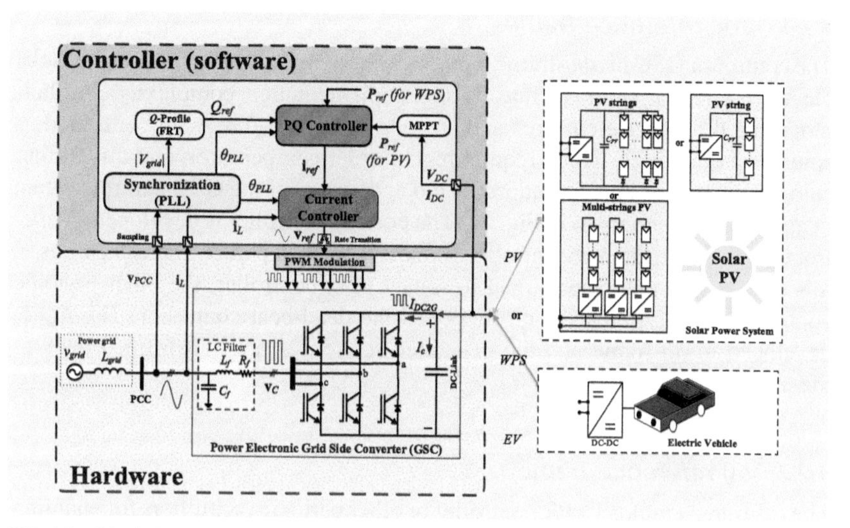

FIG. 11 Block diagram for the control of grid connected energy system.

power setpoints (PQ_{ref}), and in some cases, the DC-link voltage (V_{DC}) and the AC voltage amplitude ($|V_{grid}|$). The synchronization unit is the heart of the control system as it is responsible for determining the grid voltage phase angle and frequency, which is necessary before making the grid connection. The current and *PQ* controllers are directly influenced by the response of the synchronization method [3]; therefore, the behavior of the PLL (phase locked loop) shown in Fig. 10 is very fundamental for the overall operation of the grid connected system.

For control, the interaction between three-phase grid currents and voltages forms the basis for the overall operation of three-phase power system. Space Vector Transformation (SST) is an effective tool when dealing with the analysis and control of such system. Various SSTs exist in the literature, facilitating the simpler analysis of three-phase system under normal and abnormal grid conditions. A three-phase system can be expressed in the corresponding equivalent space phasor representation using three types of transformations: the natural *abc* reference frame transformation, the stationary $\alpha\beta$ reference frame transformation, and the synchronous reference frame *dq* transformation. The transformation from three-phase to corresponding two-phase $\alpha\beta$-reference frame and the synchronous *dq* reference frame helps in reducing the order of controller for the power electronic-based GSC; thus simplifying the design of overall GSC control system. Furthermore, the control of GSC in *dq*-frame is more straightforward as all the controlled variables become time invariant DC quantities, thereby the active and reactive powers are easily controlled in a decoupled way.

4.2.1 abc reference frame

The control of GSC in *abc*-frame requires a separate controller for each phase of the three-phase system, increasing the implementation complexity. Furthermore, the three-phase currents and/or voltages in *abc*-frame are controlled as sinusoidal quantities, which require higher order compensator, and this further complicates the design procedure. A block diagram of the GSC control system in an *abc*-frame is shown in Fig. 12. The current controller is developed in *abc*-frame with a separate subcontroller connected in each phase. The controllers in this case can be the Proportional Resonant (PR) controller; the other periodic controllers are hysteresis controllers and the dead-beat controller. The *PQ* is developed in *dq*-frame and the generated references are transformed back to the *abc*-frame.

4.2.2 αβ reference frame

The $\alpha\beta$-frame employs a PR controller or other periodic controllers for enabling the control of grid-connected system. In $\alpha\beta$-frame, the three-phase currents and/or voltages are transformed to the corresponding two-phase sinusoidal signals. Thus, for controlling these signals, the $\alpha\beta$-frame requires higher order compensators like that of the *abc*-frame. However, as opposed to the *abc*-frame, only two phases are to be controlled in $\alpha\beta$-frame, making the design of the controller comparatively simpler and less complex. The generic block diagram for the GSC controller in $\alpha\beta$-frame is depicted in Fig. 13.

Thus, the transformed $\mathbf{v}_{\alpha\beta}$ version of the three-phase voltage vector $\mathbf{v}_{abc}$ is obtained by the transformation given in Eq. (1). On the other hand, the reverse transformation is carried out using (2).

$$\mathbf{v}_{\alpha\beta} = \left[T_{\alpha\beta}\right]\mathbf{v}_{abc} \tag{1}$$

$$\mathbf{v}_{abc} = \left[T_{\alpha\beta}\right]^{-1}\mathbf{v}_{\alpha\beta} \tag{2}$$

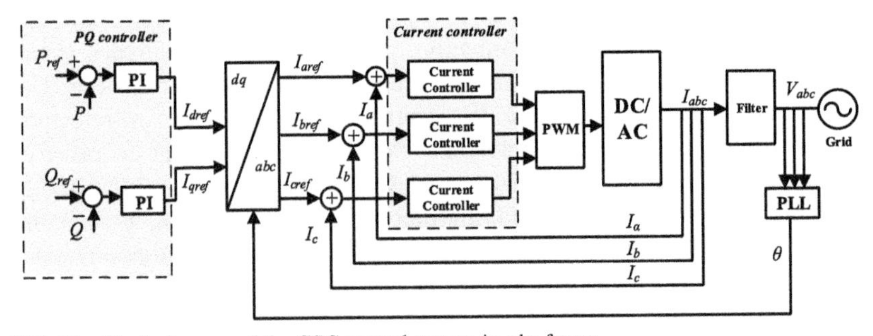

FIG. 12 Block diagram of the GSC control system in *abc*-frame.

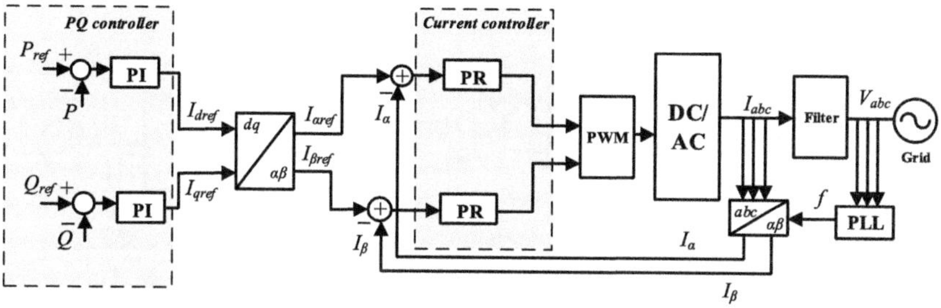

FIG. 13 Block diagram of the GSC control system in $\alpha\beta$-frame.

where,

$$\mathbf{v}_{abc} = \begin{bmatrix} v_a \\ v_b \\ v_c \end{bmatrix}, \mathbf{v}_{\alpha\beta} = \begin{bmatrix} v_\alpha \\ v_\beta \end{bmatrix} \tag{3}$$

$$[T_{\alpha\beta}] = \frac{2}{3}\begin{bmatrix} 1 & -\frac{1}{2} & -\frac{1}{2} \\ 0 & \frac{\sqrt{3}}{2} & -\frac{\sqrt{3}}{2} \end{bmatrix} \tag{4}$$

$$[T_{\alpha\beta}]^{-1} = \begin{bmatrix} 1 & -\frac{1}{2} & -\frac{1}{2} \\ 0 & \frac{\sqrt{3}}{2} & -\frac{\sqrt{3}}{2} \end{bmatrix}^T \tag{5}$$

4.2.3 dq *reference frame*

The synchronous reference dq-frame is the most important and popular frame for accurate and simplified control of power electronic converters. The problem that arises while designing the GSC controller in the *dq*-frame is the generation of undesired oscillations when a three-phase voltage or current signal containing more than one frequency is transformed with a specific speed. The fast elimination of these oscillations is the key for accurate control in *dq*-frame using a simple Proportional Integral (PI) controller. A generic block diagram of the GSC controller designed in *dq*-frame is shown in Fig. 14. The *PQ* and current controllers are developed in the *dq*-frame with PLL as the key element of the control system. The PLL allows fast and accurate synchronization of the RES/EV system with the utility grid. The grid voltage information such as the frequency, phase, and the voltage amplitude obtained through the PLL plays a vital role for enabling the accurate control of the grid-connected system.

The transformed *dq*-version of the three-phase voltage vector $\mathbf{v}_{abc}$ can be obtained by the transformation given in Eq. (6) and the reverse transformation is carried out using (7).

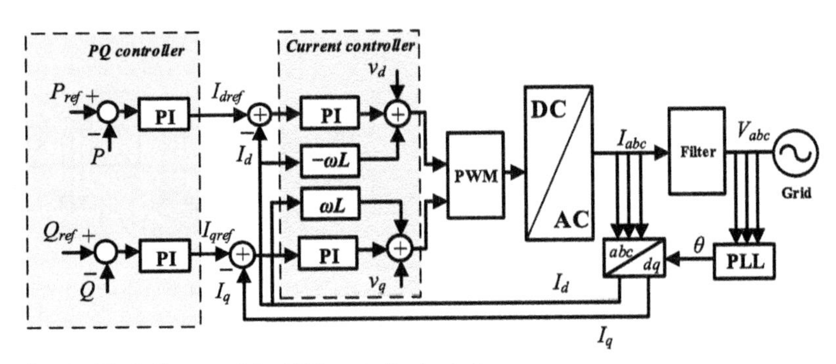

FIG. 14 Block diagram of the GSC controller in dq-frame.

$$\mathbf{v}_{dq} = \left[T_{dq}^{n/}\right]\mathbf{v}_{abc} \tag{6}$$

$$\mathbf{v}_{abc} = \left[T_{dq}^{n/}\right]^{-1}\mathbf{v}_{dq} \tag{7}$$

where,

$$\mathbf{v}_{abc} = \begin{bmatrix} v_a \\ v_b \\ v_c \end{bmatrix}, \mathbf{v}_{dq} = \begin{bmatrix} v_q \\ v_q \end{bmatrix} \tag{8}$$

$$\left[T_{dq}^{n/}\right] = \frac{2}{3}\begin{bmatrix} \cos(\theta) & \cos(\theta - 120°) & \cos(\theta + 120°) \\ \sin(\theta) & -\sin(\theta - 120°) & -\sin(\theta + 120°) \end{bmatrix} \tag{9}$$

$$\left[T_{dq}^{n/}\right]^{-1} = \begin{bmatrix} \cos(\theta) & \cos(\theta - 120°) & \cos(\theta + 120°) \\ \sin(\theta) & -\sin(\theta - 120°) & -\sin(\theta + 120°) \end{bmatrix}^T \tag{10}$$

n is the desired dq-frame to which the three-phase voltage vectors is transformed and θ is the PLL angle. For fundamental positive sequence, n=+1.

The dq-version of the voltage vectors can also be obtained directly from the $\alpha\beta$-transformed voltage vector. The $\alpha\beta$ to dq and dq to $\alpha\beta$ transformation is obtained by using (11) and (12).

$$\mathbf{v}_{dq} = \left[T_{dq}^{n}\right]\mathbf{v}_{\alpha\beta} \tag{11}$$

$$\mathbf{v}_{\alpha\beta} = \left[T_{dq}^{n}\right]^{-1}\mathbf{v}_{dq} \tag{12}$$

where,

$$\mathbf{v}_{dq} = \begin{bmatrix} v_q \\ v_q \end{bmatrix} \text{ and } \mathbf{v}_{\alpha\beta} = \begin{bmatrix} v_\alpha \\ v_\beta \end{bmatrix} \tag{13}$$

$$\left[T^n_{dq}\right] = \begin{bmatrix} cos\,(n\theta) & sin\,(n\theta) \\ -sin\,(n\theta) & cos\,(n\theta) \end{bmatrix} \tag{14}$$

$$\left[T^n_{dq}\right]^{-1} = \begin{bmatrix} cos\,(n\theta) & sin\,(n\theta) \\ -sin\,(n\theta) & cos\,(n\theta) \end{bmatrix}^T \tag{15}$$

The following section focuses on the control of converters in *dq*-frame, which is simpler and easy for the analysis and design of GSC controller for grid-connected RES and EV applications.

4.3 Controllers for the grid-side inverter

This section focuses on the AC-side control of the PV/EV grid integration using the *dq*-reference frame and a typical block diagram for such a system is presented in Fig. 13. The block diagram shows various types of subcontrol units which are used for enabling the accurate operation of the GSC controller. These controllers are the current controllers, the *PQ* controllers, and the synchronization units (PLL).

4.3.1 PQ controller

The *PQ*-controller, referred to as the outer-loop controller, is responsible for generating the reference currents ($\mathbf{i}_{ref}$) based on the setpoints defined for active and reactive powers (PQ_{ref}). The development of *PQ* controller can be enabled by either using an open-loop or a closed-loop configuration.

Conventional open-loop

In the open-loop structure, mathematical expressions are used to express the reference currents as a function of reference *PQ* power. The open-loop *PQ* controller derived using *dq*-control theory is expressed in Eq. (16).

$$\begin{bmatrix} i_d^{*+1} \\ i_q^{*+1} \end{bmatrix} = \frac{2}{3}\frac{1}{\left(v_d^{*+1}\right)^2 + \left(v_q^{*+1}\right)^2}\begin{bmatrix} v_d^{*+1} & v_q^{*+1} \\ v_q^{*+1} & -v_d^{*+1} \end{bmatrix}\begin{bmatrix} P_{ref} \\ Q_{ref} \end{bmatrix} \tag{16}$$

Where $\mathbf{v}_{dq}^{*+1}$ is the *dq*-transformed positive sequence component of grid voltage and $\mathbf{i}_{dq}^{*+1}$ is the reference *dq*-currents.

Conventional closed-loop

In the closed loop scenario, specific controllers (e.g., PI) are used in the loop to track the error between the measure and reference *PQ* powers (to zero) and current references are generated accordingly. An example of a closed-loop *PQ* controller for the fundamental positive sequence current is given in Fig. 14 and may be expressed as:

$$\begin{bmatrix} i_d^{*+1} \\ i_q^{*+1} \end{bmatrix} = \begin{bmatrix} k_p + \frac{1}{T_i}\frac{1}{s} & 0 \\ 0 & k_p + \frac{1}{T_i}\frac{1}{s} \end{bmatrix} \begin{bmatrix} \Delta P \\ \Delta Q \end{bmatrix} \tag{17}$$

where, k_p and T_i are tuning parameters for the PI controller and ΔPQ is the difference between the reference (PQ_{ref}) and measured (PQ) power. The sign of reference power decides charging or discharging for EVs.

The conventional controller allows the generation of references for fundamental positive sequence component, thus using the PV/EV as a source to fulfill the load demand and support the grid in power balance.

Advanced open-loop *PQ* controller

An advanced *PQ* controller is suggested in Ref. [37] which allows the generation of reference currents for on-purpose injection of harmonic, unbalance, and DC currents. The on-purpose injection would allow the use of PV/EV as an active power filter satisfying asymmetric and nonlinear loads, thus improving the power quality of the grid. This *PQ* control uses the expressions given in Eq. (16) for the generation of reference fundamental positive sequence currents, whereas a sequence analyser is used to provide the reference setpoints for asymmetric and nonlinear currents. The analyser acquires the measured load current and extract the required asymmetric, harmonic, and DC components and use these as a reference to PV/EV current controller so as to be injected by renewables making them act as an active power filter (compensating network harmonics, unbalance, DC offset, etc.). This adds an advantage and unique feature to grid-connected system which enables grid power quality support in addition to supplying the actual load demand. It has been shown in Ref. [37] that this could help reduce the network loses by 5.5%, improve current power quality by 85.84%, and save network capacity by reducing net energy demand on the grid (16.38%). The graphical representation of advanced *PQ* controlled is presented in Fig. 15.

4.3.2 Current controllers

The current controller is the final part of control system which enables the injection of reference current and deliver/absorb the required amount of power. The performance and response of the current controller is critical and sensitive to the grid conditions. A very conventional controller shown in Fig. 14 is suitable for tracking the reference currents but only under normal grid conditions. However, in practice, the grid supply may have harmonics, unbalanced currents, DC, and various kinds of faults. To deal with such distorted grid supply, advanced current controllers are required, which are capable of compensating grid voltage harmonic, unbalanced faults, and nonlinear load currents.

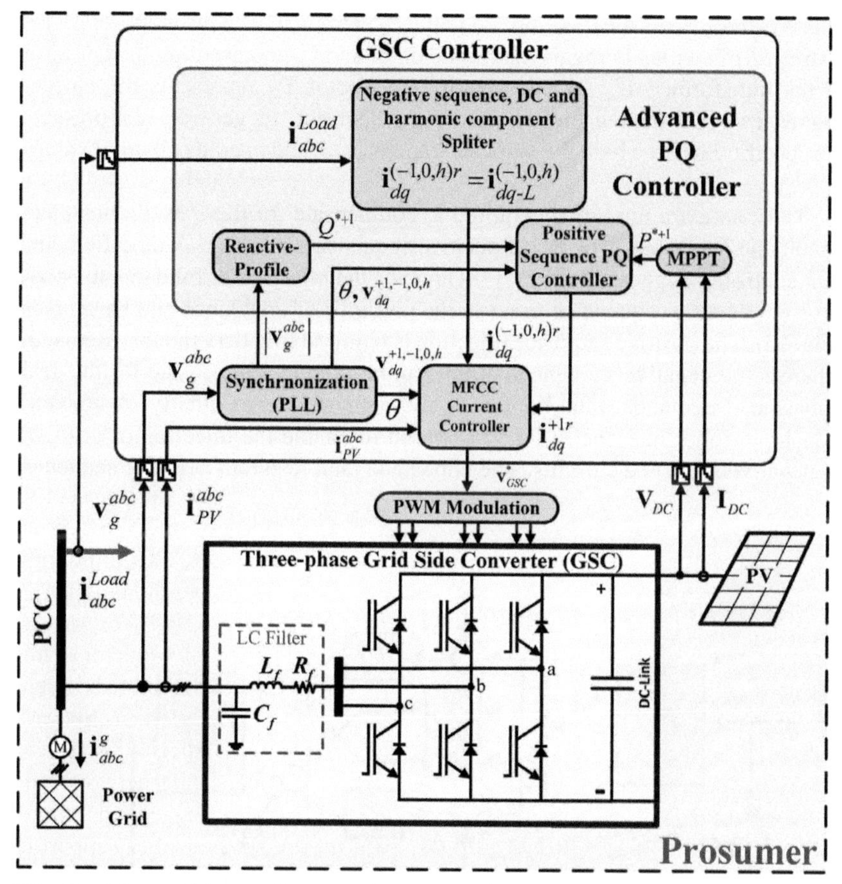

FIG. 15 Advanced PQ controller generating asymmetric and nonlinear currents [37].

Modified *dq*-current controller

The conventional controller designed in *dq*-frame fails when the grid three-phase voltages are unbalanced and/or include harmonics. This is because the *dq*-transformation results in DC quantities only if the input signal contains a single frequency (e.g., 50 Hz). The presence of unbalance gives rise to a negative sequence current (i.e., −50 Hz), which results in double frequency oscillations on positive sequence transformed vector and this jeopardize the performance of PI controller. The interaction of DC and oscillating terms in a *dq* frame is expressed in equation (18) as:

$$\mathbf{i}_{dq}^{n} = \underbrace{\mathbf{I}_{dq}^{n}}_{DC\ Term} \underbrace{\left(\sum_{m \neq n} \mathbf{I}_{dq}^{m} \left[T_{dq}^{n-m} \right] \right)}_{Oscillation Term} \tag{18}$$

where $\mathbf{i}_{dq}^{n}$ represents the transformed voltage vector in n^{th} synchronous reference frame (SRF^{n}) with n being the harmonic order and $\mathbf{I}_{dq}^{n}$ representing the DC term in the transformed $\mathbf{I}_{dq}^{n}$ voltage vector. The vector $\mathbf{I}_{dq}^{m}$ represents the current sequences other than n present in the grid current. In general, n represents the specific SRF to which the signal is transformed, and m holds all other values expect n.

Thus, an extra module is required to compensate for these oscillations and enable the accurate operation of positive sequence controller. A modified current controller suggested in Ref. [38] enables the injection of fundamental positive or negative sequence current in the presence of unbalance and harmonics (Fig. 16). The positive sequence injection is required to satisfy the load demand, whereas the negative sequence current can be used to compensate for the grid unbalanced demands, thus improving the network power quality. A conventional controller like one in Fig. 14 is used to enable the injection of positive or negative sequence currents. The unbalance (and resultant negative sequence

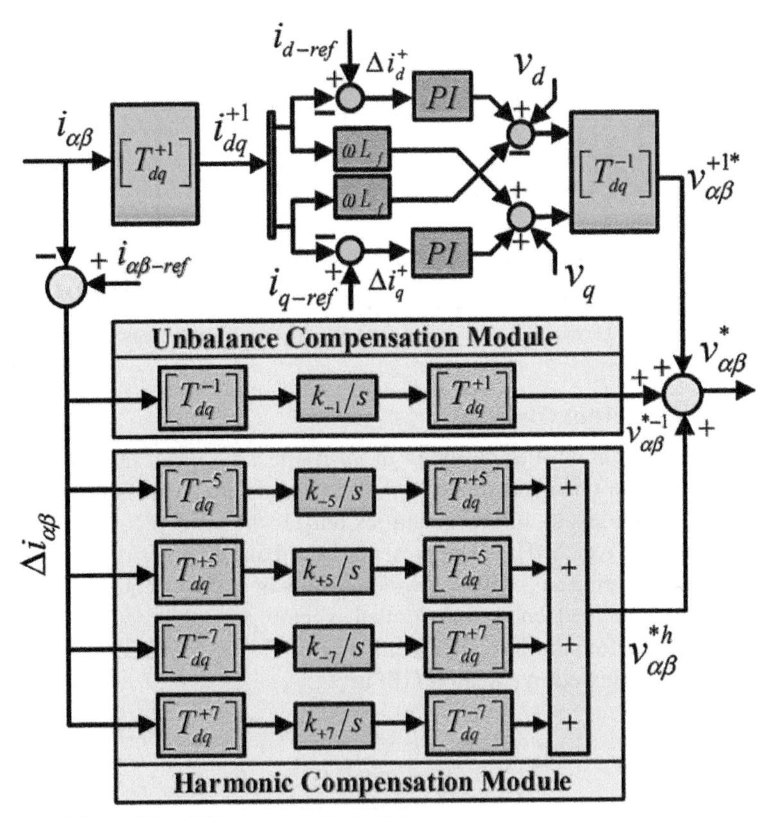

FIG. 16 The modified SRF current controller [39].

current) and voltage harmonics are compensated by introducing multi-back-to-back dq-frames and integral controllers.

The multi-functional current controller (MFCC)

It has been suggested in Ref. [15,37] that a PV system (or any RES with inverter) can be used as an active power filter which requires a more diversified and flexible current controller capable of on-purpose injecting/absorbing the asymmetric, DC, and harmonic currents from the local loads, thus enhancing the grid power quality. The reference currents provided by the advanced PQ controller of Fig. 15 can be combined with the MFCC shown in Fig. 17 to mark the on-purpose delivery of such distorted and asymmetric currents. This will make the grid-connected system more diversified and grid supporting, thus enhancing the overall operation of the power system, reducing demand on the grid, losses, and the cost of extra power conditioning devices. The MFCC utilizes dq-control theory and results in a modular controller. The P, N, DC, and h modules shown in Fig. 17 are, respectively, responsible for injecting/absorbing positive sequence, unbalance, DC, and harmonic currents. The controller takes the measured current as an input, where each module decouples and extracts the desired current component, compares it with the reference signal, and uses the PI controller to achieve the desired setpoint. The controller works in the same way when any harmonic is present in the grid voltage as well as for unbalance. The mathematical representation of MFCC for the extraction of desired $\mathbf{i}_{dq}^{n*}$ is given as (Eq. 19):

$$\mathbf{i}_{dq}^{n^*} = \left[T_{dq}^{n}\right] \underbrace{\left(\mathbf{i}_{\alpha\beta} - \sum_{m \neq n} \left[T_{dq}^{-m}\right] [F(s)] \left[T_{dq}^{m}\right] \mathbf{i}_{\alpha\beta}^{m^*} \right)}_{\mathbf{i}_{\alpha\beta}^{n^*}} \tag{19}$$

Where, the $[F(s)]$ is first order low pass filter.

The controller based on (19) has the ability to perform accurately in the presence of unbalanced faults, harmonics, and DC offset. Furthermore, it has the capability to inject on-purpose asymmetric, DC, and nonlinear harmonic currents, if required by the connected nearby loads.

4.3.3 *Synchronization techniques*

The synchronization unit, such as PLL, plays a pivotal role in the overall operation of grid-connected storage and RES. This is because the PLL is the key element of the whole control system using the PQ and current controllers in the dq-frame. The PLL allows a fast and accurate synchronization of the energy system to the utility grid. The grid voltage information such as the voltage frequency, phase, and amplitude obtained through the PLL plays a vital role for enabling the accurate control of the grid-connected RES. Consequently, the design of the PLL under normal/abnormal grid conditions is critical for the

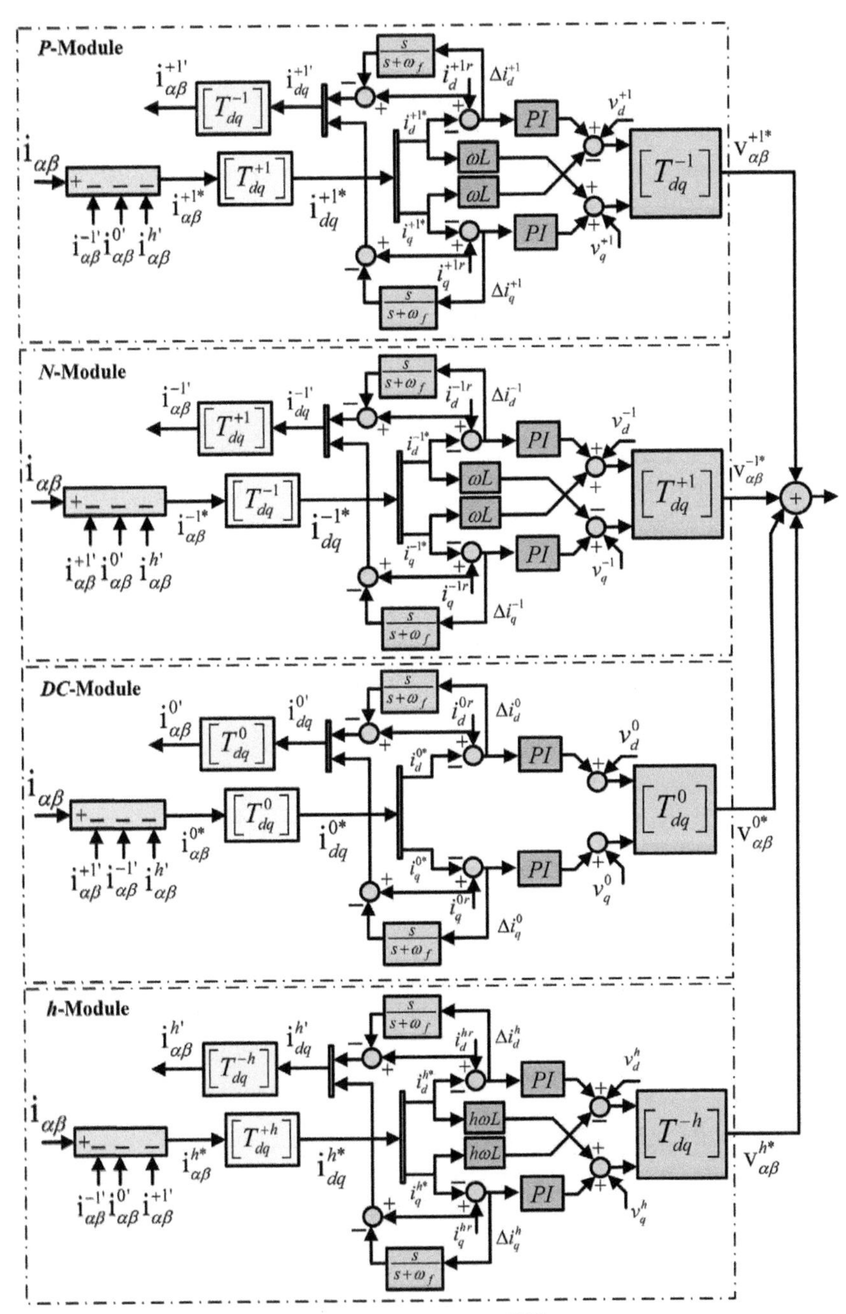

FIG. 17 The MFCC for grid-connected energy systems [37].

accurate operation of the grid-connected RES/EV. Various PLL techniques are proposed in the literature ranging from a basic to advanced techniques.

dq-PLL

A very basic and conventional PLL is the synchronous reference frame *dq*-PLL, shown in Fig. 18, which employs the fundamental frequency transformation ($n = +1$). The q-component of the resulting transformed voltage vector ($\mathbf{v}_{dq}^{+1}$) is passed through a PI controller and the phase angle and frequency information is obtained. It is worth noting that the PLL is accurate for normal grid conditions but does not work well during faults or when the grid voltage is distorted.

Decoupling network in αβ-domain PLL (DNabPLL)

The reason for inaccurate response from *dq*-PLL is because it is designed for single frequency input (i.e., for fundamental component $n = 1$). However, when the input voltage contains components other than the fundamental (such as harmonics, interharmonics, DC offset, etc.), the transformed voltage vector $\mathbf{v}_{dq}^{+1}$ in addition to DC suffers from unwanted oscillations as expressed by Eq. (18). This transformation in Eq. (18) with $n = 1$, under normal grid conditions, results in a positive sequence nonoscillating term only, that is $\mathbf{i}_{dq}^{+1}$. However, under abnormal grid conditions, the nonoscillating terms of positive $\mathrm{SRF}^{+1}\mathbf{v}_{dq}^{+1}$ is accompanied by undesired double, fundamental, and ($1 - h^{th}$) frequency oscillations because of the unbalance sequence ($m = -1$), DC offset ($m = 0$), and harmonic component ($m = h$), respectively. Likewise, if the transformation is carried out with $n = -1$, 0 *or* h, coupling oscillations are observed because of the remaining sequences present in the current. These oscillations limit the performance of PI controller as it is designed to work for DC inputs. Thus, it is important to first extract a clean positive sequence *dq*-grid voltage vector

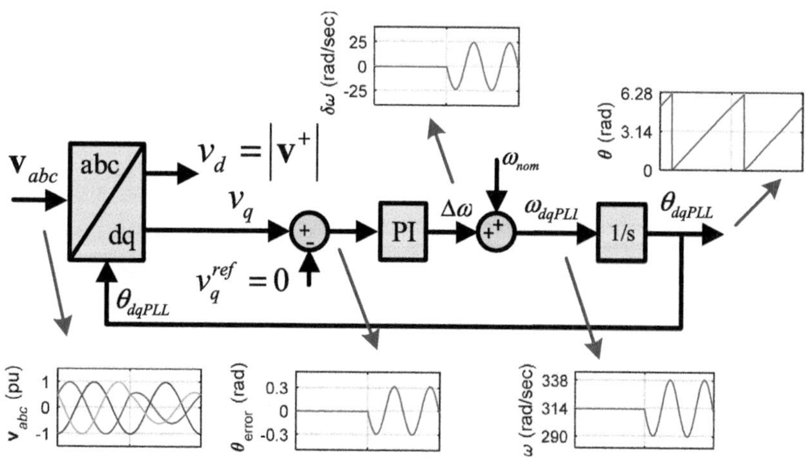

FIG. 18 The *dq-PLL* schematic diagram [40].

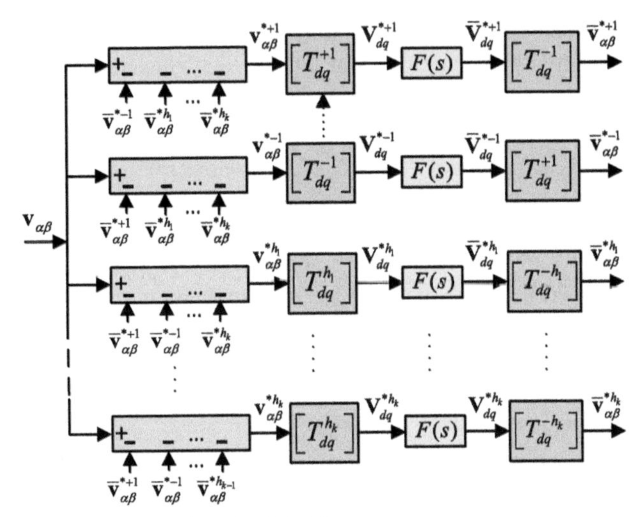

FIG. 19 Decoupling network for DNabPLL [40].

and then use it to extract the desired information. A schematic diagram of the multi-decoupling cell is given in Fig. 19, whereas its application as a PLL is depicted in Fig. 20.

The DNαβPLL [41] uses multi-harmonic decoupling cells to enable the extraction of clean fundamental positive sequence grid voltage. The PLL in each cell compensate the undesired component and extract the one that it is responsible for and at the end combine all, leading to a positive sequence dq-vector. The resulting dq^{+1} then uses an αβ-PLL [40] to extract the grid voltage phase angle and frequency. The significant advantage of PLL is that it

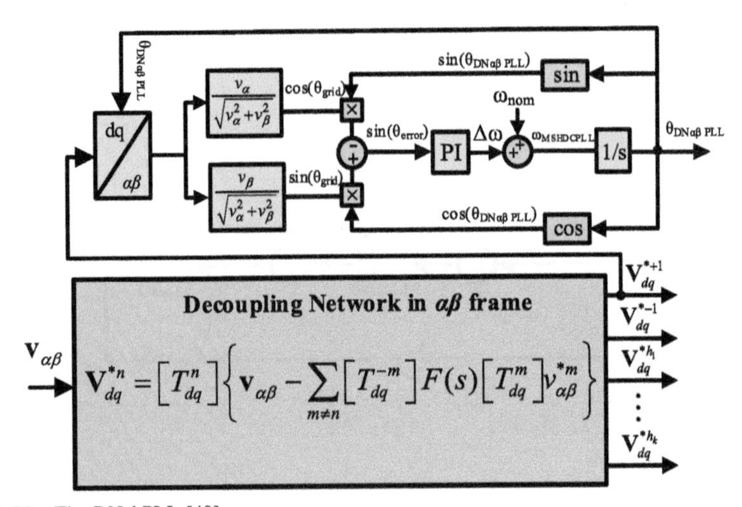

FIG. 20 The DNabPLL [40].

provides *dq*-vector for each of the component under considerations such as (+1, −1, +5, etc.) useful for feed-forward current controllers. However, this PLL requires an advance knowledge of harmonics to be compensated and additionally, it cannot cater for DC offset.

Harmonic-interharmonic DC offset PLL (HIHDOPLL)

The HIDOPLL [42] provides a less complex and advanced technique for the accurate extraction of grid voltage phase angle and frequency under harmonics, interharmonics, and DC offset. In addition, it does not require knowledge in advance for compensating grid harmonic/interharmonics. The structure of HIHDOPLL is a combination of two parts, one is the DC offset and unbalanced compensation block, and the second one is the harmonic/interharmonic compensation cell, shown in Fig. 21. The first block utilizes multi-*dq* frames to

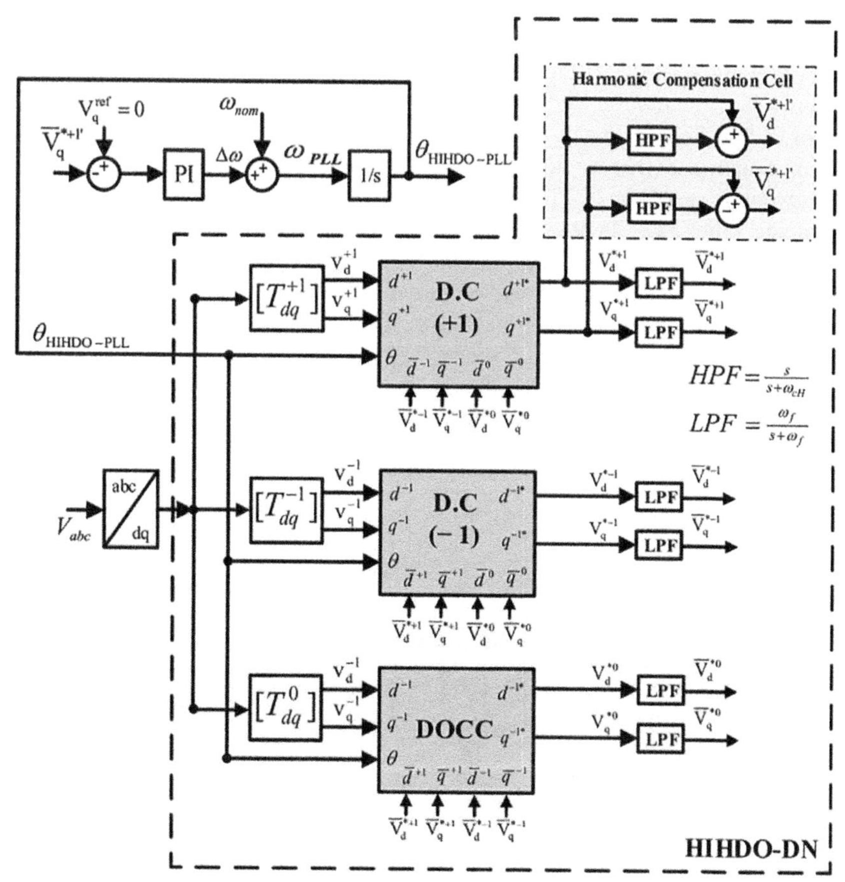

FIG. 21 The HIHDO-PLL [42].

remove the presence of undesired oscillation caused by grid voltage unbalance component (negative sequence) and DC offset. Following this, a High Pass Filter (HPF) configuration is used to provide attenuation of the harmonics and interharmonics. The cut-off frequency of the HPF plays a role in the dynamic response of the PLL. The DC offset, unbalance, and harmonic compensation are expressed mathematically by equations (20) and (21), respectively.

$$\begin{bmatrix} \mathbf{V}_{dq}^{*+1} \\ \mathbf{V}_{dq}^{*-1} \\ \mathbf{V}_{dq}^{*0} \end{bmatrix} = \left[\begin{bmatrix} \mathbf{v}_{dq}^{+1} \\ \mathbf{v}_{dq}^{-1} \\ \mathbf{v}_{dq}^{0} \end{bmatrix} - \begin{bmatrix} [0] & T_{dq}^{+1-(-1)} & T_{dq}^{+1-(0)} \\ T_{dq}^{-1-(+1)} & [0] & T_{dq}^{-1-(0)} \\ T_{dq}^{0-(+1)} & T_{dq}^{0-(-1)} & [0] \end{bmatrix} \begin{bmatrix} \overline{\mathbf{V}}_{dq}^{*+1} \\ \overline{\mathbf{V}}_{dq}^{*-1} \\ \overline{\mathbf{V}}_{dq}^{*0} \end{bmatrix} \right] \tag{20}$$

$$\mathbf{V}_{dq}^{+1\prime} = \mathbf{V}_{dq}^{*+1} - \left[T_{dq}^{-1} \right] [H] \mathbf{V}_{dq}^{*+1} \tag{21}$$

Enhanced prefiltering moving average filter type 2 PLL

The Moving Average Filter-based PLLs (MAFPLL) are also widely used for simple implementation and immunity to grid harmonics. The conventional MAFPLL suffers from slow dynamic response and inaccurate response under off-nominal grid frequencies. Several attempts have been made to improve the dynamic response of MAFPLL and one of them is the Enhanced Moving Average Filter PLL Type 2, given in Fig. 22 [43]. The PLL uses two back-to-back transformations with an intermediate MAF. This arrangement improves the dynamic response and also adds a compensation factor to improve the estimation accuracy under off-nominal frequencies. The input voltage is transferred to *dq-frame* with nominal grid frequency passed through the *dq-αβ* transformation with compensation factor added. The modified *αβ* version of the input is combined with *dq*-PLL and phase information is estimated. MAF PLLs are suitable for applications where complexity is of major concern.

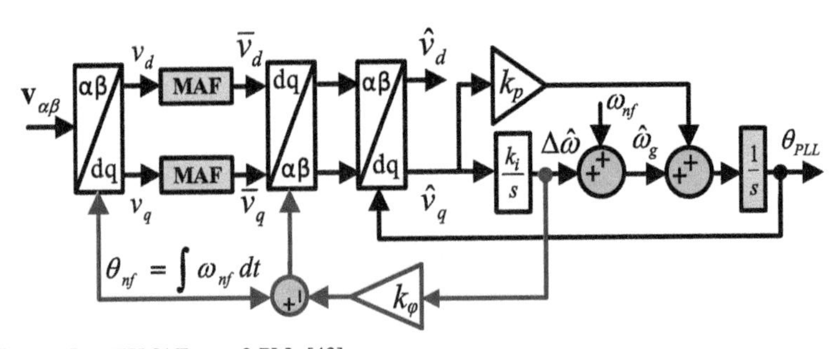

FIG. 22 The EPMAF type 2 PLL [43].

5. EV modeling and charging algorithms

5.1 EV modeling techniques

The operation of EVs and their control are influenced by the battery management system (BMS). The design of the BMS depends upon the accurate estimation of battery model and its real-time implementation. Various models of the battery exist in the literature such as chemical, electrochemical, and electrical models. The chemical models are used for manufacturing purposes. The partial differential equation based pseudo-2-dimension technique offers high modeling accuracy but presents computational complexity, thus not suitable for control applications. The electrical models, such as Equivalent Circuit Models (ECMs), are of great interest to systems engineers as they are easy to implement and used in BMS particularly for condition monitoring and control of battery. Several RC (resistor-capacitor) combinations are employed in the ECM for predicting the dynamic response of a battery. The simplest model includes one-RC element, whereas models with two or more RC elements are also presented in the literature. Increasing RC models increases the estimation accuracy but at the same time presents a high computational complexity for real-time implementation and control. The methods used for estimating ECM parameters include Kalman filter (KF) [44], Extended KF [45], Genetic Algorithm [46], and Recursive least square with multiple adaptive forgetting factors [47]. These methods provide good accuracy; however, most of these methods are very complex for embedded microcontroller of onboard BMS. There exist techniques where lookup tables containing number of parametric values are used, derived based on various operating conditions and exhaustive experiments. These methods fail when the operating conditions change, and they use lookup tables which require memory units. Recently, an online reduced complexity (ORC) technique has been proposed in Ref. [29], which offers promising estimation accuracy with a smaller number of parametric values and does not require lookup tables. The flow diagram is presented in Fig. 23 and the algorithm used trust-region optimization-based least square algorithm for estimation of parameters. The results summarized in Ref. [29] shows a significant reduction of the estimation time. For example, in the Dynamic Driving Cycle, New European Driving Cycle, and Pulsating load, the ORC techniques require 70%, 91.7%, and 92.1% less processing time as compared to other techniques. A further investigation in the laboratory showed a reduction of 96.2%, 93.2%, 88.9%, 91.3%, and 94.7% less estimation time for five different load sets. In real-time, the BMS continuously measures the terminal voltage and current and estimate the battery model parameters. Thus, an estimation technique with lower complexity and easy implementation is key for advanced operation of EVs.

5.2 Charging algorithms for EVs

EVs are increasingly employed these days as a greener technology with nearly zero direct carbon emission. However, charging of EVs still indirectly

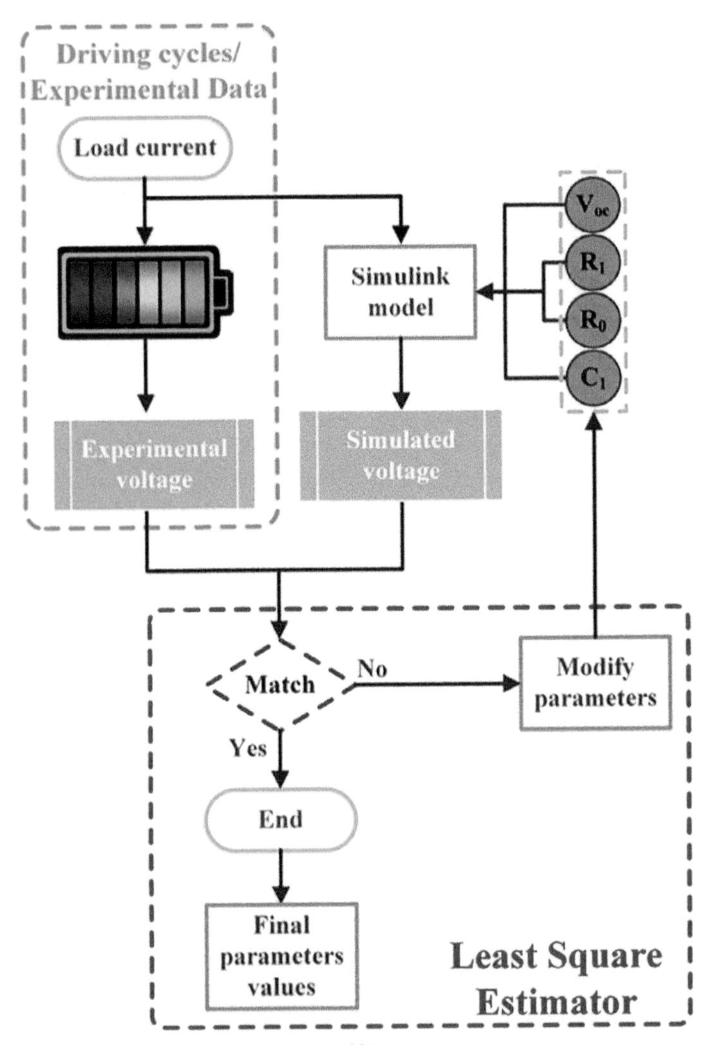

FIG. 23 Flow of ORC estimation technique [29].

consumes carbon and may end up adding to carbon production if not properly controlled. Thus, there is a need for smart charging algorithms that consider the environment and network support in addition to other objectives such as meeting drivers' requirements and being able to exploit full use of available RES for charging. Various charging algorithms exist in the literature. An ideal smart charge controller should be able to: (a) meet EV user requirements, (b) support the grid, (c) deal with intermittent nature of generation from renewables, (d) charge EVs from renewable energy, (e) increase energy autonomy, and (f) do these with minimum degradation to EV battery. The end goal is developing a control algorithm with intelligence, optimized response, and lower complexity [48]. A summary of charging algorithms available in the literature, their features, and capabilities are listed in Table 2.

6. Digital real-time simulators

The development of diversified control algorithms for RES and EV are mainly limited to their real-time operation, such as the dynamic response, implementation feasibility, computational complexity, and efficiency for certain applications. If not adequately designed, implementing such algorithms on a hardware prototype could result in failure, performance issues, hindering progress, and impeded efficiencies. Therefore, there is a need for software twinning using real-time digital simulators where the physical plant is replaced by a precisely equivalent computer model, running in real-time, and appropriately equipped with inputs and outputs (I/Os) ports that are capable of interfacing with real control systems and other equipment. Digital twinning using real-time simulators can accurately reproduce the tested system and its dynamics, together with sensors and actuators, providing comprehensive closed loop testing without the need for field testing on large-scale real systems and the subsequent expense and fault-finding issues. This approach offers significant cost and time savings that affect the product development cycle across all industries: time-to-market and system complexity [3,4]. The digital twin helps analyze behavior, performance, and prospective impact on the energy autonomy and carbon emissions. It also helps in better planning of future improvements to power system operations. It complements and improves algorithm developments and demonstration, including simulation testing, associated engineering design reviews, lab-bench testing, etc. This section focuses on the selection and use of digital real-time simulator to validate control algorithms and power system models, replicating costly and specialized equipment.

An overview of various real-time simulator is discussed in Ref. [64,65]. Most of digital simulators have common characteristics, such as (a) a host-computer is required to prepare a model (e.g., in MATLAB) and load it to the simulator for real-time operation, (b) multiple cores processor to perfume parallel operations, (c) input/output interface to connect to real-world hardware, and (d) physical or wireless communication links. The selection of a digital simulator for a particular application is critical to the performance, size, compatibility to popular software, user interface, and most importantly the cost. Some of the most popular and emerging simulators are presented in the following and summarized in Table 3.

6.1 dSPACE

dSPACEDSP Boards are very popular for implementing and validating real-time control algorithms. These are implemented for simulating stand-alone

TABLE 2 Various EV charging algorithms.

EV charging algorithms	Objectives						
	EV user requirements	*Renewable energy utilization*	*Power system support*	*Battery health*	*Energy exchanged cost*	*Dynamic tariff*	*Key highlights*
[49–51]	N	N	Y	N	N	N	EV impact on load demand and remedies
[52,53]	N	N	Y[a]	N	N	N	Centralized aggregator for EV charging
[54]	N	N	Y	N	N	Y	Differential pricing to shape load curve
[55]	N	Y	N	N	N	N	Fixed charging Demand side load profile improvement
[56]	N	N	Y	N	N	N	EV impact on voltage profile during peak hours
[57,58]	N	N	Y	N	Y	N	Mitigating effects of peak load
[59]	Y	Y	N	Y	N	N	Grid support Variable charging rate Satisfy user
[60]	Y	Y	N	N	Y	Y	Peer to peer energy trading Considering tariff

[61]	N	N	Y	N	Y	N	Optimal charging locations Reducing the negative impact of EV
[62]	N	N	Y	N	Y	Y	Customer support through price monitoring
[59]	Y	N	Y	Y	N	N	Grid support Variable charging rate Satisfy user
[63]	N	Y	N	N	N	N	50% for grid support and 50% reserved for driving

[a]*No real-time support, Y=Yes, N=No.*

TABLE 3 Overview of digital real-time simulators.

Digital technology	Hardware platform	Software compatibility	Interfacing links	Applications
dSPACE (DS1202)	NXP processor	MATLAB	Ethernet, USB, CAN, Serial, LVDS, Xilinx programmable FPGA, DIO and AIO	Robotics, electric drives control, renewable energy, vehicle engineering, or aerospace
OPAL-RT (OP5650XG)	IntelXeon processor and Xilinx FPGAs	MATLAB, RT-LAB, Hypersim	Ethernet, USB, DIO, AIO, CAN Bus, IEC61850, C37.118	Control and protection systems used in power grids, power electronics, motor drives, automotive
Speedgoat (Performance real-time target machine)	Processor and FPGA	MATLAB	Ethernet, CAN, C37.118, RS232, DIO, AIO	Power systems, smart grid, power electronics, control system, renewables, automotive
RTDS (Nova Core 1.0)	RISC processor	RSCAD Limited MATLAB	GBH port, Ethernet port, IRC ports, D/A channels, GTSYNC port	Protection and control Distributed smart grids, Education and training, Smart grids, HVDC, FACTs

controllers or systems and may be used in conjunction with actual hardware or other digital simulators (e.g., that may run a power system). In general, small budget dSPACE systems have limited processing power, but are affordable and compact. The hardware engine used by dSPACE is a processor and FPGA, and it supports BNC analog I/Os, digital I/Os, Ethernet, PCIe, and CAN communication. dSPACE units work on Windows and are compatible with MATLAB but not with all available versions, which somehow limit their performance and operation. To control and monitor a real-time operation, dSPACE units have their own GUI named as dSPACE control desk. Application areas targeted by dSPACE are real-time control, power electronics, smart algorithms, and rapid prototyping (Fig. 24).

6.2 OPAL-RT

OPAL-RT includes PC/FPGA-based real-time simulators, Hardware-in-the-Loop (HIL) testing equipment, and Rapid Control Prototyping (RCP) systems to design, test, and optimize the control and protection systems used in power grids, power electronics, motor drives, and automotive industry. They are widely used by R&D centers and universities. The OPAL-RT is primarily targeted for emulation of electric power systems and can be combined with dSPACE, for example, for combined power and control application. In terms of cost, it is relatively expensive and each CPU core needs an extra license, which further increases the cost. The OPAL-RT units use Intel processors and FPGAs and are compatible with MATLAB and controlled via RT-Lab.

FIG. 24 Front view for dSPACE MicroLab Box.

FIG. 25 Front end view of OPAL RT.

They support both Windows and Linux systems and communicate via PCIe with A/C and D/A on DSP and CAN (Fig. 25).

6.3 Speedgoat

Speedgoat offers a range of high-performance multi-core, multi-CPU target computers. Each is optimized for a different application area, from a mobile controller prototype (RCP) to multi-target rack systems for HIL applications. It uses CPUs and FPGA and is fully integrated with MATLAB Simulink (does not need extra software to control the real-time operation). The best feature is that all cores comes preactivated, so it is very effective in terms of the cost. The interface and communication links that it supports are the I/O modules, ethernet, TCP/IP, and CAN (Fig. 26).

6.4 RTDS

The RTDS unit is a very powerful machine for simulating large power systems on a multi-core system. It targets applications, such as power electronics, control systems, and smart grids. It uses NPX processors, works on Windows and

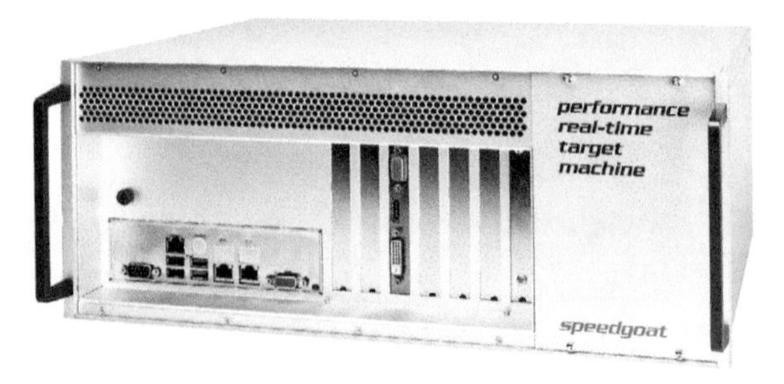

FIG. 26 Front end view of Speedgoat performance real-time simulator.

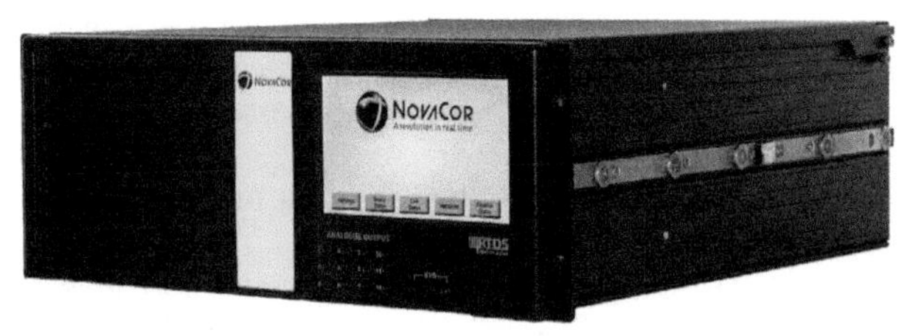

FIG. 27 RTDS Novacore hardware.

Linux, and supports ethernet. The RTDS, however, has its own simulation and model development toolbox named as RSCAD. It only supports converting very basic MATLAB models and thus is not very suitable for working with MATLAB. In addition, it is very expensive (Fig. 27).

7. Conclusions

RES and EVs are increasingly employed in order to provide green energy supply and transport systems. However, power grid faces challenges from the increased deployments of RES and EVs, such as power quality issues related to intermittent nature of renewables and uncontrolled charging of EVs. Consequently, smarter and intelligent use of distributed sources in a smart power grid is important in order to maximize the use of renewable energy and charge EVs from green energy while improving the power quality of supply and support the grid. This chapter provided a comprehensive review of global PV and EV market together with their impact on grid stability, power quality, and system cost. The chapter further provided an in-depth knowledge of control algorithms and charging techniques that can be used to control and manage energy exchange between the grid, renewables, and EVs. Finally, the use of digital real-time solutions for replicating costly and specialized equipment for indirect field testing is introduced. These solutions offer significant cost and time savings from issues affecting product development across all industries: time-to-market and system complexity.

Acknowledgment

This work is jointly supported by British Council and HAWKSBI grant funded by IUK.

References

[1] REN 21, Renewables 2021 Global Status Report, 2021, Accessed: 01/11/2021. [Online]. Available: https://www.ren21.net/reports/global-status-report/.

[2] E. Bentley, et al., On beneficial vehicle-to-grid (V2G) services, in: 2021 9th International Conference on Modern Power Systems (MPS), 16-17 June 2021, 2021, pp. 1–6, https://doi.org/10.1109/MPS52805.2021.9492671.

[3] K. Saleem, K. Mehran, Z. Ali, An improved pre-filtering moving average filter based synchronization algorithm for single-phase V2G application, in: 2020 IEEE Energy Conversion Congress and Exposition (ECCE), 11-15 Oct. 2020, 2020, pp. 4088–4093, https://doi.org/10.1109/ECCE44975.2020.9236406.

[4] IEA, Global EV Outlook 2021, IEA, Paris, 2021. Accessed: 01/11/2021. [Online]. Available: https://www.iea.org/reports/global-ev-outlook-2021.

[5] A. Tavakoli, S. Saha, M.T. Arif, M.E. Haque, N. Mendis, A.M.T. Oo, Impacts of grid integration of solar PV and electric vehicle on grid stability, power quality and energy economics: a review, IET Energy Syst. Integr. 2 (3) (2020) 243–260.

[6] US Department of Energy/PNNL/EPRI, High-resolution, time-synchronized grid monitoring devices, in: North American SynchroPhasor Initiative, NASPI-2020-TR-004, 2020. [Online]. Available: https://www.naspi.org/node/819.

[7] C.H. Dharmakeerthi, N. Mithulananthan, T.K. Saha, Impact of electric vehicle load on power system oscillatory stability, in: 2013 Australasian Universities Power Engineering Conference (AUPEC), 29 Sept.–3 Oct. 2013, 2013, pp. 1–6, https://doi.org/10.1109/AUPEC.2013.6725401.

[8] A. Tavakoli, M. Negnevitsky, D.T. Nguyen, K.M. Muttaqi, Energy exchange between electric vehicle load and wind generating utilities, IEEE Trans. Power Syst. 31 (2) (2016) 1248–1258, https://doi.org/10.1109/TPWRS.2015.2418335.

[9] C.H. Dharmakeerthi, N. Mithulananthan, T.K. Saha, Impact of electric vehicle fast charging on power system voltage stability, Int. J. Electr. Power Energy Syst. 57 (2014) 241–249, https://doi.org/10.1016/j.ijepes.2013.12.005.

[10] D. Wu, K.T. Chau, C. Liu, S. Gao, F. Li, Transient stability analysis of SMES for smart grid with vehicle-to-grid operation, IEEE Trans. Appl. Supercond. 22 (3) (2012) 5701105, https://doi.org/10.1109/TASC.2011.2174572.

[11] M. Tabari, A. Yazdani, A mathematical model for stability analysis of a dc distribution system for power system integration of plug-in electric vehicles, IEEE Trans. Veh. Technol. 64 (5) (2015) 1729–1738, https://doi.org/10.1109/TVT.2014.2336854.

[12] Z. Ali, N. Christofides, L. Hadjidemetriou, E. Kyriakides, Photovoltaic reactive power compensation scheme: an investigation for the Cyprus distribution grid, in: 2018 IEEE International Energy Conference (ENERGYCON), 3–7 June 2018, 2018, pp. 1–6, https://doi.org/10.1109/ENERGYCON.2018.8398746.

[13] S. Eftekharnejad, V. Vittal, G.T. Heydt, B. Keel, J. Loehr, Small signal stability assessment of power systems with increased penetration of photovoltaic generation: a case study, IEEE Trans. Sustain. Energy 4 (4) (2013) 960–967, https://doi.org/10.1109/TSTE.2013.2259602.

[14] F. Milano, F. Dörfler, G. Hug, D.J. Hill, G. Verbič, Foundations and challenges of low-inertia systems (invited paper), in: 2018 Power Systems Computation Conference (PSCC), 11-15 June 2018, 2018, pp. 1–25, https://doi.org/10.23919/PSCC.2018.8450880.

[15] Z. Ali, N. Christofides, L. Hadjidemetriou, E. Kyriakides, Diversifying the role of distributed generation grid-side converters for improving the power quality of distribution networks using advanced control techniques, IEEE Trans. Ind. Appl. 55 (4) (2019) 4110–4123, https://doi.org/10.1109/TIA.2019.2904678.

[16] K. Saleem, Z. Ali, K. Mehran, A single-phase synchronization technique for grid-connected energy storage system under faulty grid conditions, IEEE Trans. Power Electron. 36 (10) (2021) 12019–12032, https://doi.org/10.1109/TPEL.2021.3071418.

[17] IEEE, IEEE recommended practice for utility interface of photovoltaic (PV) systems, in: IEEE Std 929-2000, 2000, p. i, https://doi.org/10.1109/IEEESTD.2000.91304.

[18] R.L. Gonçalves, J.T. Saraiva, J.C. Sousa, V.T. Mendes, Impact of electric vehicles on the electricity prices and on the load curves of the Iberian Electricity Market, in: 2013 10th International Conference on the European Energy Market (EEM), 27-31 May 2013, 2013, pp. 1–8, https://doi.org/10.1109/EEM.2013.6607279.

[19] E. Sortomme, M.A. El-Sharkawi, Optimal scheduling of vehicle-to-grid energy and ancillary services, IEEE Trans. Smart Grid 3 (1) (2012) 351–359, https://doi.org/10.1109/TSG.2011.2164099.

[20] L. Hadjidemetriou, et al., Design factors for developing a university campus microgrid, in: 2018 IEEE International Energy Conference (ENERGYCON), 3-7 June 2018, 2018, pp. 1–6, https://doi.org/10.1109/ENERGYCON.2018.8398791.

[21] A.D. Paquette, D.M. Divan, Design considerations for microgrids with energy storage, in: 2012 IEEE Energy Conversion Congress and Exposition (ECCE), 15-20 Sept. 2012, 2012, pp. 1966–1973, https://doi.org/10.1109/ECCE.2012.6342571.

[22] A. Hirsch, Y. Parag, J. Guerrero, Microgrids: a review of technologies, key drivers, and outstanding issues, Renew. Sust. Energ. Rev. 90 (2018) 402–411, https://doi.org/10.1016/j.rser.2018.03.040.

[23] Y. Zhu, F. Wang, J. Yan, The potential of distributed energy resources in building sustainable campus: the case of Sichuan university, Energy Procedia 145 (2018) 582–585, https://doi.org/10.1016/j.egypro.2018.04.085.

[24] K.T. Akindeji, R. Tiako, I.E. Davidson, Use of renewable energy sources in university campus microgrid—a review, in: Proc. IEEE DUE, Wellington, South Africa, 25-27 March 2019, 2019, pp. 76–83.

[25] Berkely Lab (U.S. Department of Energy), Microgrids at Berkely lab, Lawrence Berkeley National Laboratory—University of California, 2019. Available: https://building-microgrid.lbl.gov/types-microgrids. (Accessed 19 December 2019).

[26] N. Hatziargyriou, Microgrids: architectures and control, Wiley/IEEE Press, Chichester, West Sussex, United Kingdom, 2014 (in English).

[27] P.G. Arul, V.K. Ramachandaramurthy, R.K. Rajkumar, Control strategies for a hybrid renewable energy system: a review, Renew. Sust. Energ. Rev. 42 (2015) 597–608, https://doi.org/10.1016/j.rser.2014.10.062.

[28] Z. Yang, C.N.-M. Ho, A review on microgrid architectures and control methods, in: 2016 IEEE 8th International Power Electronics and Motion Control Conference (IPEMC-ECCE Asia), 22-26 May 2016, 2016, pp. 3149–3156, https://doi.org/10.1109/IPEMC.2016.7512799.

[29] K. Saleem, K. Mehran, Z. Ali, Online reduced complexity parameter estimation technique for equivalent circuit model of lithium-ion battery, Electr. Power Syst. Res. 185 (2020), https://doi.org/10.1016/j.epsr.2020.106356, 106356.

[30] J.M.S. Callegari, A.F. Cupertino, V.D.N. Ferreira, H.A. Pereira, Minimum DC-link voltage control for efficiency and reliability improvement in PV inverters, IEEE Trans. Power Electron. 36 (5) (2021) 5512–5520, https://doi.org/10.1109/TPEL.2020.3032040.

[31] C. Tang, Y. Chen, Y. Chen, Y. Chang, DC-link voltage control strategy for three-phase back-to-back active power conditioners, IEEE Trans. Ind. Electron. 62 (10) (2015) 6306–6316, https://doi.org/10.1109/TIE.2015.2420671.

[32] S.R. Mohamed, P.A. Jeyanthy, D. Devaraj, M.H. Shwehdi, A. Aldalbahi, DC-link voltage control of a grid-connected solar photovoltaic system for fault ride-through capability enhancement, Appl. Sci. 9 (5) (2019), https://doi.org/10.3390/app9050952.

[33] A.S. Ahmed, B.A. Abdullah, W.G.A. Abdelaal, MPPT algorithms: Performance and evaluation, in: 2016 11th International Conference on Computer Engineering & Systems (ICCES), 20-21 Dec. 2016, 2016, pp. 461–467, https://doi.org/10.1109/ICCES.2016.7822048.

[34] C. Restrepo, C. González-Castaño, J. Muñoz, A. Chub, E. Vidal-Idiarte, R. Giral, An MPPT algorithm for PV systems based on a simplified photo-diode model, IEEE Access 9 (2021) 33189–33202, https://doi.org/10.1109/ACCESS.2021.3061340.

[35] S. Salman, X. Ai, Z. Wu, Design of a P-&-O algorithm based MPPT charge controller for a stand-alone 200W PV system, Protect. Control Modern Power Syst. 3 (1) (2018) 25, https://doi.org/10.1186/s41601-018-0099-8.

[36] B.N. Nguyen, V.T. Nguyen, M.Q. Duong, K.H. Le, H.H. Nguyen, A.T. Doan, Propose a MPPT algorithm based on thevenin equivalent circuit for improving photovoltaic system operation, Front. Energy Res. 8 (14) (2020), https://doi.org/10.3389/fenrg.2020.00014.
[37] Z. Ali, N. Christofides, L. Hadjidemetriou, E. Kyriakides, Multi-functional distributed generation control scheme for improving the grid power quality, IET Power Electron. 12 (1) (2019) 30–43. Available: https://digital-library.theiet.org/content/journals/10.1049/iet-pel.2018.5177.
[38] L. Hadjidemetriou, E. Kyriakides, F. Blaabjerg, A grid side converter current controller for accurate current injection under normal and fault ride through operation, in: IECON 2013—39th Annual Conference of the IEEE Industrial Electronics Society, 10-13 Nov. 2013, 2013, pp. 1454–1459, https://doi.org/10.1109/IECON.2013.6699347.
[39] Z. Ali, N. Chritofides, L. Hadjidemetriou, E. Kyriakides, A computationally efficient current controller for simultaneous injection of both positive and negative sequences, in: 2017 19th European Conference on Power Electronics and Applications (EPE'17 ECCE Europe), 11–14 Sept. 2017, 2017, pp. 1–10, https://doi.org/10.23919/EPE17ECCEEurope.2017.8099011.
[40] Z. Ali, N. Christofides, L. Hadjidemetriou, E. Kyriakides, Y. Yang, F. Blaabjerg, Three-phase phase-locked loop synchronization algorithms for grid-connected renewable energy systems: a review, Renew. Sust. Energ. Rev. 90 (2018) 434–452, https://doi.org/10.1016/j.rser.2018.03.086.
[41] L. Hadjidemetriou, E. Kyriakides, F. Blaabjerg, A robust synchronization to enhance the power quality of renewable energy systems, IEEE Trans. Ind. Electron. 62 (8) (2015) 4858–4868, https://doi.org/10.1109/TIE.2015.2397871.
[42] Z. Ali, N. Christofides, L. Hadjidemetriou, E. Kyriakides, Design of an advanced PLL for accurate phase angle extraction under grid voltage HIHs and DC offset, IET Power Electron. 11 (6) (2018) 995–1008, https://doi.org/10.1049/iet-pel.2017.0424.
[43] Z. Ali, N. Christofides, L. Hadjidemetriou, E. Kyriakides, Performance enhancement of MAF based PLL with phase error compensation in the pre-filtering stage, in: 2017 IEEE Manchester PowerTech, 18-22 June 2017, 2017, pp. 1–6, https://doi.org/10.1109/PTC.2017.7981086.
[44] G.L. Plett, Sigma-point Kalman filtering for battery management systems of LiPB-based HEV battery packs: part 2: simultaneous state and parameter estimation, J. Power Sources 161 (2) (2006) 1369–1384, https://doi.org/10.1016/j.jpowsour.2006.06.004.
[45] G.L. Plett, Extended Kalman filtering for battery management systems of LiPB-based HEV battery packs: part 3. State and parameter estimation, J. Power Sources 134 (2) (2004) 277–292, https://doi.org/10.1016/j.jpowsour.2004.02.033.
[46] Z. Chen, C.C. Mi, Y. Fu, J. Xu, X. Gong, Online battery state of health estimation based on genetic algorithm for electric and hybrid vehicle applications, J. Power Sources 240 (2013) 184–192, https://doi.org/10.1016/j.jpowsour.2013.03.158.
[47] V.-H. Duong, H.A. Bastawrous, K. Lim, K.W. See, P. Zhang, S.X. Dou, Online state of charge and model parameters estimation of the LiFePO4 battery in electric vehicles using multiple adaptive forgetting factors recursive least-squares, J. Power Sources 296 (2015) 215–224, https://doi.org/10.1016/j.jpowsour.2015.07.041.
[48] F.D.P. García-López, M. Barragán-Villarejo, J.M. Maza-Ortega, Grid-friendly integration of electric vehicle fast charging station based on multiterminal DC link, Int. J. Electr. Power Energy Syst. 114 (2020), https://doi.org/10.1016/j.ijepes.2019.05.078, 105341.
[49] Y. Wang, S. Huang, D. Infield, Investigation of the potential for electric vehicles to support the domestic peak load, in: Proc. IEEE IEVC, 17–19 Dec. 2014, 2014, pp. 1–8, https://doi.org/10.1109/IEVC.2014.7056124.
[50] N.B.M. Shariff, M.A. Essa, L. Cipcigan, Probabilistic analysis of electric vehicles charging load impact on residential distributions networks, in: Proc. IEEE ENERGYCON, 4–8 April 2016, 2016, pp. 1–6, https://doi.org/10.1109/ENERGYCON.2016.7513943.

[51] K. Clement-Nyns, E. Haesen, J. Driesen, The impact of charging plug-in hybrid electric vehicles on a residential distribution grid, IEEE Trans. Power Systems 25 (1) (2010) 371–380, https://doi.org/10.1109/TPWRS.2009.2036481.

[52] A.S. Masoum, S. Deilami, P.S. Moses, M.A.S. Masoum, A. Abu-Siada, Smart load management of plug-in electric vehicles in distribution and residential networks with charging stations for peak shaving and loss minimisation considering voltage regulation, IET Gener. Transm. Distrib. 5 (8) (2011) 877–888. Available: https://digital-library.theiet.org/content/journals/10.1049/iet-gtd.2010.0574.

[53] S. Deilami, A.S. Masoum, P.S. Moses, M.A.S. Masoum, Real-time coordination of plug-in electric vehicle charging in smart grids to minimize power losses and improve voltage profile, IEEE Trans. Smart Grid 2 (3) (2011) 456–467, https://doi.org/10.1109/TSG.2011.2159816.

[54] M. Nour, S.M. Said, A. Ali, C. Farkas, Smart charging of electric vehicles according to electricity price, in: Proc. IEEE ITEC, 2-4 Feb. 2019, 2019, pp. 432–437, https://doi.org/10.1109/ITCE.2019.8646425.

[55] R.M. Shukla, S. Sengupta, A.N. Patra, Smart plug-in electric vehicle charging to reduce electric load variation at a parking place, in: Proc. IEEE CCWC, 8-10 Jan. 2018, 2018, pp. 632–638, https://doi.org/10.1109/CCWC.2018.8301710.

[56] A. Neagoe-Stefana, M. Eremia, L. Toma, A. Neagoe, Impact of charging Electric Vehicles in residential network on the voltage profile using Matlab, in: Proc. IEEE International Symposium on ATEE, 7-9 May 2015, 2015, pp. 787–791, https://doi.org/10.1109/ATEE.2015.7133909.

[57] I. Sharma, C. Cañizares, K. Bhattacharya, Smart charging of pevs penetrating into residential distribution systems, IEEE Trans. Smart Grid 5 (3) (2014) 1196–1209, https://doi.org/10.1109/TSG.2014.2303173.

[58] M.F. Shaaban, M. Ismail, E.F. El-Saadany, W. Zhuang, Real-time PEV charging/discharging coordination in smart distribution systems, IEEE Trans. Smart Grid 5 (4) (2014) 1797–1807,-https://doi.org/10.1109/TSG.2014.2311457.

[59] T. Jiang, G. Putrus, Z. Gao, M. Conti, S. McDonald, G. Lacey, Development of a decentralized smart charge controller for electric vehicles, Int. J. Electr. Power Energy Syst. 61 (2014) 355–370.

[60] A. Al-Obaidi, H. Khani, H.E.Z. Farag, M. Mohamed, Bidirectional smart charging of electric vehicles considering user preferences, peer to peer energy trade, and provision of grid ancillary services, Int. J. Electr. Power Energy Syst. 124 (2021), https://doi.org/10.1016/j.ijepes.2020.106353, 106353.

[61] M. Rahman, M. Othman, H. Mokhlis, M. Muhammad, H. Bouchekara, Optimal fixed charge-rate coordination of plug-in electric vehicle incorporating capacitor and OLTC switching to minimize power loss and voltage deviation, IEEJ Trans. Electr. Electron. Eng. (2017), https://doi.org/10.1002/tee.22652.

[62] C.O. Adika, L. Wang, Smart charging and appliance scheduling approaches to demand side management, Int. J. Electr. Power Energy Syst. 57 (2014) 232–240, https://doi.org/10.1016/j.ijepes.2013.12.004.

[63] M. Singh, P. Kumar, I. Kar, Implementation of vehicle to grid infrastructure using fuzzy logic controller, IEEE Trans. Smart Grid 3 (1) (2012) 565–577, https://doi.org/10.1109/TSG.2011.2172697.

[64] L. Ibarra, A. Rosales, P. Ponce, A. Molina, R. Ayyanar, Overview of real-time simulation as a supporting effort to smart-grid attainment, Energies 10 (6) (2017) 817. Available: https://www.mdpi.com/1996-1073/10/6/817.

[65] P. Forsyth, T. Maguire, R. Kuffel, Real time digital simulation for control and protection system testing, in: 2004 IEEE 35th Annual Power Electronics Specialists Conference (IEEE Cat. No.04CH37551), 20-25 June 2004, vol. 1, 2004, pp. 329–335, https://doi.org/10.1109/PESC.2004.1355765.

Chapter 6

Adaptive control of islanded AC microgrid

Sweta Panda[a], Gyan Ranjan Biswal[a], and Bidyadhar Subudhi[b]
[a]*Department of Electrical & Electronics Engineering, VSS University of Technology, Burla, Odisha, India,* [b]*School of Electrical Sciences, Indian Institute of Technology Goa, Farmagudi, Ponda, Goa, India*

1. Introduction

A microgrid (MG) represents an amassment of distributed generation (DG) devices, load clusters, and storage devices. When storage devices combine to form a MG, it can also be called as a mini-grid even though it differs from the main grid based on various factors [1]. Control method is the most important criteria that is centralized only in traditional networks, but it can be centralized, decentralized, or decentralized in MGs [2]. The second criterion is the operating mode. MG works either in grid-tied or in stand-alone mode, but a conventional grid can only operate in grid-tied mode. On account of a traditional grid, global parameter for control methodology is frequency; however, on account of an islanded MG, it is DC link voltage [3]. MGs offer greater reliability, and power losses are also reduced [4].

MGs are basically classified as AC and DC. Most of the traditional grids are AC in nature and henceforth the emphasis is on AC MGs, which are adaptable to work in association with the main grid. Grid-tied mode and stand-alone mode are two modes in which MGs operate. The dynamic features of MG often rely largely on the grid power management in grid-tied mode, whereas MG is cut off from main grid in islanded mode and therefore needs to maintain its own frequency and voltage. The fundamental purpose of microgrid control is to ensure its economic and efficient operation. The desired highlights of the control strategy of microgrid are that output voltage and current of several DG units should monitor their reference and be able to keep the power balance, i.e., voltage and frequency variations. More difficulty arises in the control of islanded MGs due to their lower inertia, uncertainties in DG outputs, and variations in load power. Hence, islanded MGs are more challenging to control rather than grid-connected MGs.

Microgrid Cyberphysical Systems. https://doi.org/10.1016/B978-0-323-99910-6.00007-4

Over the past decade, islanded MG control is carried out using a three-layer hierarchical structure based on the response time. The primary level control is the initial level that divides the load change among different linked DGs to preserve the stability of voltage and frequency. This first level control system is generally carried out locally at every DG. After islanding, it preserves MG's voltage (v) and frequency (f) stability. It features a quick response time to keep frequency and voltage around nominal levels. The primary objectives of this level of control are preserving the stability of v and f, adequate sharing of active power (P) and reactive power (Q) among DGs, and minimizing the circulating current among DGs. For this degree of control, there is no communication system required. This level of control comprises basic hardware control and is sometimes referred to as zero-level control as it encompasses DGs internal voltage control loop (VCL) and current control loop (CCL). However, this control level results in large $v\&f$ deviations. Droop control is the most utilized primary control mechanism in inverter-based MGs.

The secondary control, which is the second layer of control, recovers the v and f variances produced by the main control operation owing to changes in loads or renewable energy sources. Depending on the power system control architecture, this control level can be implemented in one of the three ways: centralized, decentralized, or distributed control [5]. It is generally a little slower than the first control level. It determines new primary control reference values or setpoints, and the process repeats until the $v\&f$ values are set to their respective nominal levels. The MG is likewise synchronized with primary grid ensuring power quality at this control level. Low-bandwidth communication is used at this control level [6]. The tertiary control level is the top-most layer of control. It is in charge of ensuring that the MG and the primary grid have the optimum possible power flow. This level is generally slower than the secondary control level and establishes the reference values for the secondary control level.

Many secondary control techniques have been developed inside this hierarchical control configuration in order to obtain optimal power sharing as well as secondary $v\&f$management of MGs. However, there are still a number of flaws to address. A remote sensing unit monitors the voltage as well as current deviations of all DGs in the centralized secondary control technology and provides them to a central MG control system that creates controller signals and transmits them to main DG control. This approach uses a basic communication topology because information is transferred from the distant sensor block to the central controller in one way only and then from the main controller to individual DGs. This makes it much easier to implement. The disadvantage of this system is that the central controller failure interrupts the operation of the MG (single point error) and hence renders it unreliable. Another major drawback is that it needs massive communication between the central control unit and all DG units across wide geographic regions, which cannot be achieved because of greater costs. The model predictive control (MPC) [7] and optimum dispatch are other techniques utilizing this concept. MPC is a discrete method for time control, which anticipates the control and the following control sequence. The control measures include changes in the demand for load and variations in DG production. This approach is used to manage big power stations. An economical technique, employed by small MGs, is Optimum Dispatch Control [8]. It involves an offline computation of each DG unit's optimal dispatch after assessing the potential functional conditions of all DGs. However, due to the participation of many possible operational states, this results in significant complexity.

Individual DG units are controlled locally by their own controller in a decentralized control scheme. As a result, each DG receives data that is specific to it such as voltage and frequency values. Communication is streamlined because there is no requirement for a central controller and local controllers just share the information that is necessary with neighboring control units, rather than the entire information with all other local controllers. However, as the control units of all the DGs are heavily reliant on one another, totally decentralized control is not feasible. This may result in system instability. Droop control and master–slave control are two approaches for the decentralized type of control. The primary control technique and the droop control approach are identical. The master–slave control scheme, however, employs one master controller and many slave controllers. Through a communication link, the slave controllers function in accordance with the master controller's commands. In islanded mode, the controller acting as master regulates *v and f* of the system. There are two types of master control schematics: single master and multi-master control [9].

Another secondary control scheme is distributed control, which is a fusion of centralized and decentralized control. It leverages local information supplied by neighbors, such as frequency and voltage, and then exchanges information over a two-way communication link, implying some degree of centralized control. As the local controllers only share the needed information with the neighboring control units and not all the information with all of the other local controllers, this system also incorporates some decentralized control. Hence, a shorter communication link is required [10]. As a result, the MG setup is not reliant on single central controller. This lowers the chance of system failure. The main benefits of this schema are its great scalability, durability, and stability, as well as the fact that it reduces communication costs. Some of this approach include the consensus-based technique, multiagent system, feedback linearization, model predictive control (MPC), and so on. With a consensus-based method, the various DER units are converged to a single value. A global average is therefore created. This approach needs simpler communication network, as it obtains a global agreement with its neighboring units without the usage of a specialized control unit. Local agents regulate each distributed energy resource (DER) unit in this approach of multi-agent setup control. To identify system setpoints, the localized agents connect with adjacent agents for sharing data with a centralized controller. In terms of computing, this technique is more flexible and efficient.

The MPC method is discrete-time control strategy that obtains directives of control for any setup by minimizing cost functions. This cost-function is linked to the system's future performance. A cost-function is created for MPC-based approaches by integrating system state and deviance from fixed points. MPC can predict both the control sequence and the control action in the future [11]. MPC is able to handle multivariable control issues while being simple to tune.

Asymptotic convergence to *v and f* references is accomplished by the proposed controllers [12,13] utilizing feedback control and linearization scheme, indicating that the references will not be attained in a short period of time. Furthermore, despite the fact that load parameters and line impedances are unascertained in reality, a detailed model of MGs is used in development of control algorithms [14] that need accurate MG characteristics.

Although distributed proportional integral *v and f*controllers have been developed [15], the protocol's global stability cannot be guaranteed as it is based on a small-scale model. Finite-time distributed controllers using sliding mode technique are used for *v and f* control as well as power sharing [16,17]. Despite this, no link between control gain values and finite settling time has been found, making tuning a challenging process. For power sharing, a certain number of active or reactive power control methods have been proposed [18,19], but there is a necessity to have a secondary control loop to remove the *v and f* variances of proposed controls without affecting the precision of the power sharing.

Adaptive control systems have been researched extensively and given greater attention in recent decades. The fundamental applicability of adaptive control techniques is to deal with parametric unpredictability and perturbations. Adaptive control methods may undoubtedly provide long-term stability, convergence on resilience, and monitoring system dynamics. This work proposes a distributed adaptive control technique utilized at secondary control layer of MG hierarchical control system. A key contribution of this work is the development of distributed adaptive control schemes for accurate power sharing along with *v and f* control of MG.

The remaining of this work is arranged accordingly. Section 2 includes the introduction of the DG-based islanded MG model and the primary droop-based control system. This part also discusses the available secondary control techniques. The preliminary specifications of the graph theory and suggested distributed adaptive control schematic for power synchronization methods and power sharing for the AC MG system are discussed in Section 3. The findings of the simulation and the assessment are validated in Section 4. The conclusion is provided in Section 5.

2. Detailed modeling of islanded AC MG

Each DG unit in a MG setup is made up of a predominant DC supply (Photovoltaic array), a voltage source inverter (VSI), a filter, and an output connector. This setup is shown in Fig. 1. This inverter-based DG mathematical model is

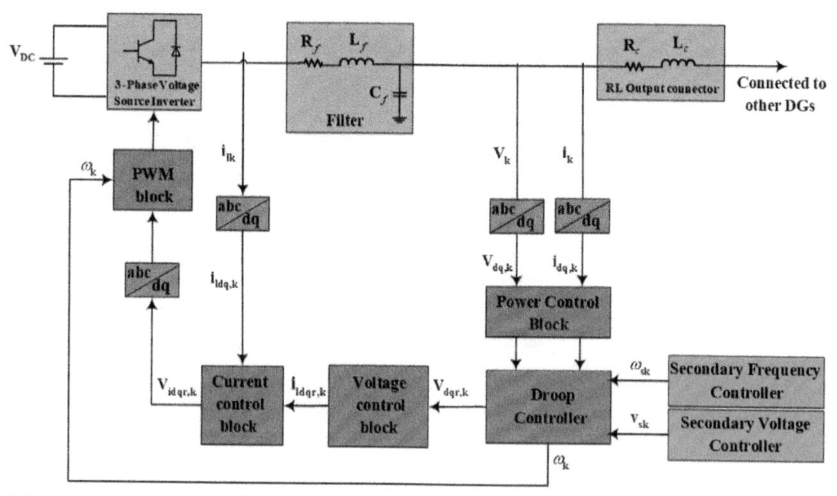

FIG. 1 Complete layout of MG system.

analyzed and presented in detail [20]. The output v *and* f of the VSI are regulated by power control loop, VCL, and CCL. In the VCL and CCL, proportional–integral (PI) controllers are utilized. As shown in Fig. 2, *P*&*Q* droop methods are utilized to describe the power controller. Each DG generally implements the primary control locally, which comprises CCL, VCL, and droop control loop.

The droop control block creates the reference input $v_{dqr,\ k}$ for voltage controller of the kth DG. The VCL block produces the reference input $i_{dqr,\ k}$for the CCL block of the kth DG, as depicted in Fig. 3. Finally, the CCL block provides the set point input for the pulse width modulation (PWM) block, as represented in Fig. 4. The VSI output is regulated by the PWM block. This control level regulates the MG's *v*&*f*, once it has been isolated from the main grid.

Droop control is the most utilized main control mechanism in inverter-based MGs. This method of control makes use of a direct relationship between Q and v, as well as P and f. The direct-quadratic (dq) reference frame is used to express the output voltage and current of MG. As the output voltage is in line with the

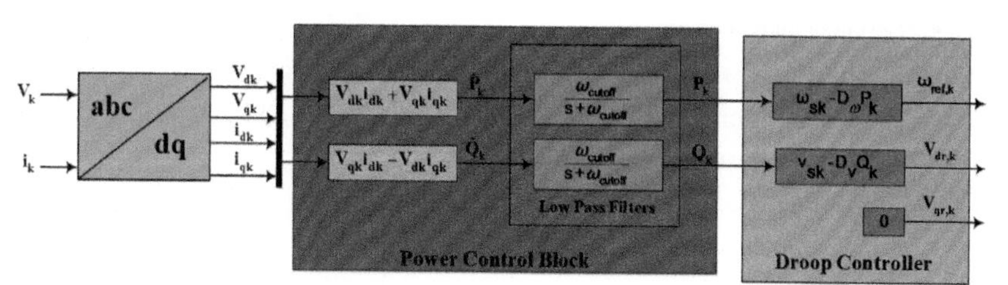

FIG. 2 Schematic outline of power controller.

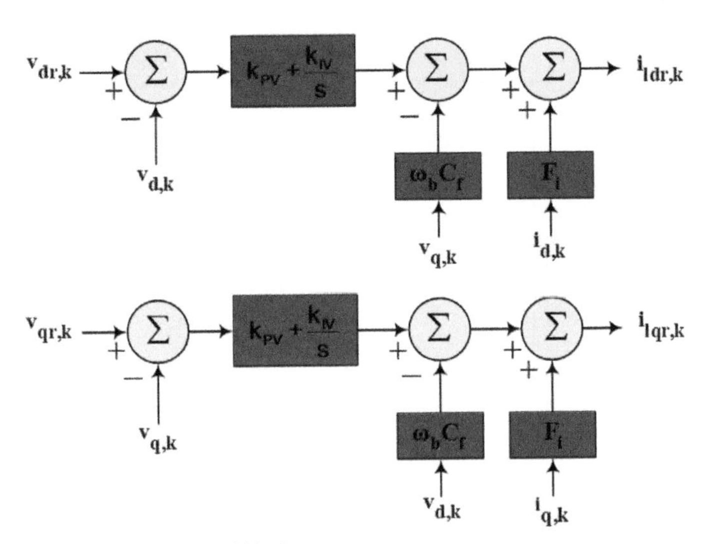

FIG. 3 Layout of voltage control block.

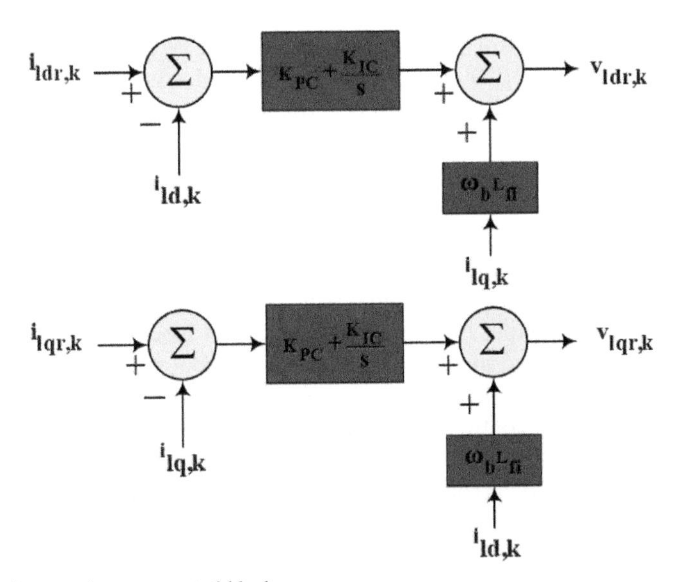

FIG. 4 Layout of current control block.

d-axis, the *q*-axis reference is considered as zero. The kth DG's *v&f* droop characteristics are given by [13]:

$$\left.\begin{aligned} \omega_k &= \omega_{sk} - D_\omega P_k \\ v_{dr,k} &= v_{sk} - D_v Q_k \\ v_{qr,k} &= 0 \end{aligned}\right\} \tag{1}$$

where ω_k is the kth DG's reference angular frequency. The reference DG output voltages are $v_{dr,\ k}$ and $v_{qr,\ k}$. The f&v droop coefficients are depicted as D_ω and D_v, respectively. The measured $P$and Q at individual DG terminals are P_k and Q_k, respectively, and are produced by a low-pass filter (LPF) with cut-off angular frequency ω_{cutoff}. Outputs of secondary control, that is, the primary control references are ω_{sk} and v_{sk}. Power controller differential equations can be represented as:

$$\dot{P}_k = -\omega_{cutoff} P_k + \omega_{cutoff} \widetilde{P}_k \tag{2}$$

$$\dot{Q}_k = -\omega_{cutoff} Q_k + \omega_{cutoff} \widetilde{Q}_k \tag{3}$$

The voltage references for the voltage controller are:

$$\left.\begin{aligned} v_{dr,k} &= v_{sk} - D_v Q_k \\ v_{qr,k} &= 0 \end{aligned}\right\} \tag{4}$$

Graphs describe the communication network of the MG. Consider an undirected graph with n-order weights $G \in (Y,E)$ with $Y = (y_1, y_2, \ldots, y_n)$ representing the node sets and $E \subseteq Y \times Y$ representing the undirected edges set. Every edge has a value $a_{kj} \geq 0$ assigned to it. If node k and node j have an edge, then $a_{kj} > 0$ and this will indicate the flow of data from DG k to DG j and vice-versa. $a_{kj} = 0$ if no connection exists between node k and node j. $A = (a_{kj})|_{k,\ j\epsilon(n\times n)}$ represents the adjacency matrix of the graph. The proposed secondary adaptive control scheme is described in the next section.

3. Proposed adaptive control scheme

From Eq. (6) representing the voltage droop control, $v_{sec,\ k}$ is to be designed. This section describes an adaptive droop control scheme design for precise Q sharing, despite unexpected uncertainties. The proposed control strategy is specified for the kth DG, $k \in \{1, \ldots, n\}$ as follows:

$$v_{sk} = \int \left(\dot{v}_{av} + \dot{v}_{aq}\right) dt \tag{5}$$

where $v_{av}$is an adaptive fixed-time reference allotment that ensures voltage management before the needed fixed-time, allowing the voltage to be driven and maintained at its nominal level despite unforeseen interruptions and v_{aq} stands for distributed adaptive droop control, which is used to precisely share reactive power despite unexpected interruptions.

Lemma 1 [21]: Consider the case of a simple scalar system.

$$\left.\begin{aligned} \dot{y}(x) &= -\alpha y^s - \beta y^t \\ y(0) &= y_0 \end{aligned}\right\} \tag{6}$$

where both α and β are greater than zero. a, b, c, and d are positive odd numbers that satisfy the requirements $s > 1$ and $t < 0$. Eq. (6) has globally fixed-time equilibrium with a settling time T limited by $T \leq T_{max}$ defined by $\frac{1}{\alpha}\frac{s}{s-1} + \frac{1}{\beta}\frac{1}{1-t}$.

Lemma 2 [22]: α needs to lie between 0 and 1 and β must be greater than 1 for satisfying the condition $y_1, y_2, \ldots, y_n \geq 0$. Hence,

$$\sum_{k=1}^{n} y_k^{\alpha}(t) \geq \left(\sum_{k=1}^{n} y_k(t)\right)^{\alpha} \tag{7}$$

$$\sum_{k=1}^{n} y_k^{\beta}(t) \geq n^{1-\beta}\left(\sum_{k=1}^{n} y_k(t)\right)^{\beta} \tag{8}$$

3.1 Adaptive voltage and reactive power control

To assure voltage management prior to the required fixed-time, an adaptive fixed-time reference allocation is used to drive and maintain nominal voltage magnitude despite unanticipated disruptions. Another distributed adaptive droop controller is designed to accurately share Q regardless of the unforeseen disruptions. Both of these adaptive controllers are designed as:

$$\dot{v}_{av} = k_{av}\left[\left(\sum_{k=1}^{n}(V_{nom} - V_{d,k})\right)^{\frac{k}{m}} + \left(\sum_{k=1}^{n}(V_{nom} - V_{d,k})\right)^{\frac{p}{q}}\right] \tag{9}$$

$$\dot{v}_{aq} = k_{aq}\left[\left(\sum_{j=1}^{n} a_{kj}(V_{d,j} - V_{d,k})\right)^{\frac{k}{m}} + \left(\sum_{j=1}^{n} a_{kj}(V_{d,j} - V_{d,k})\right)^{\frac{p}{q}}\right] \tag{10}$$

where k_{av} and k_{aq} represent the control gain of the proposed adaptive v and Q controllers, respectively. n represents the number of DGs used in the considered MG model.

The v and Q adaptive control scheme is depicted in Fig. 5.

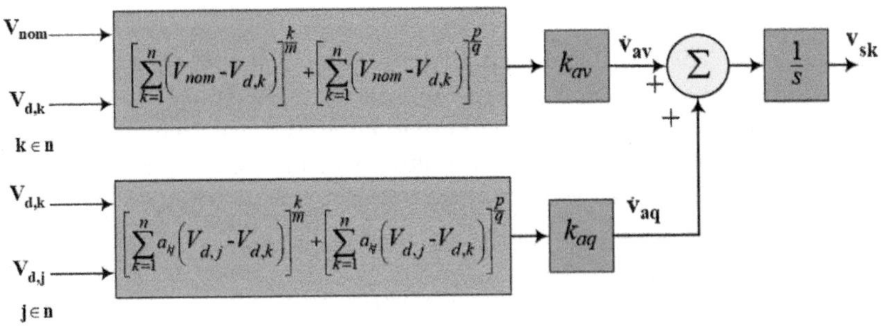

FIG. 5 Outline of proposed adaptive voltage and reactive power control block.

Two Lyapunov functions are considered for design of v and Q controllers as specified below:

$$F_v(x) = \frac{1}{4}\sum_{k=1}^{n}[y_o(t) - y_k(t)]^2 \tag{11}$$

$$F_Q(x) = \frac{1}{4}\sum_{k=1}^{n}\left(\sum_{j=1}^{n}a_{kj}\left(y_j(t) - y_k(t)\right)\right)^2 \tag{12}$$

where $y_o(t)$ is the reference value. $F_v(x)$ and $F_Q(x)$ are the Lyapunov functions considered for design of v and Q controllers, respectively. The voltage Lyapunov function's derivatives are expressed as follows:

$$\dot{F}_v(x) = \sum_{k=1}^{n}\frac{1}{4}\times 2[y_o(t) - y_k(t)][-\dot{y}_k(t)]$$

$$= -\frac{1}{2}\sum_{k=1}^{n}[y_o(t) - y_k(t)]\left[k_{av}\left(\sum_{k=1}^{n}y_o(t) - y_k(t)\right)^{\frac{k}{m}} + k_{av}\left(\sum_{k=1}^{n}y_o(t) - y_k(t)\right)^{\frac{p}{q}}\right]$$

$$= -\frac{1}{2}k_{av}\sum_{k=1}^{n}[y_o(t) - y_k(t)]^{1+\frac{k}{m}} - \frac{1}{2}k_{av}\sum_{k=1}^{n}[y_o(t) - y_k(t)]^{1+\frac{p}{q}}$$

$$\dot{F}_v(x) = -\frac{1}{2}k_{av}\sum_{k=1}^{n}\left[(y_o(t) - y_k(t))^2\right]^{\frac{1}{2}\left(1+\frac{k}{m}\right)} - \frac{1}{2}k_{av}\sum_{k=1}^{n}\left[(y_o(t) - y_k(t))^2\right]^{\frac{1}{2}\left(1+\frac{p}{q}\right)}$$

$$\dot{F}_v(x) = -\frac{1}{2}\times 2k_{av}\sum_{k=1}^{n}\frac{1}{2}\left[(y_o(t) - y_k(t))^2\right]^{\frac{k+m}{2m}} - 2k_{av}\sum_{k=1}^{n}\frac{1}{4}\left[(y_o(t) - y_k(t))^2\right]^{\frac{p+q}{2q}} \tag{13}$$

Using Lemma 2 in Eqs. (7) and (8),

$$\dot{F}_v(x) \le -\frac{k_{av}}{n^{\frac{m-k}{2m}}}[F_v(x)]^{\frac{k+m}{2m}} - 2k_{av}[F_v(x)]^{\frac{p+q}{2q}}$$

Thus, the Lyapunov function verifies $\dot{y}(x) = -\alpha y^s - \beta y^t$.
Here, $\alpha = \frac{k_{av}}{n^{\frac{m-k}{2m}}}$, $\beta = 2k_{av}$, $s = \frac{k+m}{2m}$, and $t = \frac{p+q}{2q}$.

Theorem 1 If control laws are implemented according to Eqs. (9) and (10), for precise reactive power sharing is obtained earlier than the requisite fixed-time.

Proof:

Similar to proposed adaptive voltage control, the Lyapunov function as specified in Eq. (12).

The Lyapunov function's derivative is expressed as follows:

$$\dot{F}_Q(x) = \sum_{k=1}^{n} \frac{1}{4} \times 2 \sum_{j=1}^{n} a_{kj}\left(y_j(t) - y_k(t)\right) \times \left[-a_{kj}\dot{y}_k(t)\right]$$

$$= -\frac{1}{2}\sum_{k=1}^{n}\sum_{j=1}^{n} a_{kj}a_{kj}\left(V_{d,j} - V_{d,k}\right)\left[k_{aq}\left(\sum_{j=1}^{n} a_{kj}\left(V_{d,j} - V_{d,k}\right)\right)^{\frac{k}{m}} + k_{aq}\left(\sum_{j=1}^{n} a_{kj}\left(V_{d,j} - V_{d,k}\right)\right)^{\frac{p}{q}}\right]$$

Let $\sum_{j=1}^{n} a_{kj} = c_k$ and $f_{iv} = \sum_{j=1}^{n} a_{kj}\left(V_{d,j} - V_{d,k}\right)$ for simplicity.

$$\dot{F}_Q(x) = -\frac{1}{2}k_{aq}\sum_{k=1}^{n} c_k\left[(f_{iv})^{1+\frac{k}{m}} + (f_{iv})^{1+\frac{p}{q}}\right]$$

$$= -\frac{1}{2}k_{aq}\sum_{k=1}^{n} c_k(f_{iv})^{\frac{k+m}{m}} - \frac{1}{2}k_{aq}\sum_{k=1}^{n} c_k(f_{iv})^{\frac{p+q}{q}}$$

$$= -2k_{aq}\sum_{k=1}^{n} c_k\frac{1}{4}\left[(f_{iv})^2\right]^{\frac{k+m}{2m}} - 2k_{aq}\sum_{k=1}^{n} c_k\frac{1}{4}\left[(f_{iv})^2\right]^{\frac{p+q}{2q}}$$

$$= -2k_{aq}\sum_{k=1}^{n} c_k\frac{1}{4}\left[(f_{iv})^2\right]^{\frac{k+m}{2m}} - 2k_{aq}\sum_{k=1}^{n} c_k\frac{1}{4}\left[(f_{iv})^2\right]^{\frac{p+q}{2q}}$$

$$\dot{F}_Q(x) = -2k_{aq}\sum_{k=1}^{n} c_k\left[F_Q(x)\right]^{\frac{k+m}{2m}} - 2k_{aq}\sum_{k=1}^{n} c_k\left[F_Q(x)\right]^{\frac{p+q}{2q}} \tag{14}$$

Using Lemma 2 in Eqs. (7) and (8),

$$\dot{F}_Q(x) \leq -\frac{2k_{aq}}{n^{\frac{m-k}{2m}}}\sum_{k=1}^{n} c_k\left[F_Q(x)\right]^{\frac{k+m}{2m}} - 2k_{aq}\sum_{k=1}^{n} c_k\left[F_Q(x)\right]^{\frac{p+q}{2q}}$$

Thus, the Lyapunov function verifies $\dot{y}(x) = -\alpha y^{\frac{a}{b}} - \beta y^{\frac{c}{d}}$.
Here, $\alpha = \frac{2k_{aq}}{n^{\frac{m-k}{2m}}}$, $\beta = 2k_{aq}$, $s = \frac{k+m}{2m}$, and $t = \frac{p+q}{2q}$
Settling time is given by:

$$T_v = \frac{1}{\alpha}\frac{k+m}{k-m} + \frac{1}{\beta}\frac{2q}{q-p} \tag{15}$$

Thus, the simplified form of the proposed controller is derived as:

$$v_{sk} = \int \left[k_{av} \left(\sum_{k=1}^{n} (V_{nom} - V_{d,k}) \right)^{\frac{k}{m}} + k_{av} \left(\sum_{k=1}^{n} (V_{nom} - V_{d,k}) \right)^{\frac{p}{q}} + k_{aq} \left(\sum_{j=1}^{n} a_{kj} (V_{d,j} - V_{d,k}) \right)^{\frac{k}{m}} + k_{aq} \left(\sum_{j=1}^{n} a_{kj} (V_{d,j} - V_{d,k}) \right)^{\frac{p}{q}} \right] dt \tag{16}$$

3.2 Adaptive frequency and active power control

ω_{sk} is to be designed from Eq. (1), which represents the frequency droop control. A distributed adaptive droop control is devised in this section to enable frequency management and precise active power sharing despite unforeseen disruptions.

The proposed adaptive control for the kth DG, $k \in \{1, \ldots, n\}$ is represented as follows:

$$\omega_{sk} = \int (\dot{\omega}_{a\omega} + \dot{\omega}_{ap}) dt \tag{17}$$

where $\omega_{a\omega}$ represents an adaptive fixed-time reference allotment for assuring frequency management before the needed fixed-time so that the frequency magnitude is driven and held at its nominal level despite unforeseen interruptions and ω_{ap} stands for distributed adaptive control, which is designed for precisely sharing active power.

First of all, reference allotment is carried out to assure frequency management, before the required time, so that nominal frequency magnitude is maintained despite unexpected disruptions. Another distributed adaptive droop controller is designed to accurately share P regardless of the unforeseen disruptions. Both of these adaptive controllers are designed as:

$$\dot{\omega}_{a\omega} = k_{a\omega} \left(\sum_{k=1}^{n} (\omega_{nom} - \omega_k) \right)^{\frac{k}{m}} + k_{a\omega} \left(\sum_{k=1}^{n} (\omega_{nom} - \omega_k) \right)^{\frac{p}{q}} \tag{18}$$

$$\dot{\omega}_{ap} = k_{ap} \left(\sum_{j=1}^{n} a_{kj} (\omega_j - \omega_k) \right)^{\frac{k}{m}} + k_{ap} \left(\sum_{j=1}^{n} a_{kj} (\omega_j - \omega_k) \right)^{\frac{p}{q}} \tag{19}$$

The ω and P adaptive control scheme is depicted in Fig. 6.

Here, we consider the same Lyapunov functions as used in v and Q control given in Eqs. (11) and (12). The frequency Lyapunov function's derivative is expressed as follows:

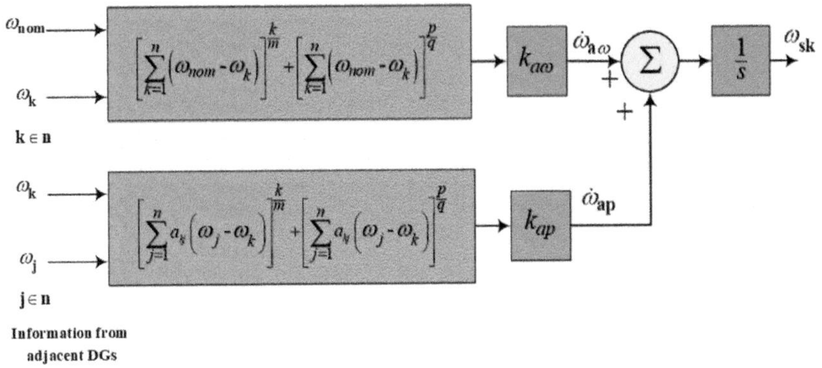

FIG. 6 Outline of proposed adaptive frequency and active power control block.

$$\dot{F}_{\omega}(x) = \sum_{k=1}^{n} \frac{1}{4} \times 2[y_o(t) - y_k(t)][-\dot{y}_k(t)]$$

$$= -\frac{1}{2}\sum_{k=1}^{n}[y_o(t) - y_k(t)]\left[k_{a\omega}\left(\sum_{k=1}^{n} y_o(t) - y_k(t)\right)^{\frac{k}{m}} + k_{a\omega}\left(\sum_{k=1}^{n} y_o(t) - y_k(t)\right)^{\frac{p}{q}}\right]$$

$$= -\frac{1}{2}k_{a\omega}\sum_{k=1}^{n}[y_o(t) - y_k(t)]^{1+\frac{k}{m}} - \frac{1}{2}k_{a\omega}\sum_{k=1}^{n}[y_o(t) - y_k(t)]^{1+\frac{p}{q}}$$

$$\dot{F}_{\omega}(x) = -\frac{1}{2}k_{a\omega}\sum_{k=1}^{n}\left[(y_o(t) - y_k(t))^2\right]^{\frac{1}{2}\left(1+\frac{k}{m}\right)} - \frac{1}{2}k_{a\omega}\sum_{k=1}^{n}\left[(y_o(t) - y_k(t))^2\right]^{\frac{1}{2}\left(1+\frac{p}{q}\right)}$$

$$\dot{F}_{\omega}(x) = -\frac{1}{2} \times 2k_{a\omega}\sum_{k=1}^{n}\frac{1}{2}\left[(y_o(t) - y_k(t))^2\right]^{\frac{k+m}{2m}} - 2k_{a\omega}\sum_{k=1}^{n}\frac{1}{4}\left[(y_o(t) - y_k(t))^2\right]^{\frac{p+q}{2q}} \tag{20}$$

Using Lemma 2 in Eq. (7),

$$\dot{F}_{\omega}(x) \le -\frac{k_{a\omega}}{n^{\frac{m-k}{2m}}}[F_{\omega}(x)]^{\frac{k+m}{2m}} - 2k_{a\omega}[F_{\omega}(x)]^{\frac{p+q}{2q}}$$

Thus, the Lyapunov function verifies $\dot{y}(x) = -\alpha y^s - \beta y^t$. Here, $\alpha = \frac{k_{a\omega}}{n^{\frac{m-k}{2m}}}$, $\beta = 2k_{a\omega}$, $s = \frac{k+m}{2m}$, and $t = \frac{p+q}{2q}$

Theorem 2 If control protocol is implemented according to Eqs. (19) and (21), for precise active power sharing is obtained earlier than the requisite fixed-time.

Proof:

Lyapunov function is identical to that of the adaptive Q regulation provided by Eq. (12). The Lyapunov function's derivative considered for P is expressed as follows:

$$\dot{F}_P(x) = \sum_{k=1}^{n} \frac{1}{4} \times 2 \sum_{j=1}^{n} a_{kj}\left(y_j(t) - y_k(t)\right) \times \left[-a_{kj}\dot{y}_k(t)\right]$$

$$= -\frac{1}{2} \sum_{k=1}^{n} \sum_{j=1}^{n} a_{kj}a_{kj}(\omega_j - \omega_k)\left[k_{aq}\left(\sum_{j=1}^{n} a_{kj}(\omega_j - \omega_k)\right)^{\frac{k}{m}}\right.$$

$$\left. + k_{aq}\left(\sum_{j=1}^{n} a_{kj}(\omega_j - \omega_k)\right)^{\frac{p}{q}}\right]$$

Let $c_k = \sum_{j=1}^{n} a_{kj}$ and $f_{i\omega} = \sum_{j=1}^{n} a_{kj}(\omega_j - \omega_k)$ for simplicity.

$$\dot{F}_P(x) = -\frac{1}{2} k_{ap} \sum_{k=1}^{n} c_k\left[(f_{i\omega})^{1+\frac{k}{m}} + (f_{i\omega})^{1+\frac{p}{q}}\right]$$

$$= -\frac{1}{2} k_{ap} \sum_{k=1}^{n} c_k (f_{i\omega})^{\frac{k+m}{m}} - \frac{1}{2} k_{ap} \sum_{k=1}^{n} c_k (f_{i\omega})^{\frac{p+q}{q}}$$

$$= -2k_{ap} \sum_{k=1}^{n} c_k \frac{1}{4}\left[(f_{i\omega})^2\right]^{\frac{k+m}{2m}} - 2k_{ap} \sum_{k=1}^{n} c_k \frac{1}{4}\left[(f_{i\omega})^2\right]^{\frac{p+q}{2q}}$$

$$= -2k_{ap} \sum_{k=1}^{n} c_k \frac{1}{4}\left[(f_{i\omega})^2\right]^{\frac{k+m}{2m}} - 2k_{ap} \sum_{k=1}^{n} c_k \frac{1}{4}\left[(f_{i\omega})^2\right]^{\frac{p+q}{2q}}$$

$$\dot{F}_P(x) = -2k_{ap} \sum_{k=1}^{n} c_k [F(x)]^{\frac{k+m}{2m}} - 2k_{ap} \sum_{k=1}^{n} c_k [F(x)]^{\frac{p+q}{2q}} \tag{21}$$

Using Lemma 2 in Eq. (12),

$$\dot{F}_P(x) \le -\frac{2k_{ap}}{n^{\frac{m-k}{2m}}} \sum_{k=1}^{n} c_k [F(x)]^{\frac{k+m}{2m}} - 2k_{ap} \sum_{k=1}^{n} c_k [F(x)]^{\frac{p+q}{2q}}$$

Thus, the Lyapunov function verifies $\dot{y}(x) = -\alpha y^s - \beta y^t$.
Here, $\alpha = \frac{2k_{ap}}{n^{\frac{m-k}{2m}}}$, $\beta = 2k_{ap}$, $s = \frac{k+m}{2m}$, and $t = \frac{p+q}{2q}$
Settling time is given by:

$$T_\omega = \frac{1}{\alpha} \frac{k+m}{k-m} + \frac{1}{\beta} \frac{2q}{q-p} \tag{22}$$

4. Simulation results and analysis

The proposed control scheme's efficacy is validated by a simulation analysis that compares it to the conventional secondary control and is used to determine its effectiveness. Some works that deal with v and f regulation and power sharing will be compared in this work. This simulation work takes into account the v and Q controllers proposed by Bidram et al. [14] and the f and P controllers proposed by Dehkordi et al. [16]. Reference v&f tracking and resilience against load power variations are compared.

The simulation test system used is depicted in Fig. 7 [23]. It is a low-voltage stand-alone MG with five DGs and five loads coupled via a communication network represented in a non-directional graph with adjacent matrices as follows:

$$A = \begin{bmatrix} 0 & 1 & 0 & 0 & 1 \\ 1 & 0 & 1 & 0 & 0 \\ 0 & 1 & 0 & 1 & 0 \\ 0 & 0 & 1 & 0 & 1 \\ 1 & 0 & 0 & 1 & 0 \end{bmatrix}$$

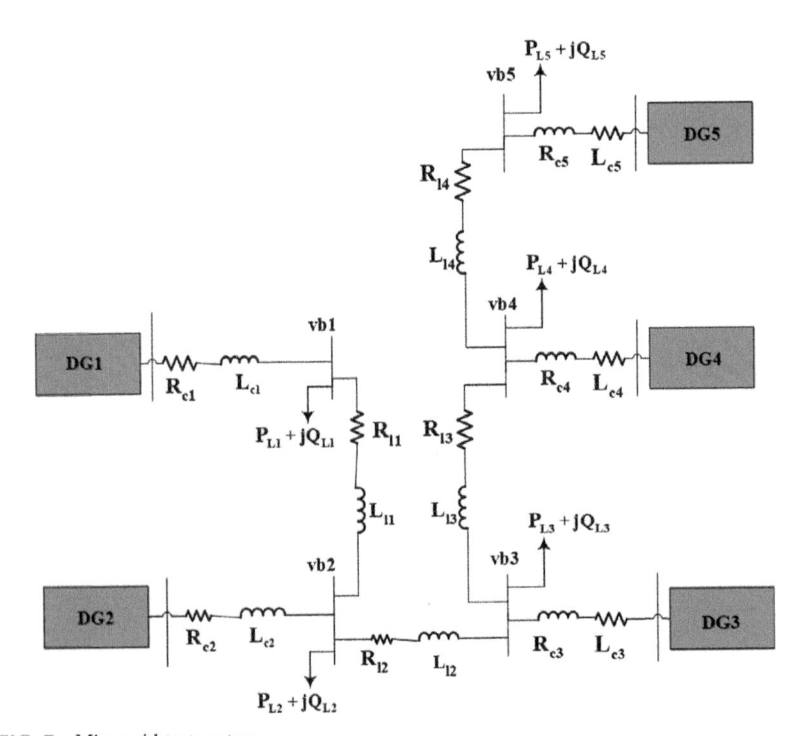

FIG. 7 Microgrid test system.

TABLE 1 Droop coefficients and controller gains.

Parameter	DG1, DG2, and DG3	DG4 and DG5
V_{nom}	380 V	380 V
ω_{nom}	314.16 rad/s	314.16 rad/s
D_ω	1.06×10^{-4}	1.46×10^{-4}
D_v	5.2×10^{-3}	4.8×10^{-3}
R	0.03 Ω	0.03 Ω
L	0.35 mH	0.35 mH
R_f	0.1 Ω	0.1 Ω
L_f	1.35 mH	1.35 mH
C_f	50 μF	50 μF
K_{PV}	0.1	0.5
K_{IV}	420	390
K_{PC}	15	10.5
K_{IC}	20,000	16,000
ω_{cutoff}	31.41 rad/s	31.41 rad/s

The simulations are carried out with the MG structure shown in Section 2. Some of the MG system parameters have been taken from the work done by Shrivastava et al. [5]. Droop coefficients and controller gains are outlined in Table 1. Tables 2 and 3 summarize the line parameters and load parameters, respectively.

In addition, 50 Hz for frequency and 380 V for voltage magnitude are used as reference values. The following are control parameters for proposed controllers: $k_{av}=1500$, $k_{a\omega}=1500$, $k_{aq}=200$, $k_{ap}=200$, $\frac{k}{m}=1.5$, and $\frac{p}{q}=0.5$. The voltage as well as current controller gains are adjusted to large values for rejection of high frequency disturbances along with damping of filters. Four steps are considered in this simulation study as depicted in Table 4.

TABLE 2 Line parameters.

Parameters	Z_{line12}	Z_{line23}	Z_{line34}	Z_{line45}
R_l	0.2 Ω	0.15 Ω	0.2 Ω	0.15 Ω
L_l	300 μH	1850 μH	300 μH	1850 μH

TABLE 3 Load parameters.

Parameters	Load 1	Load 2	Load 3	Load 4	Load 5
R	30 Ω	40 Ω	55 Ω	45 Ω	60 Ω
L	47 μH	64 μH	64 μH	54 μH	75 μH

TABLE 4 Steps considered in simulation work.

Simulation time range (t)	Steps
0–0.25 s	Primary control is only activated
0.25 s	Secondary adaptive v&Q synchronization scheme and f&P synchronization scheme are activated
0.5 s	Load 1 and Load 5 are increased

The adaptive v&f control presented in Figs. 8 and 9 recovers the reference v&f while rejecting the voltage disturbance caused by variations in load power. Fixed settling time set for voltage controller is $T_v = 1.73 \times 10^{-3}$ s and frequency controller is $T_\omega = 1.35 \times 10^{-3}$ s. The actual settling times for both the controllers are same as that of the fixed allotted time, thus validating the proposed controller design. In terms of load power variation control, the simulation results show that the proposed control technique outperforms the traditional controllers. At $t = 0.25$ s, the output voltage and frequency of DG 1 vary much more in conventional method as compared to the proposed method. The proposed frequency controller is significantly more resilient as no or very few variations

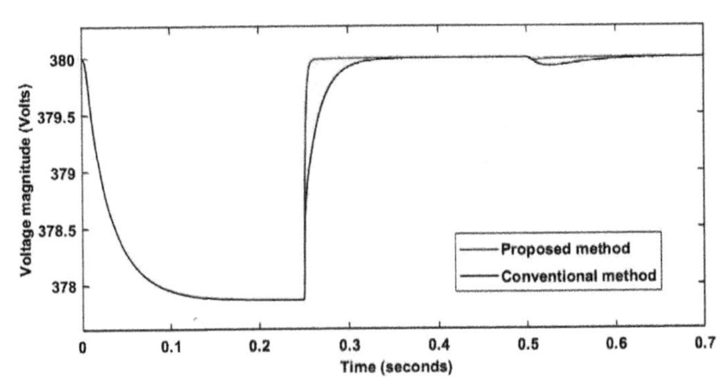

FIG. 8 Terminal voltage magnitude comparison of DG 1.

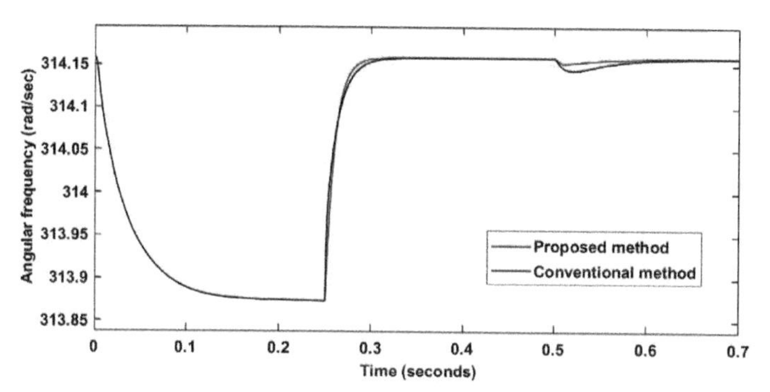

FIG. 9 Angular frequency comparison of DG 1.

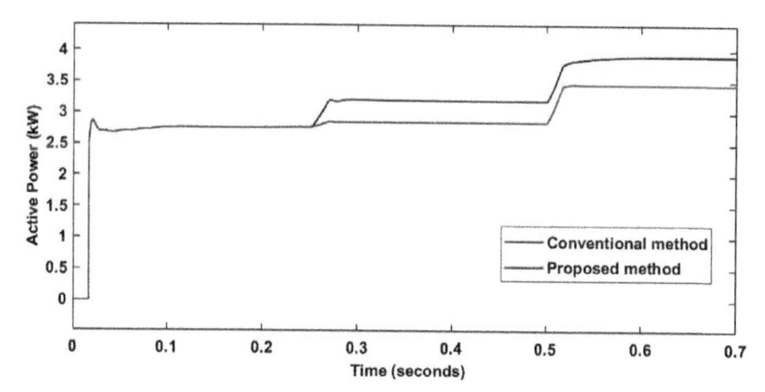

FIG. 10 Active power comparison of DG 1.

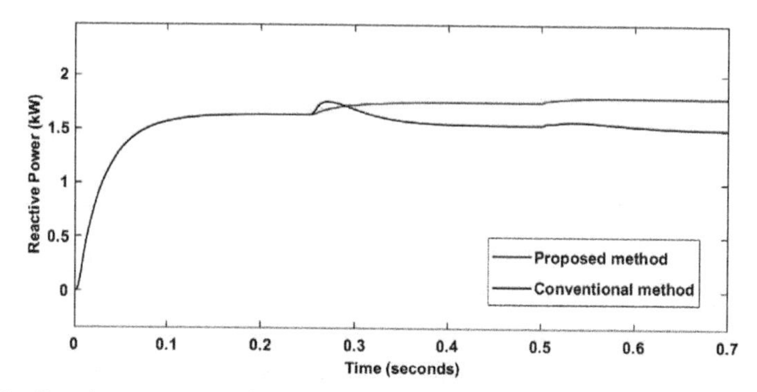

FIG. 11 Reactive power comparison of DG 1.

occurred during load power changes. The conventional controllers are slower in their dynamics and more responsive to system conditions when the demand for load power varies.

The differences in power sharing are further reduced as seen in Figs. 10 and 11 and a correct management of power flow among DGs is achieved without affecting v or f control at the desired settling time.

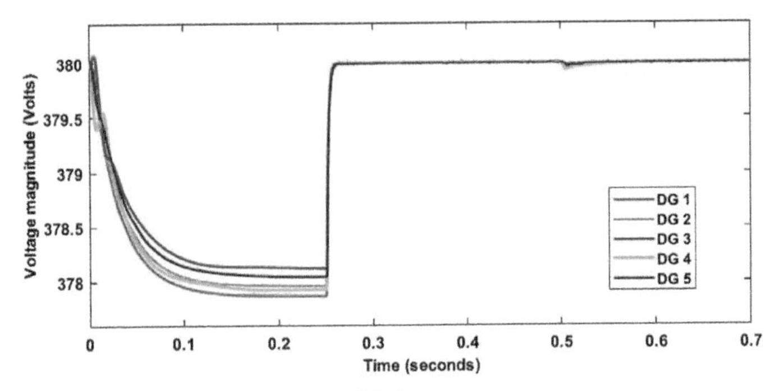

FIG. 12 Terminal voltage magnitudes of all DGs.

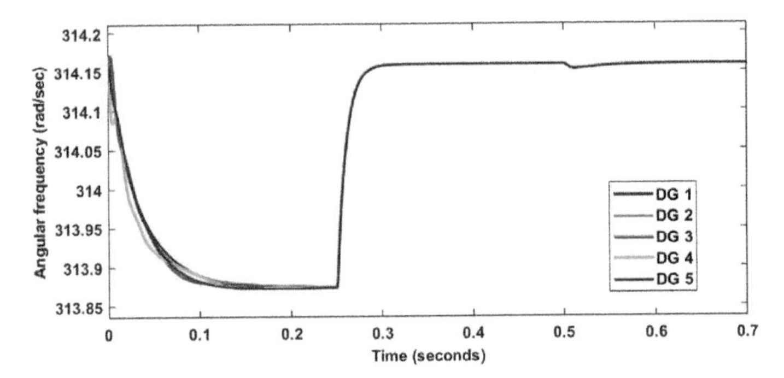

FIG. 13 Angular frequency of all DGs.

The output voltages of all the DGs converge to their nominal value after activation of adaptive control scheme at $t=0.25$ s as depicted in Fig. 12. The angular frequency output of all the DGs converge to their nominal value after activation of adaptive control scheme at $t=0.25$ s as depicted in Fig. 13.

5. Conclusion

In this paper, a stand-alone AC MG with adaptive *v&f* control and a *P&Q* sharing mechanism is proposed. In comparison to the existing traditional control approaches, the proposed methodology offers several advantages. To begin with, it ensures exact power distribution, frequency control, and reference tracking regardless of system disruptions. Second, the suggested control method is very simple and basic. Finally, compared to traditional controllers, the suggested controller is more resilient. The suggested controller's performance has been compared to that of a well-tuned conventional controller, and it has been determined that the former provides more reliable results than the latter. The proposed controller is less sensitive to system changes as compared to the conventional method.

References

[1] Y. Levron, J.M. Guerrero, Optimal power flow in microgrids with energy storage, IEEE Trans. Power Syst. 28 (3) (2013) 3226–3234.

[2] T.L. Vandoorn, B. Meersman, J.D.M.D. Kooning, L. Vandevelde, Analogy between conventional grid control and islanded microgrid control based on a global DC-link voltage droop, IEEE Trans. Power Deliv. 27 (3) (2012) 1405–1414.

[3] A. Hirsch, Y. Parag, J. Guerrero, Microgrids: a review of technologies, key drivers, and outstanding issues, Renew. Sust. Energ. Rev. 90 (2018) 402–411.

[4] B.S. Hartono, Budiyanto, R. Setiabudy, Review of microgrid technology, in: International Conference on QiR, Yogyakarta, 2013, pp. 127–132.

[5] S. Shrivastava, B. Subudhi, S. Das, Distributed voltage and frequency synchronisation control scheme for islanded inverter-based microgrid, IET Smart Grid 1 (2) (2018) 48–56.

[6] T.L. Nguyen, Q. Tran, R. Caire, C. Gavriluta, V.H. Nguyen, Agent based distributed control of islanded microgrid—real-time cyber-physical implementation, in: 2017 IEEE PES Innovative Smart Grid Technologies Conference Europe (ISGT-Europe), Torino, 2017, pp. 1–6.

[7] J. Lai, X. Lu, X. Yu, W. Yao, J. Wen, S. Cheng, Gossip-based distributed active load voltage control for low-voltage microgrids, in: 2017 Chinese Automation Congress (CAC), Jinan, 2017, pp. 3830–3835.

[8] F. Mumtaz, I. SafakBayram, Planning, operation, and protection of microgrids: an overview, Energy Procedia 107 (2017) 94–100.

[9] B. Satish, S. Bhuvaneswari, Control of microgrid—a review, in: 2014 International Conference on Advances in Green Energy (ICAGE), Thiruvananthapuram, 2014, pp. 18–25.

[10] M.S. Mahmoud, S. AzherHussain, M.A. Abido, Modeling and control of microgrid: an overview, J. Franklin Inst. 351 (5) (2014) 2822–2859.

[11] M. Yazdanian, A. Mehrizi-Sani, Distributed control techniques in microgrids, IEEE Trans. Smart Grid 5 (6) (2014) 2901–2909.

[12] N.M. Dehkordi, N. Sadati, M. Hamzeh, Fully distributed cooperative secondary frequency and voltage control of islanded microgrids, IEEE Trans. Energy Convers. 32 (2) (2017) 675–685.

[13] X. Wu, C. Shen, R. Iravani, A distributed, cooperative frequency and voltage control for microgrids, IEEE Trans. Smart Grid 9 (4) (2018) 2764–2776.

[14] A. Bidram, A. Davoudi, F.L. Lewis, A multiobjective distributed control framework for islanded AC microgrids, IEEE Trans. Industr. Inform. 10 (3) (2014) 1785–1798.

[15] J.W. Simpson-Porco, Q. Shafiee, F. Dörfler, J.C. Vasquez, J.M. Guerrero, F. Bullo, Secondary frequency and voltage control of islanded microgrids via distributed averaging, IEEE Trans. Ind. Electron. 62 (11) (2015) 7025–7038.

[16] N.M. Dehkordi, N. Sadati, M. Hamzeh, Distributed robust finite-time secondary voltage and frequency control of islanded microgrids, IEEE Trans. Power Syst. 32 (5) (2017) 3648–3659.

[17] X. Lu, X. Yu, J. Lai, Y. Wang, J.M. Guerrero, A novel distributed secondary coordination control approach for islanded microgrids, IEEE Trans. Smart Grid 9 (4) (2018) 2726–2740.

[18] H. Mahmood, D. Michaelson, J. Jiang, Reactive power sharing in islanded microgrids using adaptive voltage droop control, IEEE Trans. Smart Grid 6 (6) (2015) 3052–3060.

[19] M. Ramezani, S. Li, Y. Sun, Combining droop and direct current vector control for control of parallel inverters in microgrid, IET Renew. Power Gener. 11 (1) (2017) 107–114.

[20] N. Pogaku, M. Prodanovic, T.C. Green, Modeling, analysis and testing of autonomous operation of an inverter-based microgrid, IEEE Trans. Power Electron. 22 (2) (2007) 613–625.

[21] Z. Zuo, B. Tian, M. Defoort, Z. Ding, Fixed-time consensus tracking for multiagent systems with high-order integrator dynamics, IEEE Trans. Autom. Control 63 (2) (2018) 563–570.

[22] Z. Zuo, Nonsingular fixed-time consensus tracking for second order multi-agent networks, Automatica 54 (2015) 305–309.

[23] S. Panda, G.R. Biswal, B. Subudhi, Distributed adaptive control of an islanded microgrid, Proceedings of the 2021 International Symposium of Asian Control Association on Intelligent Robotics and Industrial Automation (IRIA), Goa, 2021, pp. 234–239.

Chapter 7

Proactive defense system for enhanced resiliency of power grids with microgrids

Najda Vadakkeyveettil Mohamed Anwar, Gopakumar Pathirikkat, and Sunitha Rajan
Department of Electrical Engineering, National Institute of Technology Calicut, Kerala, India

1. Introduction

Renewable energy resource integration to power system led by global green energy targets initiated as several small-scale energy resources in distribution sector transfigured to the high penetration of renewable energy in terms of large wind farms and solar parks in transmission level. Emerging deregulated power market facilitates the consumers to be stakeholders having flexible power management strategies. Advancement in self-sufficient microgrids, which maintain an active link with main power grid, is a perfect example for the same. However, uncertainties instigated by variable and intermittent renewable sources (RES) and forecasting errors are stumbling the prospect of widespread microgrid dominant with RES in view of stability challenges within main power grid [1–3]. Besides, the policies adopted for deregulation and restructuring of the power grids could radically influence resiliency of the whole power grid. The market-oriented strategies and policies framed for stakeholders open up both opportunities and challenges to energy management [4]. While limited reactive power support being one of such challenges, microgrid as an aggregator offering ancillary services can assist for better control in evolving grid.

Even though numerous researches and developments are transpiring to overcome the aforementioned issues, disturbance that propagates in these deregulated market-oriented grids are evolving as new challenges [5,6]. Many recent researches substantiate the significance of defending or mitigating such wide area propagating disturbances that could instigate wide area power quality issues [7,8]. This research outlines the disturbance propagation phenomenon due to uncertainties associated with RES in a deregulated power grid, which could be aggravated by disturbance triggered uncontrolled wide area microgrid islanding.

Microgrid Cyberphysical Systems. https://doi.org/10.1016/B978-0-323-99910-6.00008-6

Based on the studies conducted, a proactive defense strategy is presented, which utilizes microgrids as an aggregator of wide area ancillary support, which in turn limits its uncontrolled islanding. The defense strategy prevents manifestation of wide spreading disturbances into major power quality issues, thereby enhancing the reliability, resiliency, and self-healing capability of grid.

2. Dependencies and energy management in deregulated grid

Current practice of integration of RES to grid is in the form of grid following inverter-based systems extracting maximum power (MPPT) to cope up with the global carbon emission reduction targets, leading to variable power injection. Even with availability of several forecasting techniques, the prediction of renewable power output is still erroneous. The forecast error makes RES non-dispatchable, and corresponding uncertainty in their output is usually defined in terms of forecast error. The corresponding variation in inverter output can be modeled using the probability density functions (PDF) of RES and the energy conversion systems. The PDF of this variation is represented by $f_{FE}(\mu_{\Delta P}, \sigma^2_{\Delta P}, \gamma_{\Delta P}, \kappa_{\Delta P}, t)$, where μ, σ^2, γ, and κ are the mean, variance, skewness, and kurtosis of the distribution [7]. These uncertainties appear in the overall power offered or drawn by the microgrid under prevailing market conditions.

The competitive market-oriented operation has drastically changed the operational behavior of different entities of the system. Several services that were ample in conventional grid became scarce in deregulated grid, a few being the frequency regulation reserve and reactive power. Literature has emphasized the significance of availability of fast and flexible reserve resources across the grid for the resilient, reliable, and secure operation of evolving deregulated grid with distributed microgrids dominated with RES. Usual resiliency enhancement measures spread across different phases of the event, and state of the grid along with the role of microgrid in each of the phases are outlined in Fig. 1. The lack of inertia driven large ROCOF, frequency nadir, and accompanied risk factors in the evolving grid together with the methods to alleviate them are well dealt with in literature. The distributed RES following MPPT injects power to the grid at unity power factor meeting local loads forming almost self-sufficient microgrids in terms of active power requirement. Relatively low active power imbalance in microgrids will be dealt with import from or export to the grid while operating in grid-connected mode. In their grid-connected mode of operation, the converters of these distributed resources operate in grid following mode and hence the entire microgrid depends on the grid for its reactive power requirement. In addition, resources under the microgrid aggregators are following the resiliency-based scheduling during proactive phase of resiliency enhancement, so as to achieve defensive islanding of these microgrid with minimum curtailment of their non-critical loads during preventive phase. These microgrids also act as resilient resources during the recovery phase by providing demand response, re-dispatching the resources, non-critical load curtailment, etc. The resiliency requirements encourage the steady state operation of grid

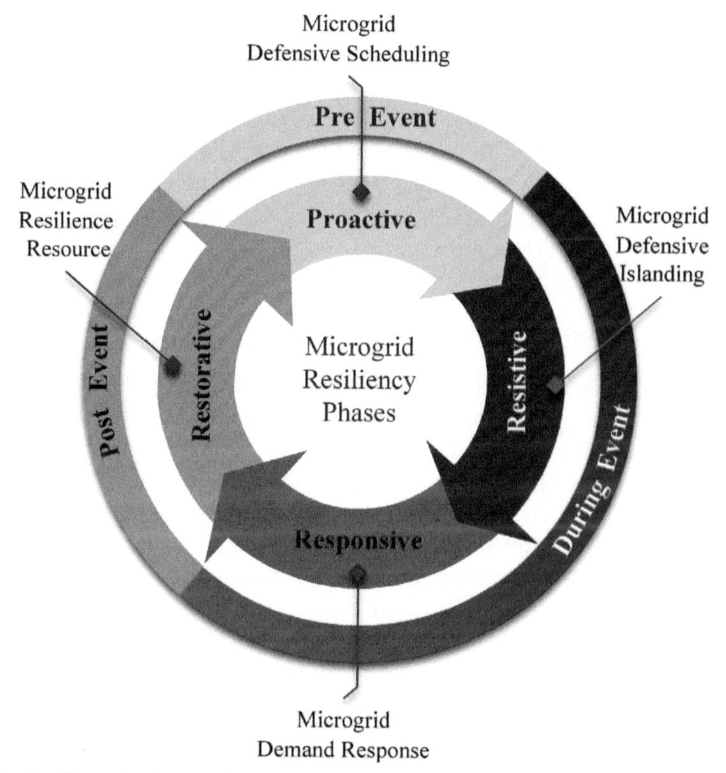

FIG. 1 Resiliency implementation phases.

in terms of forming several grid-connected microgrids in sub-transmission and distribution level. Hence, aggregated load at power system bus is becoming more reactive in nature. Moreover, reactive power is required for maintaining the lightly loaded interconnections involving long transmission lines that are essential for resilient, reliable, and secure operation of a highly variable and uncertain deregulated grid.

Reactive power facilitates the required active power flow and maintains the bus voltage within the desired limits so as to have secure operation of power system equipments and loads. The risk factors associated with scarcity of reactive power and allied voltage fluctuation and collapse, in deregulated grid with distributed microgrid dominant with RES, is attaining research focus [2,9–11]. The reactive power support provided by the converter-operated distributed renewable generation is at the expense of curtailment of available active power from them following MPPT and hence in contrast with the emission reduction policies. Accordingly, FERC order no: 888 or similar policies and regulatory steps identify the reactive power as an important ancillary service in future grid in place of obligatory reactive power requirement by policy regulations, being against the best interest of competitive market as well as the emission reduction targets [12].

The recent technological achievements and the cost benefits attained have assisted in the availability of distributed fast and flexible resources (FFR) like battery energy storage system throughout the grid, capable of providing dynamic reactive power support along with distributed microgrid aggregators [1,5,8,13,14]. Inline with this, several market models that deal with reactive power as ancillary service in day ahead energy market and real time balancing market are reported in literature aiding toward a deregulated grid dominated with microgrids [15–17].

2.1 Market model for procurement of reactive power

The localized nature of reactive power leads to promoting the locally available reactive power through market incentives. Researches advocate local or zonal reactive power markets formulated on schemes like voltage control area (VCA) to ensure deliverability of resources [15,18]. One of such reactive power procurement market models formulated as bilevel optimization problem is shown by Eq. (1) [15,18]. Market gets cleared when the amount to be paid by the balancing authority (BA) operating the market is minimized, given by Eq. (1a), while the profit obtained by the reactive power resource aggregators concurrently ensuring the bilateral contracts is maximized, given by Eq. (1b). In deregulated market model involving several microgrids, each microgrid aggregator will be acquiring reactive power from distributed resources coming under respective microgrid in local market and dealing with balancing authority as single entity of reactive power [4].

$$
\left.\begin{array}{l}
Minimize : PP_{ISO} = \sum\limits_{i \in G} c_i Q_{gi} + CF_{vi} + CF_{resv_i} \\
\text{such that :} \\
P_{gi} - P_{di} = \sum\limits_{j \in N} |V_i||V_j| Y_{ij} cos(\theta_{ij} + \delta_j - \delta_i) \\
Q_{gi} - Q_{di} = -\sum\limits_{j \in N} |V_i||V_j| Y_{ij} sin(\theta_{ij} + \delta_j - \delta_i) \\
P_{gi}^{min} \le P_{gi} \le P_{gi}^{max} \quad \forall i \in G_{NC} \\
P_{gi}^{mincontract} \le P_{gi} \le P_{gi}^{maxcontract} \quad \forall i \in G_{BC} \\
S_l^{min} \le S_l \le S_l^{max} \quad \forall l \in L \\
|V_i^{min}| \le |V_i| \le |V_i^{max}| \quad \forall i \in N \\
\sum\limits_{j \in N} \dfrac{X_{ij} P_{ij}^2}{|V_j|^2} \le Q_{vi} \le \sum\limits_{j \in N} \dfrac{\left(V_{ref}^2 - |V_i|^2 + 1\right)}{X_{ij}}
\end{array}\right\} \quad (1a)
$$

$$\left.\begin{aligned}
&\text{and subject to :}\\
&\text{Maximize : } GP_i = c_i Q_{gi} - CF_i\left(Q_{gi}\right) \quad \forall i \in G\\
&\text{such that :}\\
&Q_{gi}^{\min} \le Q_{gi} \le Q_{gi}^{\max} \quad \forall i \in G\\
&\text{where,}\\
&\mathrm{CF}_{\mathrm{vi}} = (OC_i + LOC_i.Q_{vi})Q_{vi}.RRF_i\\
&CF_{resv_i} = LOC_i\left(Q_{gi}^{\max} - Q_{gi}\right)^2
\end{aligned}\right\} \qquad (1b)$$

where i and j are the indices for set of all buses N, l index for set of all lines L, G_{BC} and G_{NC} for set of all reactive power sources including microgrid aggregators involved and not involved in bilateral contracts, respectively, X and Y are the reactance and admittance matrix, respectively. c_i is the price signal issued to ith reactive power resource or microgrid aggregator, OC and LOC are operational cost and lost opportunity cost included in the bid submitted by aggregators to BA. RRF_i is the reactive power relevance factor of ith reactive power resource or microgrid aggregator in providing reactive power to weaker section of the grid. P, Q, and S are active, reactive, and apparent power, "g" stands for generation, and "d" for load. The associated zonal or local partitioning of market drew on VCA is discussed in the following section.

2.2 Voltage control area formation

Several techniques of partitioning the system to VCA based on the electrical distance of load buses from voltage controlling buses as well as voltage sensitivity, based on $\partial Q/\partial V$ and $\partial V/\partial Q$, respectively, from the Jacobean matrix, bus admittance matrix, heuristic methods, Monte Carlo simulation, and so forth are reported in the literature [12,18]. Among them, the partitioning method based on voltage and apparent power coupling shown in Eq. (2) seems promising to the future uncertain deregulated grid, as real time operation of active and reactive power balancing markets are interdependent [12,15]. Voltage apparent power coupled-VCA derived from bus admittance matrix given in Eq. (2) [12] aids in partitioning the BA into local or zonal market for reactive power procurement as discussed in Section 2.1.

$$\left.\begin{array}{l}\begin{bmatrix} V_{load} \\ I_{gen} \end{bmatrix} = \begin{bmatrix} Y_{ll}^{-1} & -Y_{ll}^{-1}Y_{\mathrm{lg}} \\ Y_{gl}Y_{ll}^{-1} & Y_{gg} - Y_{gl}Y_{ll}^{-1}Y_{\mathrm{lg}} \end{bmatrix} \begin{bmatrix} I_{load} \\ V_{gen} \end{bmatrix} \\ V_{load} = M_{ll}V_{ll}^{*-1}S_l^* + N_{gl}V_{gg}^{*\ -1}S_g^* = V_{load}^l + V_{load}^g \\ \text{where} \\ Y_{BUS} = \begin{bmatrix} Y_{ll} & Y_{\mathrm{lg}} \\ Y_{gl} & Y_{gg} \end{bmatrix} \\ M_{ll} = Y_{ll}^{-1}Y_{\mathrm{lg}}Y_{gg}^{-1}Y_{gl} \\ N_{gl} = Y_{gl}^{-1} - Y_{ll}^{-1}Y_{\mathrm{lg}}Y_{gg}^{-1} \\ S_g^* = V_{gg}^* I_{gen} \end{array}\right\} \quad (2)$$

M_{ll} and N_{gl} gives relative electrical distance between load buses among themselves and between generator buses and load buses as complex measures, respectively. These measures can be utilized to define VCA by means of any clustering methods.

As partitioning of VCA are based on voltage apparent power coupling, the bilateral contracts existing among different local balancing areas under hierarchically upper BA or with microgrid aggregators and load serving entities of different balancing areas highly influence their formation. Moreover, in a microgrid dominant highly competitive deregulated grid, the microgrid aggregators can bid their resources in several markets operating at different time scales like day ahead energy market, real time ancillary service market for frequency regulation in different time scale, as contingency reserve, reactive power ancillary service, black start reserves, etc. Stakeholders' willingness to bid in any of these markets depends on their profit optimization in different time scale. Hence, the market clearing decides the availability of reactive power stakeholders at different locations. As the voltage control areas are defined hierarchically based on the electrical distance of load buses from voltage controlling buses as well as voltage sensitivity, these VCAs may overlap the grid regions operated by different BAs interconnected with long transmission lines. This market scenario as well as existing special protection schemes as adopted in Hydro-Québec network [19] to effectively utilize reactive power compensators may result in dependency of region of grid on reactive power available through long transmission line, in order to facilitate active power flow and maintain the voltage profile [18]. The voltage control strategy that could be deployed in such deregulated grid is discussed in the following section.

2.3 Automatic voltage control (AVC) in emerging grid

Deregulated market policies encourage stake holders to offer ancillary or energy services while maximizing their social welfare. Accordingly, utility scale fast acting reserves based on energy resources, conventional generators, microgrid aggregators, reactive power aggregators coordinating several distributed reactive power source, etc. can offer their service in reactive power market. In most grids, the reactive power/voltage set points of these aforesaid resources were determined by optimal power flow during scheduling process. Conventionally, these setpoints were varied manually based on operator's discretion, predefined rules formulated from past experience, etc. during normal operation or contingency [20,21]. As renewable energy penetration increases above 30%, the uncertainties in generation becomes the critical parameter determining the voltage stability rather than the large system loads and thus becomes the major source of voltage fluctuations and power quality issues [2,5]. Hence, automatic voltage controls like three-level hierarchical control implemented in France and Italy, OPF-based method in PJM as well as method based on adaptive zoning implemented in China are becoming quite important in deregulated grid with distributed RES [3,6,22]. Most of the AVC techniques implemented and presented in literature make use of voltage measurements at pilot buses as disturbance to formulate the control input setpoints for reactive power resources. These control set points were derived from multi criterion, multi objective optimization problem. Some of the several objectives in literature are minimization of voltage variations at critical buses, minimizing the ∞-norm of pilot bus voltage deviation vector, minimizing the reactive power output of resources so as to maximize the reserve, cost factors such as loss minimization, number of switching operations of discrete devices, minimal deviation from market cleared or OPF based setpoints etc., could be achieved through methods like model predictive control [6,23–26]. In all these methods, voltage measurements at pilot buses are treated as disturbance that is approximated as a measure of reactive generation or load disturbance in linearized decoupled Q-V model of the system. One of such models [22] is shown in Eq. (3) as,

$$\left.\begin{aligned} \text{Minimize}: \Delta|V_{load}| &= K_1 \Delta Q_{load} - K_2 \Delta u \\ \Delta u &= \begin{bmatrix} \Delta Q_{Qg} & \Delta V_{Qc} \end{bmatrix}_{q\times 1} \end{aligned}\right\} \tag{3}$$

where K_1 and K_2 are the sensitivity matrices derived from the susceptance matrix of the system. u is the control input that forms the reactive power or voltage reference depending on the type of the device providing the regulation service.

3. Renewable uncertainty instigated disturbance propagation

The well-known active power P, reactive power Q, and voltage V interdependency shown in Eq. (4) is the basis of short-term voltage stability analysis of grid in terms of P-V, Q-V, or P-Q curve [27].

$$V_l = \sqrt{\frac{E_{th}^2}{2} - Q_l X_{th} \pm \sqrt{\frac{E^4}{4} - X_{th}^2 P_l^2 - X_{th} E_{th}^2 Q}} \tag{4}$$

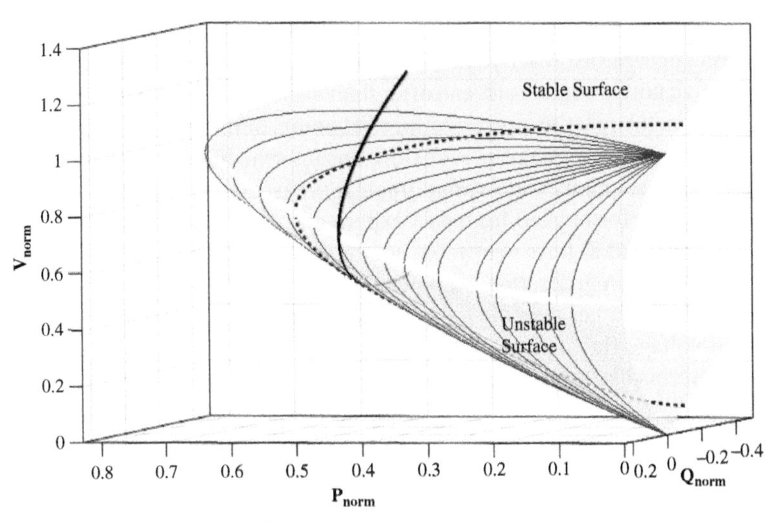

FIG. 2 *P-Q-V* surface at system bus.

where E_{th} and X_{th} are the Thevenin equivalent voltage source and impedance of the grid at the concerned load bus "l" with load $P_{l+j}Q_l$. It clearly shows that the voltage at any bus in the grid is highly sensitive to the active power (P) and reactive power (Q) flow toward the bus and the interdependency is shown by the *P-Q-V* surface in Fig. 2. If the load behavior is to maintain constant power factor, the bus voltage varies along the constant power factor lines shown in thin, red ones along the stable surface. Similarly, if load draws constant real power, the bus voltage varies along thick, black, and similar lines along the surface depending on the condition. Likewise, dotted blue line shows the bus voltage variation for constant reactive power load.

The intermittency in active load/generation of microgrid at PCC owing to uncertainty of distributed RES operating at UPF will result in variation of net effective real power requirement at PCC of microgrid. This uncertainty in active power at PCC affects the real power flow through the lines connected to PCC. This variation in real power flow causes voltage variation at PCC as per Eq. (4) along the voltage surface presented in Fig. 2. The inherent nature of transmission line offers dynamic reactive power to restore the voltage resulting in variation in reactive power flow through connected lines ensuing further voltage variations. Even though the distributed fast and flexible active power reserves compensate the uncertain real power from microgrids, the reactive power flow disturbance instigated by the real power uncertainty at microgrid PCC propagates through the grid visible in the spatial voltage profile. It has been shown that unlike active power disturbance, reactive power disturbance propagates in an "echelon" structure such that disturbance amplifies and attenuates while propagating through weakly and strongly coupled buses with reactive power sources, respectively [28]. The lack of distributed fast acting resources to provide reactive power support will result in amplified propagation of voltage disturbance to weaker part of the grid resulting in unintentional islanding of microgrids in weaker section of grids aggravating the disturbance and leading to long-term voltage stability problem. The unintentional islanding of microgrid results in load curtailment in grid-dependent microgrids affecting the power system resiliency. When a portion of BA depend on reactive power available through long transmission line, the reverberation characteristics of long transmission line defined by Sonia Leva et al. in terms of space vectors [29] also highly affects the disturbance seen and response provided to mitigate the spread of voltage disturbance on the other side of the line.

4. Enhancing the power grid resiliency against long-term voltage instability

As mentioned in Section 2, power system resiliency is an evolving concept and measures to quantify it is not yet standardized. Literature is reported with a wide variety of matrices, key idea being the relationship between the area of real and desired performance curves [30]. Roles of microgrids in enhancing resiliency of power grid are in the form of defensive islanding of microgrids planned in proactive phase and implemented in resistive phase (as shown in Fig. 1), emergency demand response from microgrid in responsive phase and microgrid as resilience resource in local or community-based aspects in restorative phase. The prevailing market condition can lead the microgrid aggregators to facilitate reactive power support by coordinating its converter connected RES, energy storages, demand responses, etc.

In a deregulated grid with several microgrid aggregators capable of providing the reactive power support along with other utility scale fast acting reactive power resources could alleviate the spread of the uncertainty-induced voltage disturbance. The uncertainty-instigated voltage disturbance can be well within the permissible limits in premises to the source but escalates toward weaker part of the grid following echelon structure and eventually lead to voltage instability. The widespread availability of reactive power support from microgrid aggregators and their effective coordination with wide area control system (WACS) is vital in enhancing the resiliency of power grid against voltage instability in proactive phase.

5. Proactive defense methodology for enhancing power system resiliency

The wide area spread of disturbance induced by uncertain microgrid being a prime concern in emerging deregulated grid, a proactive defense methodology to mitigate the spread of disturbance is presented in this section. The methodology is devised to function at regional system protection centers (SPC). The propagation characteristics of reactive power disturbance instigated by uncertainty at microgrid PCC vary spatially based on spatial characteristics of the grid as well as temporally based on the temporal characteristics of uncertainty as explored in Section 3 and is visible in spatial and temporal voltage profile of the grid. The spread of this reactive power disturbance can be effectively mitigated by coordinated control of the several distributed reactive power resources like fast acting BESS-based resources, microgrid aggregators, conventional reactive power sources, etc. throughout the grid. The presented proactive defense methodology to enhance power system resiliency against voltage instability by mitigating the spread of reactive power disturbance relies on the time-synchronized measurement of voltage across the grid and coordinated control of distributed reactive power sources. Hence, PMU-based wide area measurement system (WAMS) and wide area control of reactive power resources forms the backbone of the presented proactive defense strategy.

The bottom layer of the methodology involves data collecting, updating and pre-processing at various time scales from the physical grid and virtual deregulated market. The market data get updated in different time scales like 24-h, 1-h, 5-min, etc. as and when market clearing of day ahead energy market, intra-day, and real time ancillary or balancing markets happens. Network data get updated based on the breaker and relay status from WAMS as and when related event happens. This real-world data collection at base layer are being utilized for several other control and monitory applications carried out at SPC. In the presented proactive defense strategy, these data are pre-processed and it enables to partition the grid into VCAs and select pilot buses based on sensitivity data as explained in Section 2.2. The next layer of the methodology involves the real time synchronized measurement of voltage profile of the grid, features of which acts as the main input to the methodology. These real-world data collection and real time measurement sections form the base layer of the proactive defense strategy. The feature extraction from the real time measurement of voltage profile of the grid happening at the data analysis and synthesis layer gives awareness about spatial and temporal characteristics of the disturbance. The situational awareness from data analysis layer together with the pre-processed data from bottom most layer provides necessary information required in the next layer to devise a decision logic. The presented disturbance mitigation strategy is formulated meeting several objectives while maintaining different constraints and limits. Besides, the methodology considers the highly nonlinear nature of emerging deregulated grid where reactive power for certain area is made available through long transmission line, which decides the disturbance propagation characteristics that are not considered by the reactive power regulation strategies so far. The proactive defense strategy-based decision logic helps to choose effective sources among the several distributed reactive power aggregators including the microgrids for mitigating the spread of uncertainty-induced voltage disturbance. These situational awareness layer and decision logic form the heart of the presented proactive defense strategy to mitigate

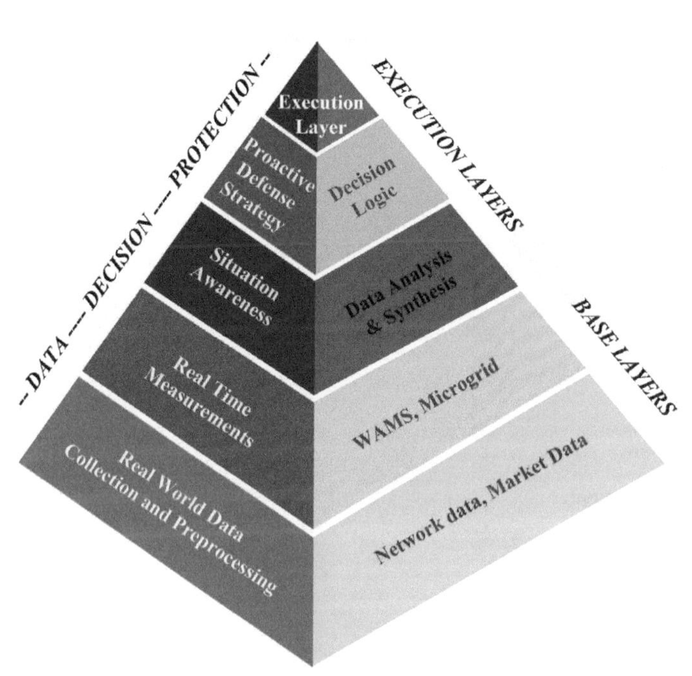

FIG. 3 Presented proactive defense methodology against disturbance propagation.

the voltage disturbance spread, initiated by the uncertainty at microgrid PCC, which otherwise would have amplified across the weaker section of grid adversely affecting the power system resiliency. This strategy can be a part of hierarchical AVC, gaining importance in emerging deregulated grid due to the reactive power regulation issues.

Finally, the execution layer implements the strategy through WACS, giving setpoint update to the most appropriate distributed fast acting reactive power resources like utility scale BESS, STATCOM, microgrid aggregators, and so on. WAMS-based data acquisition and real time measurement in base layer and WACS in execution layer are implemented through wired communication like PLC and fiber optics or through wireless communications, making use of protocols like IEEE 802.22, LTE, LTE-A, etc. complying with synchro phasor standard IEEE C37.118. These layers of the presented proactive defense strategy to enhance the resiliency of the grid is illustrated in the Fig. 3.

6. Implementation of presented methodology

Wide area propagation of uncertainty-induced disturbance is stumbling the prospect of widespread renewable dominant microgrid in the evolving deregulated power system. The propagation of these disturbance can be mitigated

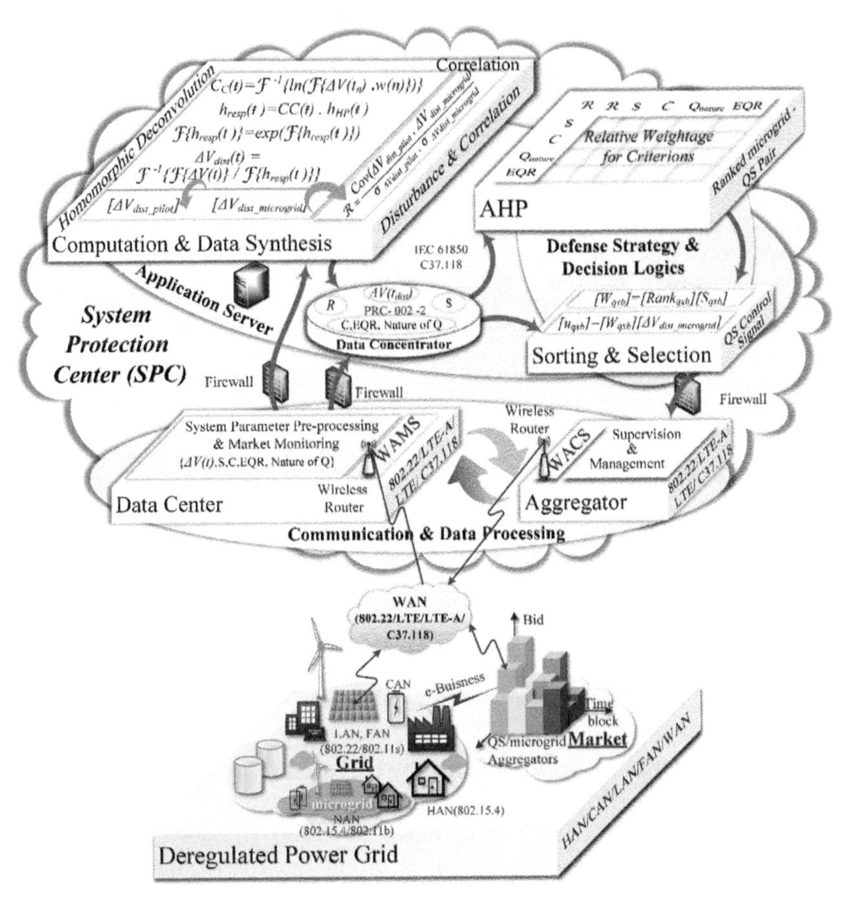

FIG. 4 Implementation of presented proactive defense methodology against disturbance propagation.

effectively through the proactive defense scheme. The different steps involved in the implementation of presented proactive defense methodology at regional SPC for resiliency enhancement against the generation uncertainty in deregulated power system with several grid supporting and grid dependent microgrids is shown in Fig. 4. The implementations happen with the interaction between the physical and cyber layer of the emerging deregulated grid.

As discussed in Section 2.3, most AVC techniques consider the voltage variation at candidate buses as the disturbance against which the set points of coordinated reactive power sources are obtained as control input to minimize the disturbance effect. The assumption of bus voltage variation at any bus as the linear combination of reactive power disturbance at that bus and system response to that disturbance including the control measures based on the Q-V decoupled linear model of the system as described in Eq. (3) in

Section 2.3 is utilized to develop the optimal control input in AVC is implemented and presented in literature. In system where the dominant uncertainty is related to load, the time frame related to the regulation requirement justifies this assumption. Emerging deregulated grid is dominated with resources like distributed RES, utility scale converter based fast acting flexible resources, smart and flexible microgrid aggregators, etc. The dynamics related to generation uncertainty caused by intermittent RES and the related response from the aforesaid resources happen at lower time frame in future grid and hence nonlinearities are crucial in the control implementation time frame. Moreover, the system depending on reactive power through the long transmission line are influenced by the reverberation characteristics of long line and will be highly nonlinear in nature. Hence, control strategies ignoring the nonlinearities of the system completely won't be effective for mitigating the spread of the disturbance in future grid, which in turn affects the resiliency.

The voltage variation at a particular bus is the convolution of voltage variation due to disturbance at same bus (forcing function) and the system response there due to the disturbance as well as the other disturbances that happened at some other locations. The heart of the presented proactive defense strategy is to deconvolute the voltage variation at candidate bus ($\Delta V(t)$) as voltage variation due to actual disturbance there, denoted as $\Delta V_{dist}(t)$ and system response ($h_{resp}(t)$). Then, develop the control setpoints for reactive power source to mitigate the propagating disturbance by making use of characteristics of $\Delta V_{dist}(t)$ and $h_{resp}(t)$.

The e-Business based market clearing for reactive power procurement as described in Section 2.1 happens at cyber layer of the system. This market clearing parameters along with network parameters from WAMS and state estimator are exploited to partition the system into different VCA and select pilot bus as described in Section 2.2. This system partitioning is crucial to implement the presented proactive defense for enhancing the resiliency at physical layer of the emerging grid. The PMU-based real time voltage measurement at pilot buses and RES PCCs in a VCA happens at physical layer. These are transmitted to regional SPC over cyber layer following any of the protocols like IEEE 802.22, LTE, LTE-A, etc. complying with synchro phasor standard IEEE C37.118. These form the base layer of the methodology and happen at data center and will be stored in the data concentrator as illustrated in Fig. 4.

The decisive part of the methodology is to utilize actual voltage disturbance $\Delta V_{dist}(t)$ caused by microgrid PCC and pilot bus, other than the mixture $\Delta V(t)$ of several disturbance and their response to derive the control signal for obtaining the reactive power regulation to mitigate the spread of the voltage disturbance. The homomorphic deconvolution based on complex cepstral transform decomposes the signal into its forcing function and system response without the prior knowledge of any of the component, either the forcing function or the system response, by means of liftering. In homomorphic deconvolution employing cepstral analysis, the nonlinear convolved signal in time domain is transformed into linear quefrency domain where it is liftered (linear filtering in quefrency domain) to high quefrency (high energy forcing function) and low quefrency part (system response) and then transformed back into time domain. The beauty of complex cepstral analysis is its ability to recover the signal in time domain without loss of information as the phase information is always retained. Hence, this tool widely used in speech, seismic, vibration analysis, etc. proves to be vital in evolving grid where WAMS-based big data are effectively utilized to develop the control applications. As a first step to the future of power system applications employing cepstral analysis, the presented proactive defense strategy decisively utilizes this homomorphic deconvolution to decompose the voltage variation signal at candidate buses, the pilot bus $\Delta V_{pilot}(t)$ and the PCC of microgrids $\Delta V_{microgrid}(t)$, into voltage variation due to disturbance there and system response to it as given by Eq. (5).

$$\left.\begin{aligned}
&\Delta V(t) = \left[\Delta V_{pilot}(t)\Delta V_{microgrid}(t)\right]\\
&C_C(t) = \mathcal{F}^{-1}\{\ln(\mathcal{F}\{\Delta V(t_n).w(n)\})\}\\
&h_{resp}(t) = C_C(t).h_{LP}(t)\\
&\mathcal{F}\{h_{resp}(t)\} = \exp\left(\mathcal{F}\{h_{resp}(t)\}\right)\\
&\Delta V_{dist}(t) = \mathcal{F}^{-1}\{\mathcal{F}\{\Delta V(t_n)\}/\mathcal{F}\{h_{resp}(t)\}\}\\
&\Delta V_{dist}(t) = \left[\Delta V_{dist_{pilot}}(t)\Delta V_{dist_{microgrid}}(t)\right]
\end{aligned}\right\} \tag{5}$$

where $\mathcal{F}, \mathcal{F}^{-1}, C_C(t), h_{LP}(t)$ represents Fourier transform, inverse Fourier transform, complex cepstral transform, and impulse response of low pass filter at quefrency domain, respectively. If there are "b" uncertain microgrids based on market condition in a VCA, then $\Delta V_{microgrid}(t)$ is a column vector of size b at each time instant.

Correlation coefficients ($\mathcal{R}_{b\times 1}$) of voltage variation due to the actual disturbance at pilot bus and PCCs of microgrids ($\Delta V_{dist_{pilot}}(t)$ and $\Delta V_{dist_{microgrid}}(t)$) as in Eq. (6) gives the relative influence of uncertainty at ith uncertain microgrid at pilot bus. These computations were carried out in application server of regional SPC and forms the data analysis and synthesis layer of the presented methodology shown in Figs. 3 and 4.

$$\mathcal{R}_{b\times 1} = \frac{\operatorname{cov}\left(\Delta V_{dist_{microgrid}}(t), \Delta V_{dist_{pilot}}(t)\right)}{\sigma_{\Delta V_{dist_{microgrid}}(t)}\sigma_{\Delta V_{dist_{pilot}}(t)}} \tag{6}$$

This correlation coefficient vector $\mathcal{R}_{b\times 1}$, sensitivity matrix $S_{q\times b}$ among q reactive power sources including the microgrid aggregators offering reactive power reserve and b uncertain microgrids in the VCA, nature of reactive power resources and the pricing signal of q reactive power sources for reactive resource offered at different time frame $C_{q\times 1}$ and the effective reactive power reserve (EQR) expected from them during worst case contingency are the criterion, whose relative importance can be decided by reactive power aggregator for the welfare of both aggregator and FFR, to find the effective reactive power source and renewable resource pair in alleviating the disturbance spread and thus enhancing resiliency. Analytic hierarchical process (AHP) is an effective multi-criterion decision making tool where the relative significance of the criterion could be decided and varied as and when required. Hence, the presented strategy formulates the ranking matrix $Rank_{q\times b}$ for reactive power source and uncertain microgrids in the VCA based on their effectiveness in mitigating the disturbance while meeting these criteria, employing AHP. This rank matrix together with the sensitivity matrix $S_{q\times b}$ forms the weighing matrix $W_{q\times b}$ in Eq. (7) that acts as the decision logic, determining the setpoint control signals $u_{q\times 1}$ for q reactive sources. The data analysis and synthesis layer and the decision logic layer are implemented at regional SPC application server, shown in Fig. 4, complying with access security protocols and at data rate requirements with NREL PRC-002-02.

$$\left[W_{q\times b}\right] = \left[Rank_{q\times b}\right]\left[S_{q\times b}\right] \tag{7}$$

$$\left[u_{q\times 1}\right] = \left[W_{q\times b}\right]\left[\Delta V_{dist_{microgrid}b}\right] \tag{8}$$

The WACS-based execution happens through cyber layer in which the reactive power aggregator communicates this setpoint control signal $u_{q\times 1}$ to local controller at aggregator offering reactive power service. This WACS-based implementation of presented proactive defense strategy in the physical grid through cyber layer is illustrated in Fig. 4.

7. Case studies and discussions

Case studies are conducted to investigate the microgrid uncertainty instigated wide area disturbance propagation and influence of the reactive power support on this propagation characteristic. The study is carried out on modified version of New York-New England (NY-NE) 16 machine 68 bus system where the RES is integrated into the system [31]. RES locations are considered as microgrid locations for our case study and details are given in Table 1. Investigation shows

TABLE 1 Details of modified NY-NE system.

Bus locations	Microgrid	Conventional generator
NE area	26, 57, 59, and 68	1–9
NY area	17, 33, and 53	10–13
Other areas	18, 41, and 42	14–16

the spread of disturbance caused by the uncertainty during lack of reactive power support from the resources. Later part of the study reveals the effect of concentrated and distributed reactive power support in mitigating the spread of disturbance. Power flow characteristics of the system is in such a way that the NE area depends on NY area for reactive power support through long transmission line and hence the disturbance propagation is influenced by the reverberation characteristics of the long line. Study period lasts from 95 to 150 s, where uncertainty period is 50 s starting from 100 s before which system is operating under steady state without any disturbance. We consider a 35% uncertainty in the active power injected by a grid supporting microgrid at bus 26, due to the uncertainties associated with the RES in that microgrid, while other microgrids are operating in dispatchable mode. Investigation is carried out in MATLAB/SIMULINK platform and has four main parts as follows:

- Case 1: Investigation of propagation characteristics of disturbance
- Case 2: Investigation about the influence of concentrated reactive power support on disturbance propagation
- Case 3: Investigation about the effectiveness of distributed reactive power support on disturbance propagation
- Case 4: Influence of nonlinear long line reverberation characteristics on all the above three cases

7.1 Case 1: Investigation of propagation characteristics of disturbance

Disturbance propagation induced by uncertainty of microgrid at bus 26 shown in Fig. 5 is analyzed in this case. This uncertainty affects the real and reactive power flow through the connected line and eventually influences the reactive power support from area-2 of studied system through long transmission line. This influence is also depicted in Fig. 5. Spatial characteristics of disturbance propagation over interconnected grid is clearly visible from the effect uncertainty causes on the other part of the system as in Fig. 5. As the support from area-2 is the main source of reactive power in area-1, the disturbance in this support affects the available reactive power at other buses, which could be depicted from their voltage profile shown in Fig. 6.

7.2 Case 2: Investigation about the influence of concentrated reactive power support on disturbance propagation

Conventionally, as a measure against voltage instability, reactive support is provided at weaker bus of the area. As a next case, we analyze the influence of reactive power support provided at bus 64, which is the weakest bus in the area as illustrated in previous researches related to voltage stability. The influence again is visible in voltage profile of the grid and is captured in Fig. 7. Now,

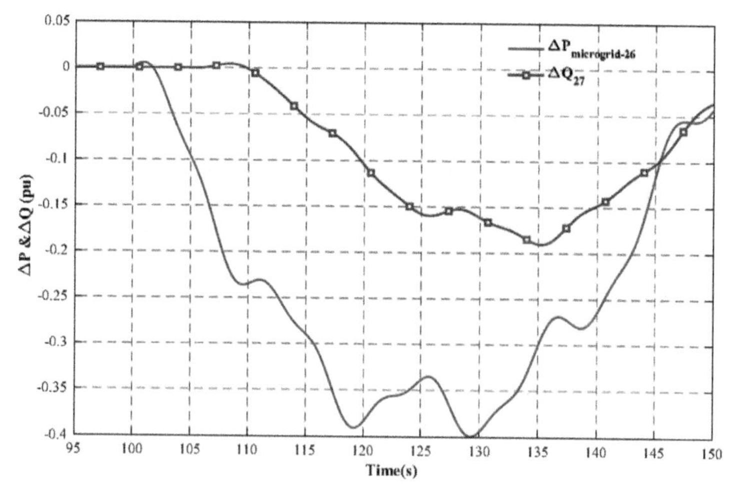

FIG. 5 Uncertainty in active power from microgrid dominant with RES at bus 26 and its effect on Q support from NY area on bus 27.

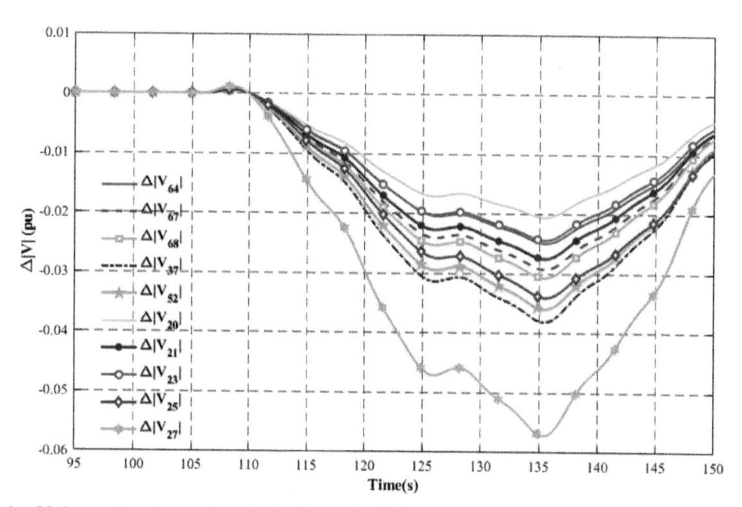

FIG. 6 Voltage disturbance in selected buses in NE area instigated by uncertainty in microgrid at bus 26.

we analyze the case where the reactive power is provided at the strongest bus in the area, i.e., the pilot bus. The influence is shown in Fig. 8. Comparison of Figs. 7 and 8 clearly reveals the incapability of the conventional reactive power support provided at weaker bus against voltage instability in mitigating the propagation of disturbance. It also emphasizes the importance of pilot bus-based AVC in enhancing the resiliency by mitigating the disturbance spread in the evolving deregulated grid. When reactive power support is provided at

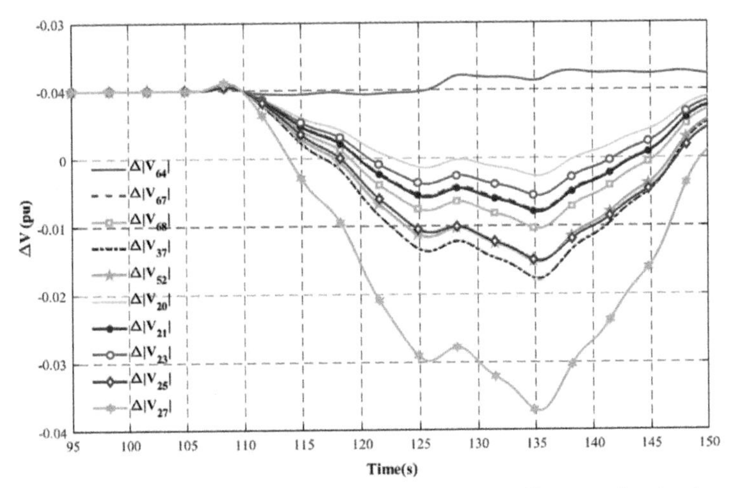

FIG. 7 Voltage disturbance in selected buses in NE area instigated by uncertainty in microgrid at bus 26 with reactive power support at weaker bus 64.

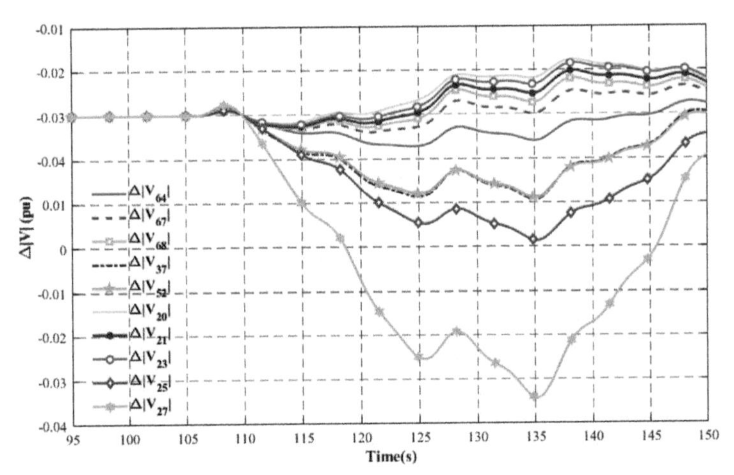

FIG. 8 Voltage disturbance in selected buses in NE area instigated by uncertainty in microgrid at bus 26 with reactive power support at pilot bus 68.

the weakest bus, its voltage profile alone improved, while the support at pilot bus improved the voltage profile of considerable number of buses other than those highly sensitive to the disturbance source. Fig. 9 shows the effectiveness of support provided at the disturbance source in improving the voltage profile thereby enhancing the resiliency. The study reveals the importance of identifying the source effective in mitigating the spread of the disturbance based on which the presented proactive disturbance strategy ranks the reactive power sources and disturbance source pair available in the area in the order of their ability in mitigating the disturbance.

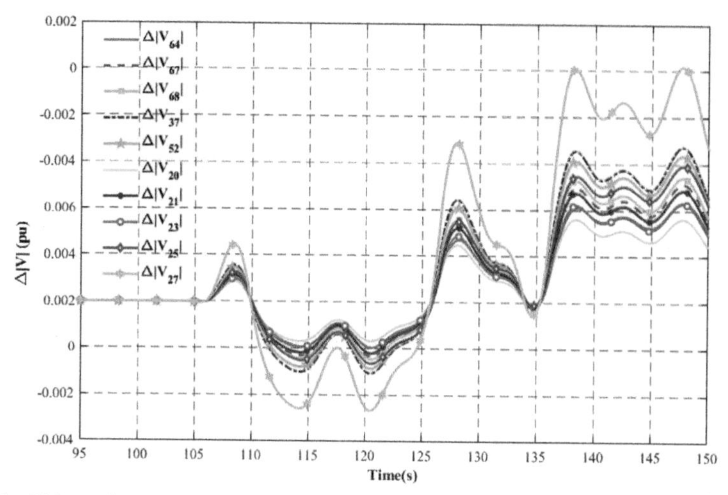

FIG. 9 Voltage disturbance in selected buses in NE area instigated by uncertainty in microgrid at bus 26 with reactive power support at bus 27.

7.3 Case 3: Investigation about the effectiveness of distributed reactive power support on disturbance propagation

The effectiveness of distributed reactive power support at three different buses in deficient area is clearly captured in Fig. 10 compared to the support provided at pilot bus in Fig. 9. Provision of reactive power support at pilot bus is the best case other than the provision of support against the disturbance at source itself in case of concentrated reactive power support. In actual power grid, there will be more than one disturbance source at a time. Investigation so far reveals that identifying each of the disturbance source and finding out the best source that could mitigate the disturbance is quiet important in enhancing the power system resiliency. It also puts emphasis on the importance of providing the control signal to the effective source according to the disturbance it has to mitigate.

Based on the investigation results, the presented proactive control strategy is devised to ensure that the aforesaid two objectives are met so that uncertainty-instigated disturbance spread throughout the system is effectively mitigated enhancing the power system resiliency.

7.4 Case 4: Influence of nonlinear long line reverberation characteristics on all the above three cases

In order to quantify and compare the influence of long line propagation characteristics on the voltage disturbance in all the three cases, we consider the

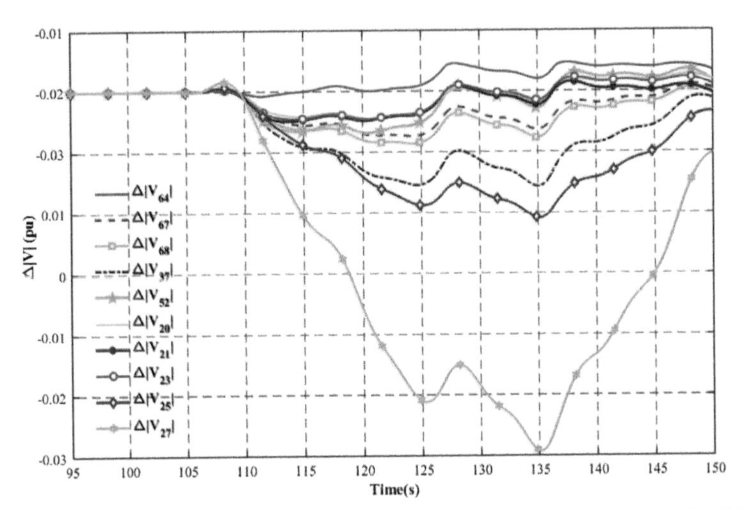

FIG. 10 Voltage disturbance in load buses instigated by uncertainty in microgrid at bus 26 with distributed reactive power support at buses 24, 60, and 64.

energy content (ξ_i) and peak variation ($\Delta Peak_i$) of the voltage disturbance signal at ith bus given by Eqs. (9) and (10) [7] and are summarized in Table 2.

$$\xi_i = \int_{t_0}^{T} |\Delta V_i| dt \tag{9}$$

$$\Delta Peak_i = \begin{cases} -\min(\Delta V_i); \text{if} |\min(\Delta V_i)| > |\max(\Delta V_i)| \\ \max(\Delta V_i); \text{otherwise} \end{cases} \tag{10}$$

Table 2 clearly shows the availability of reactive power in longer lines to mitigate the disturbance than from comparatively shorter ones in case 1 and case 2, when Q aid is not provided or when Q aid provided is insufficient. The ξ of voltage disturbance at bus 23 is 0.5 when long transmission line connecting area 1 and area 2 is 200 km compared to 0.3 when the line is 400 km long, indicating the Q aid given by longer transmission line. This indicates that the long line is providing dynamic reactive power support in case 1 and case 2 and as a result uncertainty-induced disturbance in area 1 spread toward area-2. When the disturbance induced by long transmission line is well mitigated by Q aid from distributed source in area 1 itself, dynamic Q support from long line is no longer required that helps in mitigating the spread of disturbance to the other area. As Table 2 reveals, the ξ_i of bus voltage variations are almost same irrespective of line length in case 3, indicating that area-1 is independent of dynamic Q support provided by area-2 during the time of disturbance. Hence, validating the ability of distributed Q sources to mitigate wide area spread of disturbance induced by the uncertainty.

TABLE 2 Features of voltage disturbance at different bus locations during case 1, case 2, and case 3.

Q aid	Line length (km)	Bus no	64	67	68	37	52	20	21	23	25	27
–	200	ξ	0.41	0.56	0.62	0.96	0.85	0.28	0.50	0.39	0.77	2.14
		$\Delta peak$	−2.72	−3.20	−3.36	−4.17	−3.94	−2.25	−3.01	−2.67	−3.72	−6.26
	300	ξ	0.36	0.50	0.55	0.84	0.75	0.24	0.44	0.35	0.68	1.89
		$\Delta peak$	−2.56	−3.01	−3.16	−3.92	−3.71	−2.11	−2.83	−2.51	−3.50	−5.89
	400	ξ	0.32	0.44	0.48	0.74	0.67	0.22	0.39	0.30	0.60	1.67
		$\Delta peak$	−2.40	−2.83	−2.97	−3.68	−3.48	−1.98	−2.66	−2.36	−3.29	−5.53
At B64	200	ξ	0.00	0.20	0.27	0.50	0.42	0.10	0.21	0.16	0.41	1.46
		$\Delta peak$	−0.06	−1.98	−2.27	−3.07	−2.81	−1.41	−1.99	−1.72	−2.77	−5.21
	300	ξ	0.00	0.18	0.24	0.44	0.37	0.09	0.18	0.14	0.36	1.28
		$\Delta peak$	−0.06	−1.83	−2.11	−2.87	−2.62	−1.31	−1.85	−1.60	−2.59	−4.88
	400	ξ	0.00	0.16	0.21	0.39	0.32	0.08	0.16	0.12	0.32	1.13
		$\Delta peak$	−0.05	−1.70	−1.97	−2.67	−2.44	−1.22	−1.73	−1.49	−2.42	−4.56
At B64, B59, B24	200	ξ	0.00	0.01	0.01	0.04	0.01	0.00	0.00	0.00	0.07	0.56
		$\Delta peak$	−0.04	−0.42	−0.49	−0.93	−0.35	−0.23	−0.25	−0.23	−1.19	−3.31
	300	ξ	0.00	0.01	0.01	0.03	0.01	0.00	0.00	0.00	0.06	0.50
		$\Delta peak$	−0.04	−0.38	−0.44	−0.83	−0.35	−0.23	−0.25	−0.23	−1.09	−3.07
	400	ξ	0.00	0.01	0.01	0.03	0.00	0.00	0.00	0.00	0.05	0.44
		$\Delta peak$	−0.04	−0.36	−0.42	−0.75	−0.34	−0.22	−0.24	−0.23	−1.00	−2.84

$\Delta peak$ is in 10^{-2} pu.

8. Conclusions

This chapter investigates the opportunities in deregulated power grid to curb the challenges posed by renewable energy-dominant microgrids. In accordance with this, a proactive defense strategy to attenuate the wide area spread of disturbance instigated by uncertainty associated with renewable dominant microgrid is presented. Case studies are conducted to analyze the wide area propagation phenomenon of the disturbance instigated by the uncertainty. The influence of various conditions of reactive power support provided by the system, which in turn decided by the policies and market clearing conditions in a deregulated grid, is also analyzed. Observations from the investigations emphasize the need for AVC in evolving deregulated grid for mitigating the spread of the disturbance, thereby enhancing the resiliency of power grid. Propagation characteristics of disturbance evident from investigation under different conditions of reactive power regulation aided in developing the presented proactive defense strategy to enhance the power system resiliency in view of evolving deregulated market. The proactive defense strategy utilizes microgrid aggregators along with other reactive power resources to mitigate the disturbance spread induced by microgrid uncertainty itself and hence, aids in better prospects toward a future smart grid with large-scale microgrids.

References

[1] L. Meng, J. Zafar, S.K. Khadem, A. Collinson, K.C. Murchie, F. Coffele, G.M. Burt, Fast frequency response from energy storage systems—a review of grid standards, projects and technical issues, IEEE Trans. Smart Grid 11 (2) (2020) 1566–1581.

[2] B. Qi, K.N. Hasan, J.V. Milanovic, Identification of critical parameters affecting voltage and angular stability considering load-renewable generation correlations, IEEE Trans. Power Syst. 34 (4) (2019) 2859–2869, https://doi.org/10.1109/TPWRS.2019.2891840.

[3] H. Sun, Q. Guo, J. Qi, V. Ajjarapu, R. Bravo, J. Chow, Z. Li, R. Moghe, E. Nasr-Azadani, U. Tamrakar, G.N. Taranto, R. Tonkoski, G. Valverde, Q. Wu, G. Yang, Review of challenges and research opportunities for voltage control in smart grids, IEEE Trans. Power Syst. 34 (4) (2019) 2790–2801, https://doi.org/10.1109/TPWRS.2019.2897948.

[4] J. Hu, G. Yang, C. Ziras, Aggregator operation in the balancing market through network-constrained transactive energy, IEEE Trans. Power Syst. 34 (5) (2019) 4071–4080.

[5] S. Alzahrani, R. Shah, N. Mithulananthan, Exploring the dynamic voltage signature of renewable rich weak power system, IEEE Access 8 (2020) 216529–216542, https://doi.org/10.1109/ACCESS.2020.3041410.

[6] M.A. Elizondo, N. Samaan, Y.V. Makarov, J. Holzer, M. Vallem, R. Huang, B. Vyakaranam, X. Ke, F. Pan, Literature survey on operational voltage control and reactive power management on transmission and sub-transmission networks, in: IEEE Power and Energy Society General Meeting, 2018 January, 2018, pp. 1–5, https://doi.org/10.1109/PESGM.2017.8274068.

[7] N.V. Mohamed, G. Pathirikkat, S. Rajan, Modelling of inter-area angular dynamics and proactive defence strategy in deregulated power grids with large-scale integration of renewable energy sources, IET Gener. Transm. Distrib. 14 (15) (2020) 2940–2950, https://doi.org/10.1049/iet-gtd.2019.1868.

[8] A. Nikoobakht, J. Aghaei, M. Shafie-Khah, J.P.S. Catalao, Allocation of fast-acting energy storage systems in transmission grids with high renewable generation, IEEE Trans. Sustain. Energy 11 (3) (2020) 1728–1738, https://doi.org/10.1109/TSTE.2019.2938417.
[9] S. Lin, Y. Lu, M. Liu, Y. Yang, S. He, H. Jiang, SVSM calculation of power system with high wind-power penetration, IET Renew. Power Gener. 13 (8) (2019) 1391–1401, https://doi.org/10.1049/iet-rpg.2018.6144.
[10] E. Vittal, M. O'Malley, A. Keane, A steady-state voltage stability analysis of power systems with high penetrations of wind, IEEE Trans. Power Syst. 25 (1) (2010) 433–442, https://doi.org/10.1109/TPWRS.2009.2031491.
[11] X. Xu, S. Chen, Power system voltage stability evaluation considering renewable energy with correlated variabilities, IEEE Trans. Power Syst. 33 (3) (2018) 3236–3245.
[12] D. Jay, K.S. Swarup, Isoperimetric clustering-based network partitioning algorithm for voltage-apparent power coupled areas, IET Gener. Transm. Distrib. 13 (22) (2019) 5109–5116, https://doi.org/10.1049/iet-gtd.2019.0115.
[13] A. Jalali, M. Aldeen, Short-term voltage stability improvement via dynamic voltage support capability of ESS devices, IEEE Syst. J. 13 (4) (2019) 4169–4180, https://doi.org/10.1109/JSYST.2018.2882643.
[14] M. Padhee, A. Pal, C. Mishra, K.A. Vance, A fixed-flexible BESS allocation scheme for transmission networks considering uncertainties, IEEE Trans. Sustain. Energy 11 (3) (2020) 1883–1897, https://doi.org/10.1109/TSTE.2019.2946133.
[15] D. Jay, K.S. Swarup, Game theoretical approach to novel reactive power ancillary service market mechanism, IEEE Trans. Power Syst. 36 (2) (2021) 1298–1308, https://doi.org/10.1109/TPWRS.2020.3019786.
[16] E. Lobato Miguélez, F.M. Echavarren Cerezo, L. Rouco Rodrguez, On the assignment of voltage control ancillary service of generators in Spain, IEEE Trans. Power Syst. 22 (1) (2007) 367–375, https://doi.org/10.1109/TPWRS.2006.888984.
[17] J. Zhong, K. Bhattacharya, J. Daalder, Reactive power as an ancillary service: issues in optimal procurement, in: PowerCon 2000–2000 International Conference on Power System Technology, Proceedings, Vol. 2(2), 2000, pp. 885–890, https://doi.org/10.1109/ICPST.2000.897138.
[18] J. Zhong, E. Nobile, A. Bose, K. Bhattacharya, Localized reactive power markets using the concept of voltage control areas, IEEE Trans. Power Syst. 19 (3) (2004) 1555–1561, https://doi.org/10.1109/TPWRS.2004.831656.
[19] E. Ghahremani, A. Heniche, M. Perron, M. Racine, S. Landry, H. Akremi, A detailed presentation of an innovative local and wide-area special protection scheme to avoid voltage collapse: from proof of concept to grid implementation, IEEE Trans. Smart Grid 10 (5) (2018) 5196–5211, https://doi.org/10.1109/TSG.2018.2878980.
[20] H. Ma, D.J. Hill, Adaptive coordinated voltage control - part I: basic scheme, IEEE Trans. Power Syst. 29 (4) (2014) 1546–1553, https://doi.org/10.1109/TPWRS.2013.2293577.
[21] H.M. Ma, K.T. Ng, K.F. Man, A multiple criteria decision-making knowledge-based scheme for real-time power voltage control, IEEE Trans. Industr. Inform. 4 (1) (2008) 58–66, https://doi.org/10.1109/TII.2008.919320.
[22] H.Y. Su, F.M. Kang, C.W. Liu, Transmission grid secondary voltage control method using PMU data, IEEE Trans. Smart Grid 9 (4) (2018) 2908–2917, https://doi.org/10.1109/TSG.2016.2623302.
[23] H. Ma, D.J. Hill, A fast local search scheme for adaptive coordinated voltage control, IEEE Trans. Power Syst. 33 (3) (2018) 2321–2330, https://doi.org/10.1109/TPWRS.2017.2748149.
[24] Z. Tang, E. Ekomwenrenren, J.W. Simpson-Porco, E. Farantatos, M. Patel, H. Hooshyar, Measurement-based fast coordinated voltage control for transmission grids, IEEE Trans. Power Syst. 8950 (December) (2020) 1–14, https://doi.org/10.1109/TPWRS.2020.3045379.

[25] Z. Tang, D.J. Hill, T. Liu, H. Ma, Hierarchical voltage control of weak subtransmission networks with high penetration of wind power, IEEE Trans. Power Syst. 33 (1) (2017) 187–197, https://doi.org/10.1109/tpwrs.2017.2700996.

[26] W. Yan, W. Cui, W.J. Lee, J. Yu, X. Zhao, Pilot-bus-centered automatic voltage control with high penetration level of wind generation, IEEE Trans. Ind. Appl. 52 (3) (2016) 1962–1969, https://doi.org/10.1109/TIA.2015.2511166.

[27] T. Van Cutsem, C. Vournas, Transmission system aspects, Voltage Stability of Electric Power Systems, in:, M.A. Pai (Ed.), The Springer International Series in Engineering and Computer Science, Springer, 2008.

[28] J. Thorp, D. Schulz, M. Ilić-Spong, Reactive power-voltage problem: conditions for the existence of solution and localized disturbance propagation, Int. J. Electr. Power Energy Syst. 8 (2) (1986) 66–74, https://doi.org/10.1016/0142-0615(86)90001-3.

[29] S. Leva, A.P. Morando, Analysis of physically symmetrical Lossy three-phase transmission lines in terms of space vectors, IEEE Trans. Power Deliv. 21 (2) (2006) 873–882.

[30] M. Panteli, P. Mancarella, D.N. Trakas, E. Kyriakides, N.D. Hatziargyriou, Metrics and quantification of operational and infrastructure resilience in power systems, IEEE Trans. Power Syst. 32 (6) (2017) 4732–4742, https://doi.org/10.1109/TPWRS.2017.2664141.

[31] A. Adrees, J.V. Milanović, P. Mancarella, Effect of inertia heterogeneity on frequency dynamics of low-inertia power systems, IET Gener. Transm. Distrib. 13 (14) (2019) 2951–2958, https://doi.org/10.1049/iet-gtd.2018.6814.

Chapter 8

Adaptive controller-based shunt active power filter for power quality enhancement in grid-integrated PV systems

Jitendra Kumar Sao[a], Pravat Kumar Ray[b], Ram Dayal Patidar[a], and Sushree Diptimayee Swain[a]

[a]*OPJU Raigarh (C.G.), Raigarh, India,* [b]*Department of Electrical Engineering, National Institute of Technology, Rourkela, India*

1. Introduction

The use of nonlinear loads such as computers, printers, rectifiers, switched-mode power supply, variable speed drivers, and other converters is increasing continuously in both domestic and industrial purposes. These nonlinear loads are responsible for harmonics generation because they draw nonsinusoidal current from the source [1]. The presence of harmonics degrades the power quality of the electrical power system. Low-power quality yields problems such as overheating, instability, interference in the communication line, and damaging of sensitive equipment. These concerning power quality problems can be mitigated by compensating generated harmonics. Voltage and current harmonics problems can be compensated by using series and shunt filters, respectively [2]. These filters may be passive or active, but passive filters have serious disadvantages such as selective compensation, resonance problem, and bulky in size [3]. Hence, active power filter (APF) became more popular for harmonics compensation. When an APF is connected with the load in shunted form, it is known as shunt active power filter (SAPF). SAPF is a power conditioning device which improves power quality by providing current harmonics compensation. SAPF consists of a voltage source inverter (VSI), which injects filter current at the point of common coupling (PCC) in the system to compensate generated harmonics. Injected current is the same as harmonics current but opposite in phase so it can cancel the harmonics [4].

Microgrid Cyberphysical Systems. https://doi.org/10.1016/B978-0-323-99910-6.00010-4

Renewable energy sources are clean and create less pollution, hence effective use of these sources is one of the major thrust areas for researchers. There are many renewable energy sources, but photovoltaic (PV) energy source is most popular due to its advantages over other sources, such as low maintenance and lesser implementation cost. Hence, PV-integrated SAPF is used in this chapter. Solar panels convert sunlight into electrical energy due to the PV effect [5]. But like other renewable energy sources, PV energy source also lacks in its efficiency. As solar panels have nonlinear *I-V* characteristics, hence they are unable to provide constant voltage or constant current. This low efficiency due to nonlinear characteristics can be increased by maximum power point tracking (MPPT) technique. Maximum power is drawn from the solar panel only at a particular voltage and current value. Many methods of MPPT are suggested by researchers [6, 7]. Among them, we have used incremental conductance (IncCond)-based MPPT in this chapter. PV system is integrated with VSI of SAPF, which provides reference DC-link capacitor voltage. This DC-link capacitor voltage (V_{dc}) is controlled by conventional proportional-integral (PI) controller [8].

SAPF consists of a VSI along with passive $R - L$ filter and DC-link capacitor C_{dc}. This SAPF, as the name specifies, is connected in shunted form with grid and nonlinear load at the PCC and provides compensating current [9]. It has advantages over passive filter because it provides better performance in both steady-state and transient conditions. The performance of SAPF depends on the reference current generation methods used. For reference current generation, there are many methods suggested by researchers such as instantaneous reactive power theory or *p-q* theory (IRPT) [10], synchronous reference frame, and unit vector technique [11]. In all the above-mentioned techniques, IRPT is used in this chapter because of its simplicity, satisfactory performance, and ease of calculation. Also, IRPT tracks instantaneous values of system parameters and works well in three-phase three-wire or three-phase four-wire system. There are mainly four kinds of controlling involved in designing of SAPF: (1) controlling of current references generation or harmonics extraction; (2) controlling of DC-link capacitor voltage (V_{dc}); (3) controlling of switching pulses; and (4) controlling of synchronization. The performance of SAPF depends on the effectiveness of these controlling algorithms. To track the reference current, direct current controller or indirect current controller (ICC) technique can be used, which provides the actual reference current [12]. These compensating reference currents are converted to compensating reference voltage through suitable transformation, which is used to generate switching pulse for insulated-gate bipolar transistors (IGBTs) of VSI using various pulse width modulation (PWM) techniques such as space-vector PWM, carrier-based PWM (CBPWM), etc. [13]. Unipolar CBPWM technique is used in this chapter because of simple design, less calculation, and less cost requirements. Performance of SAPF also depends on DC voltage stabilization across the DC-link capacitor of VSI hence proper tuning of PI controller is required to stabilize the DC-link capacitor voltage.

PI controller can be tuned by various optimization methods for giving the best performance for particular load conditions. However, these optimization methods have limitations such as adaptability, robustness, that is, if load or system parameter changes, tuned gains of PI controller may not provide optimum performance also, PI controller is not robust to faster transient response. Hence to eliminate the above-mentioned DC voltage stabilization techniques, ICC-based model reference adaptive control (ICC-MRAC) has been proposed in this chapter. In this technique, DC-link capacitor voltage is stabilized through a reference system, which sends the required response as a command signal. This model reference adaptive controller (MRAC) is superior to conventional in terms of its flexibility, adaptability, and robustness because this adaptive controller adapts the controller gain as per the changes of load or system parameters [14, 15].

The main aim of any SAPF is to make source current harmonics free and in phase with source voltage, for that controller of SAPF plays a vital role hence a robust, adaptive, and reliable MRAC is proposed in this chapter, which is superior to conventional PI controller. This is evident by simulation results and again verified by experimental validation.

This chapter is organized in the following sections. First in this section, brief introduction about harmonics problem and their compensation methods is described. In Section 2, SAPF configuration with reference estimation technique and incremental conductance-based MPPT is discussed. Section 3 described the modeling of three-phase three-wire SAPF. Conventional PI controller and proposed MRAC for DC-link voltage regulation are described in Sections 4 and 5, respectively. In Sections 6 and 7, simulation and experimental results for both methods are discussed, and finally conclusion of this chapter is given in Section 8.

2. Filter configuration

PV-integrated SAPF configuration is shown in Fig. 1. In this configuration, SAPF is connected in parallel with three-phase grid supply and three-phase bridge rectifiers at PCC. This SAPF provides filter current, which compensates generated harmonics by nonlinear loads.

SAPF is a VSI having a DC-link capacitor (C_{dc}). The operation of this VSI is controlled by switching pulses, which are generated by PWM hysteresis band current controller. IRPT, which is also known as *p-q* theory, is one of the most common techniques of reference current generation, hence used for our work. This generated reference current is compared with actual sensed source current for VSI switching pulse generation. A DC voltage source is required to provide reference DC voltage, which regulates DC-link capacitor voltage. This reference DC voltage is provided by the PV system connected with VSI. To draw the maximum power from PV system, MPPT is needed as discussed here.

2.1 MPPT algorithm

PV system delivers maximum power at a particular value of current and voltage. MPPT is required to maximize the efficiency of the PV system. There are many

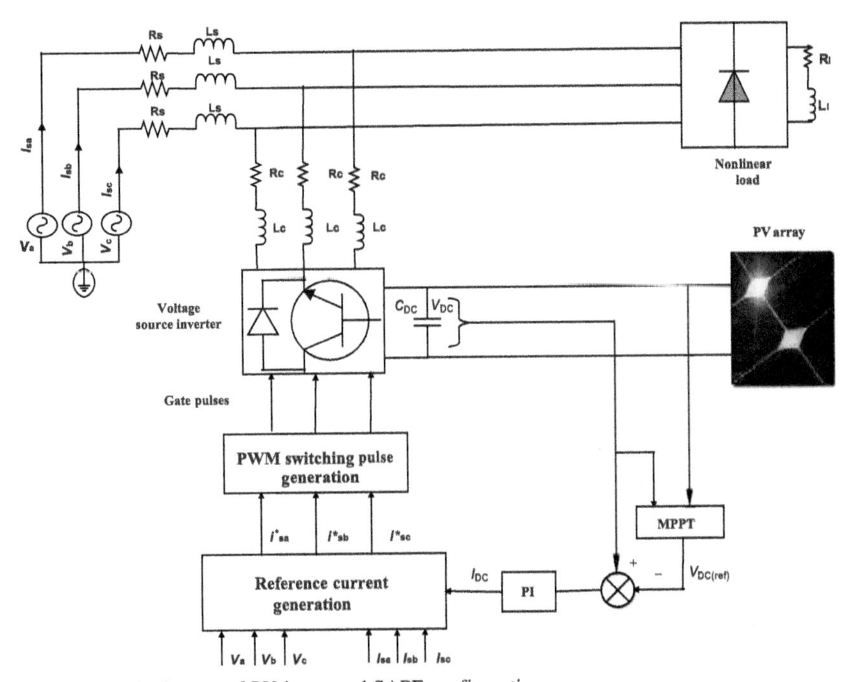

FIG. 1 Block diagram of PV-integrated SAPF configuration.

methods suggested by researchers to track the maximum power point of PV systems such as hill climbing, perturbation and observation (P&O), incremental conductance-based MPPT, artificial neural network, and fuzzy logic-based method. These all methods differ in their performance, speed of convergence, popularity, and complexity. In this chapter, one of the most popular and efficient methods of MPPT, incremental conductance-based MPPT is considered for our experimentation [16, 17].

2.2 Reference current estimation

The most popular and simple method to estimate reference currents is IRPT, which was proposed by Japanese electrical engineer and Professor Hirofumi Akagi in 1996 [18]. Proposed reference current generation *p-q* algorithm is shown in Fig. 2. This method is preferred over others because of its simplicity, satisfactory performance, and ease of calculation.

3. Modeling of three-phase three-wire SAPF

Circuit configuration of three-phase three-wire grid system with three-phase SAPF is presented in Fig. 3 and Fig. 4 represents the small-signal equivalent of three-phase SAPF and represented by Eqs. (1)–(3) as follows:

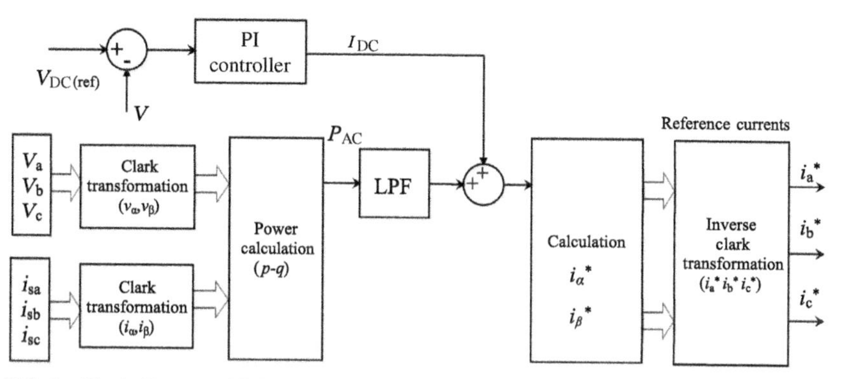

FIG. 2 Block diagram of IRPT.

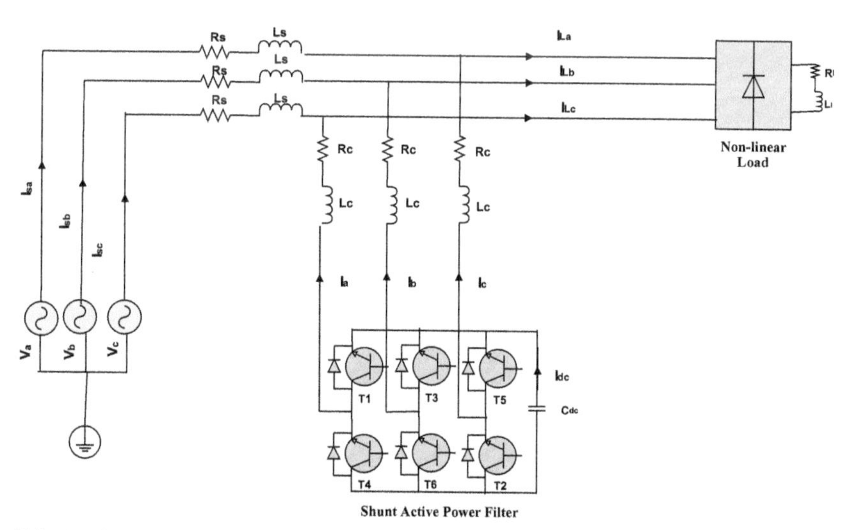

FIG. 3 Circuit configuration of proposed three-phase three-wire gird system with three-phase SAPF.

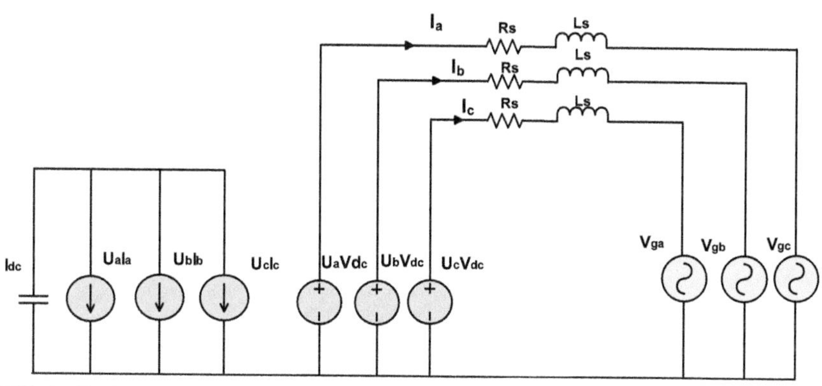

FIG. 4 Equivalent circuit of three-phase SAPF.

$$\frac{dI_a}{dt} = \frac{(u_a - u_b)}{L_f} \cdot V_{dc} + \frac{R_f}{L_f} \cdot I_c \tag{1}$$

$$\frac{dI_b}{dt} = \frac{(u_b - u_c)}{L_f} \cdot V_{dc} - \frac{R_f}{L_f} \cdot (I_b + I_c) \tag{2}$$

$$\frac{dI_c}{dt} = \frac{(u_c - u_a)}{L_f} \cdot V_{dc} - \frac{R_f}{L_f} \cdot (I_c + I_a) \tag{3}$$

Three-phase three-wire PV-integrated balanced power system is used in this chapter, hence sum of current and voltages of all the three phases will be equal to zero as presented in Eqs. (4), (5), respectively.

$$I_a + I_b + I_c = 0 \tag{4}$$

$$V_{ga} + V_{gb} + V_{gc} = 0 \tag{5}$$

where, u_a, u_b, u_c are the switching sequences of IGBTs of phases a, b, c, respectively, then,

$$\left.\begin{aligned} u_a - u_b &= d_x \\ u_b - u_c &= d_y \\ u_c - u_a &= d_z \end{aligned}\right\} \tag{6}$$

where d_x, d_y, d_z are the switching state functions.

Substituting Eq. (6) into Eqs. (1)–(3), we get

$$\frac{dI_a}{dt} = \frac{d_x}{L_f} \cdot V_{dc} + \frac{R_f}{L_f} \cdot I_c \tag{7}$$

$$\frac{dI_b}{dt} = \frac{d_y}{L_f} \cdot V_{dc} + \frac{R_f}{L_f} \cdot I_a \tag{8}$$

$$\frac{dI_c}{dt} = \frac{d_x}{L_f} \cdot V_{dc} + \frac{R_f}{L_f} \cdot I_b \tag{9}$$

Equation of DC voltage is obtained by applying Kirchhoff's current law (KCL) on the DC side of equivalent circuit of SAPF as shown in Eq. (10):

$$\frac{dV_{dc}}{dt} = \frac{1}{C_{dc}}(u_a \cdot I_a + u_b \cdot I_b + u_c \cdot I_c) \tag{10}$$

State-space representation of Eqs. (7)–(10) is shown in Eq. (11):

$$\frac{d}{dt}\begin{bmatrix} I_a \\ I_b \\ I_c \\ V_{dc} \end{bmatrix} = \begin{bmatrix} 0 & 0 & \frac{R_f}{L_f} & \frac{d_x}{L_f} \\ \frac{R_f}{L_f} & 0 & 0 & \frac{d_y}{L_f} \\ 0 & \frac{R_f}{L_f} & 0 & \frac{d_z}{L_f} \\ \frac{u_a}{V_{dc}} & \frac{u_b}{V_{dc}} & \frac{u_c}{V_{dc}} & 0 \end{bmatrix} \cdot \begin{bmatrix} I_a \\ I_b \\ I_c \\ V_{dc} \end{bmatrix} \quad (11)$$

By applying inverse Clarke's transform, the dynamic transfer function of SAPF can be represented by Eq. (12):

$$\frac{d}{dt}\begin{bmatrix} I_\alpha \\ I_\beta \\ V_{dc} \end{bmatrix} = \begin{bmatrix} \frac{R_f}{L_f} & -\omega & \frac{d_\alpha}{L_f} \\ \omega & \frac{R_f}{L_f} & \frac{d_\beta}{L_f} \\ \frac{u_\alpha}{C_{dc}} & \frac{u_\beta}{C_{dc}} & 0 \end{bmatrix} \cdot \begin{bmatrix} I_\alpha \\ I_\beta \\ V_{dc} \end{bmatrix} \quad (12)$$

4. PI controller for DC voltage stabilization

A PI controller is used for the stabilization of actual DC-link capacitor voltage (V_{dc}) such that it follows the DC reference voltage ($V_{dc(ref)}$). PI controller is responsible for peak value of reference currents, settling time, speed of response, and DC-power losses ($p_{dc(loss)}$). Difference between DC reference voltage ($V_{dc(ref)}$) and actual DC-link capacitor voltage (V_{dc}) is considered as error voltage. This error voltage is minimized using PI controller. For proper operation, this PI controller should be tuned properly. In PI controller technique, a closed loop is needed with PI controller and open-loop gain of SAPF as shown in Fig. 5. For modeling, we need transfer function of DC-link capacitor, VSI, and PI controller as discussed in the following sections.

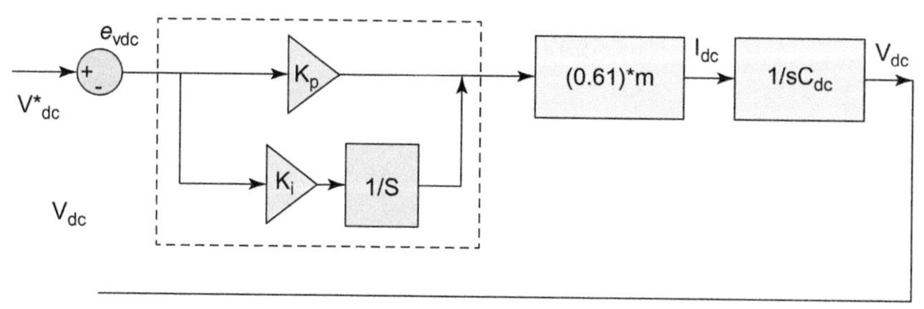

FIG. 5 Block diagram of closed-loop PI controller-based SAPF.

4.1 Transfer function for DC-link capacitor

From the small-signal equivalent model of SAPF as shown in Fig. 4, DC-link capacitor current is given by

$$I_{\mathrm{dc}} = C_{\mathrm{dc}} \cdot \frac{\mathrm{d}V_{\mathrm{dc}}}{\mathrm{d}t} \tag{13}$$

where

C_{dc} = DC-link capacitance;
V_{dc} = DC-link capacitor voltage; and
I_{dc} = DC-link capacitor current.

By applying Laplace transform on Eq. (13), we get

$$I_{\mathrm{dc}} = s \cdot C_{\mathrm{dc}} \cdot V_{\mathrm{dc}} \tag{14}$$

Hence, transfer function of DC-link capacitor is given by Eq. (15)

$$\therefore (\mathrm{TF})_{C_{\mathrm{dc}}} = \frac{I_{\mathrm{dc}}}{V_{\mathrm{dc}}} = s \cdot C_{\mathrm{dc}} \tag{15}$$

4.2 Transfer function of VSI

Transfer function for VSI can be modeled by using small-signal equivalent model of SAPF as shown in Fig. 4. In Fig. 4, power transfer from AC side to VSI and DC side to VSI is equal, then equating both powers as shown in Eqs. (16), (17)

$$V_{\mathrm{dc}} \cdot I_{\mathrm{dc}} = V_{\mathrm{ac}} \cdot I_{\mathrm{ac}} \tag{16}$$

$$V_{\mathrm{dc}} \cdot I_{\mathrm{dc}} = \begin{bmatrix} V_{\mathrm{ca}} \\ V_{\mathrm{cb}} \\ V_{\mathrm{cc}} \end{bmatrix} \begin{bmatrix} I_{\mathrm{ca}} & I_{\mathrm{cb}} & I_{\mathrm{cc}} \end{bmatrix} \tag{17}$$

Clarke's transformation is given by Eq. (18):

$$\begin{bmatrix} A_\alpha \\ A_\beta \end{bmatrix} = [K] \cdot \begin{bmatrix} A_{\mathrm{a}} \\ A_{\mathrm{b}} \\ A_{\mathrm{c}} \end{bmatrix} \tag{18}$$

where K is the transformation matrix of Clarke's transformation as follows:

$$K = \begin{bmatrix} \frac{2}{3} & \frac{-1}{3} & \frac{-1}{3} \\ 0 & \frac{1}{\sqrt{3}} & \frac{-1}{\sqrt{3}} \\ \frac{1}{3} & \frac{1}{3} & \frac{1}{3} \end{bmatrix} \tag{19}$$

Hence from Eqs. (17), (18), we can derive Eq. (20)

$$\therefore V_{\text{dc}} \cdot I_{\text{dc}} = \left[K^{-1}\right]\begin{bmatrix} V_\alpha \\ V_\beta \end{bmatrix}\left[K^{-1}\right]^T\left[I_{\text{c}\alpha}\ I_{\text{c}\beta}\right] \tag{20}$$

where α, β are the axis after the Clarke's transformation and $I_{\text{c}\alpha}, I_{\text{c}\beta}$ are the two-phase line currents in α, β axis.

$\because$ modulation index (m) is represented by Eq. (21):

$$m = \frac{2\sqrt{2}}{\sqrt{3}} \cdot \frac{V_{\text{Line}}}{V_{\text{dc}}} \tag{21}$$

Hence, after Clarke transformation, we get

$$V_{\text{dc}} \cdot I_{\text{dc}} = \frac{\sqrt{3} \cdot m \cdot V_{\text{dc}}}{2\sqrt{2}}\left[I_{\text{c}\phi}\right] \tag{22}$$

where $I_{\text{c}\phi}$ represents compensating current and $\phi = \alpha, \beta$

$$\therefore (\text{TF})_{\text{VSI}} = \frac{I_{\text{c}\phi}}{I_{\text{dc}}} = \frac{2\sqrt{2}}{\sqrt{3} \cdot m} \tag{23}$$

4.3 Transfer function of PI controller

Block diagram of PI controller is given in Fig. 5. From Fig. 5, transfer function of PI controller can be represented by Eq. (24):

$$I_{\text{ac}} = K_{\text{p}}\widetilde{V}_{\text{dc}} + K_{\text{i}}\int \widetilde{V}_{\text{dc}}\text{d}t \tag{24}$$

By applying Laplace transform on Eq. (24), we get

$$\frac{I_{\text{ac}}}{\widetilde{V}_{\text{dc}}} = \left(\frac{sK_{\text{p}} + K_{\text{i}}}{s}\right) \tag{25}$$

After obtaining individual transfer function for DC-link capacitor, VSI, and PI controller as shown in Eqs. (15), (23), (25), respectively, we can find the closed-loop transfer function from closed-loop block diagram of PI controller as shown in Fig. 5. Solving the closed-loop block diagram of transfer function of SAPF is given by Eq. (26):

$$\frac{V_{\text{dc}}}{V_{\text{dc}}^*} = \frac{\sqrt{3}m}{2\sqrt{2}C_{\text{dc}}}\left(\frac{sK_{\text{p}} + K_{\text{i}}}{s^2 + s\left(\frac{\sqrt{3}mK_{\text{p}}}{2\sqrt{2}C_{\text{dc}}}\right) + \left(\frac{\sqrt{3}mK_{\text{i}}}{2\sqrt{2}C_{\text{dc}}}\right)}\right) \tag{26}$$

where

m = modulation index;
C_{dc} = DC capacitance;
K_p = proportional constant;
K_i = integral constant;
V_{dc}^* = reference DC voltage; and
V_{dc} = DC-link capacitor voltage.

and modulation index (m) is given by Eq. (27):

$$m = \frac{2\sqrt{2} \cdot V_{peak}}{V_{dc}} \tag{27}$$

where V_{peak} = peak line voltage.

∵ The standard second-order control system transfer function $G(s)$ is shown by

$$G(s) = \frac{\omega_n^2}{s^2 + 2 \cdot \xi \cdot \omega_n \cdot s + \omega_n^2} \tag{28}$$

where

ω_n = natural undamped frequency of the system and
ζ = damping ratio.

Comparing Eqs. (26), (28), we get

$$K_p = \frac{4\sqrt{2}\xi\omega_n C_{dc}}{\sqrt{3}m} \tag{29}$$

$$\omega_n^2 = \frac{\sqrt{3}mK_i}{2\sqrt{2}C_{dc}} \tag{30}$$

$$K_i = \frac{2\sqrt{2}C_{dc}\omega_n^2}{\sqrt{3}m} \tag{31}$$

Since here $m = 0.83$ and $C_{dc} = 2200\ \mu F$, we get

$$\xi = \frac{\log M_p}{\sqrt{\Pi^2 + \log M_p^2}} \tag{32}$$

where M_p = 30% overshoot and natural undamped frequency of the system is shown in Eq. (33):

$$\omega_n = \frac{4}{\xi \cdot t_s} \tag{33}$$

where settling time $t_s = 6\,s$.

From Eqs. (29), (31), we get the values of proportional (K_p) constant and integral (K_i) constant, respectively, as follows:

$$K_p = \frac{4\sqrt{2} \cdot \xi \cdot \omega_n \cdot C_{dc}}{\sqrt{3} \cdot m} = 0.0058 \tag{34}$$

$$K_i = \frac{2\sqrt{2} \cdot \omega_n^2 \cdot C_{dc}}{\sqrt{3} \cdot m} = 0.0036 \tag{35}$$

5. MRAC for DC voltage stabilization

The MRAC is superior to the conventional PI controller in terms of its adaptability, flexibility, and robustness. When there is a change in load or system parameters, then conventional PI controller with fixed gain is unable to provide optimum performance but in MRAC gain is adaptive, which changes automatically to adapt to the changes in the system hence provides better performance and stability than conventional PI controller. In MRAC, manual control of gains is not required as it is required in PI controller. Fig. 6 represents the block diagram of MRAC.

Simplifying the block diagram of MRAC shown in Fig. 6, we get the transfer function of SAPF as shown in Eq. (36):

$$\frac{V_{dc}}{I_{c\phi}} = \frac{\sqrt{3} \cdot m}{2\sqrt{2} \cdot s \cdot C_{dc}} \tag{36}$$

where

m = modulation index = 0.83 and
C_{dc} = 2200 μF.

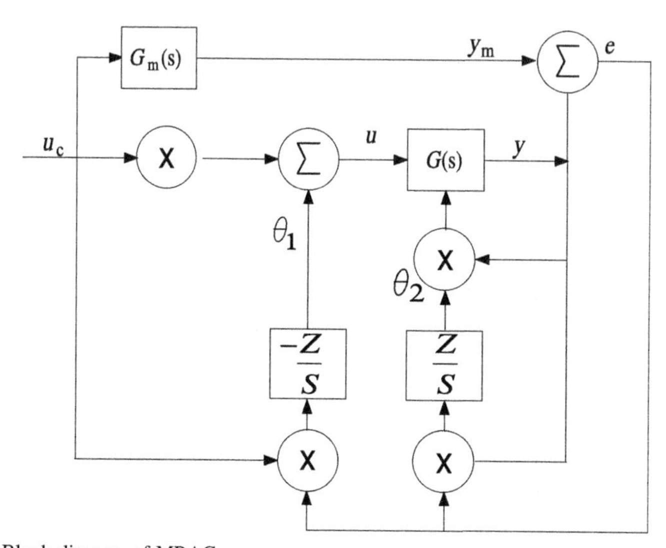

FIG. 6 Block diagram of MRAC.

Hence, open-loop transfer function (OLTF) of SAPF is represented by

$$\therefore (\mathrm{TF})_{\mathrm{SAPF}} = \frac{236.627}{s} \tag{37}$$

State-space equation of the actual plant is shown by

$$\frac{\mathrm{d}y}{\mathrm{d}t} = -ay + bu \tag{38}$$

Applying the Laplace transformation on Eq. (38), we get

$$s \cdot y(s) = -a \cdot y(s) + b \cdot u(s) \tag{39}$$

$$(s + a) \cdot y(s) = b \cdot u(s) \tag{40}$$

$$G(s) = \frac{y(s)}{u(s)} = \frac{b}{s + a} \tag{41}$$

where $G(s)$ is the transfer function of actual plant.

By comparing Eqs. (37), (38), we get the values of the constants of the transfer function of actual plant as shown in Eqs. (42), (43), respectively,

$$a = 0 \tag{42}$$

$$b = 231.05 \tag{43}$$

where

$y(s) = V_{\mathrm{dc}}$ = output of the actual plant and
$u(s) = I_{\mathrm{dc}}$ = input to the actual plant.

we know that

$$u(t) = \theta_1 u_{\mathrm{c}} - \theta_2 y \tag{44}$$

where u_{c} is the control input to the MRAC block, which has to be controlled. So, the state-space equation of the reference plant model is shown in Eq. (45):

$$\frac{\mathrm{d}y}{\mathrm{d}t} = -a_{\mathrm{m}} y_{\mathrm{m}} + b_{\mathrm{m}} u_{\mathrm{c}} \tag{45}$$

By applying Laplace transform on Eq. (45), we get

$$G_{\mathrm{m}}(s) = \frac{y_{\mathrm{m}}(s)}{u_{\mathrm{c}}(s)} = \frac{b_{\mathrm{m}}}{(s + a_{\mathrm{m}})} \tag{46}$$

where

$G_{\mathrm{m}}(s)$ = transfer function of a reference plant model;
$a_{\mathrm{m}}, b_{\mathrm{m}}$ = constants of transfer function of reference plant model, as shown in Eqs. (47), (51), respectively,

$$\because a_m = a + b\theta_2 \tag{47}$$

$$a_m = 231.05\theta_2 \tag{48}$$

and

$$\theta_2 = \frac{a_m - a}{b} \tag{49}$$

$$\theta_2 = \frac{a_m}{231.05} \tag{50}$$

Similarly,

$$b_m = b\theta_1 \tag{51}$$

$$b_m = 231.05\theta_1 \tag{52}$$

and

$$\theta_1 = \frac{b_m}{b} \tag{53}$$

$$\theta_1 = \frac{b_m}{231.05} \tag{54}$$

Hence, transfer function of MRAC model plant is obtained and shown in Eq. (55):

$$G_m(s) = \frac{y_m}{u_c} = \frac{b_m}{s + a_m} = \frac{b\theta_1}{s + a + b\theta_2} = \frac{231.05\theta_1}{s + 231.05\theta_2} \tag{55}$$

The candidate Lyapunov function is shown in Eqs. (56)–(58):

$$V(t, \theta_1, \theta_2) = \frac{1}{2}\left[e^2 + \frac{1}{b\gamma}(b\theta_2 + a - a_m)^2 + \frac{1}{b\gamma}(b\theta_1 - b_m)^2\right] \tag{56}$$

$$\frac{dV}{dt} = \frac{de}{dt} + \frac{1}{\gamma}(b\theta_2 + a - a_m)\left(\frac{d\theta_2}{dt}\right) + \frac{1}{\gamma}(b\theta_1 - b_m)\left(\frac{d\theta_1}{dt}\right) \tag{57}$$

$$\begin{aligned}\frac{dV}{dt} &= -a_m e^2 + \frac{1}{\gamma}(b\theta_2 + a - a_m)\left(\frac{d\theta_2}{dt} - \gamma y e\right)\\ &\quad + \frac{1}{\gamma}(b\theta_1 - b_m)\left(\frac{d\theta_1}{dt} + \gamma u_c e\right)\end{aligned} \tag{58}$$

Eq. (58) suggests the adaptation laws as shown in Eq. (59):

$$\theta_2 = \frac{\gamma y e}{s} \tag{59}$$

$$\theta_1 = \frac{-\gamma u_c e}{s} \tag{60}$$

Applying Eqs. (59), (60) into Eq. (58), we get

$$\frac{dV}{dt} = -a_m e^2 \quad (61)$$

$$\frac{dV}{dt} = -231.05\theta_2 e^2 \quad (62)$$

where z is the adaptation gain of the controller γ, $e = y - y_m$, y is the output of actual plant, y_m is the output of reference model plant, and $u_c = 160$ is the DC reference voltage.

$$y_m = \frac{b\theta_1 u_c}{p + a + b\theta_2} = \frac{231.05\theta_1 u_c}{p + 231.05\theta_2} \quad (63)$$

$$\gamma = z \quad (64)$$

and assuming adaptation law as shown in Eq. (65):

$$z = \frac{231.05*3}{s + 231.05} \quad (65)$$

6. Simulation results

PV-integrated SAPF is modeled in the MATLAB/Simulink environment for both proposed MRAC and conventional PI controller. Then, the performance of both the controllers has been studied under different conditions as mentioned in the following sections.

6.1 Case I: Balanced supply with steady-state load condition

First, both controllers have been implemented for bridge rectifiers with resistive-inductive load for three-phase supply. For this case, we have considered $R = 100\,\Omega$ and $L = 50\,\text{H}$. For experimentation, initially, SAPF is turned off and turned on at 0.1 s. Figs. 7A and 8A represent the waveform of three-phase source current, filter current, and DC-link capacitor voltage for conventional PI controller and proposed MRAC, respectively. In both waveforms, the source current is nonsinusoidal until 0.1 and became sinusoidal after that. Because SAPF provides compensating filter current, which suppresses harmonics when SAPF is turned on at 0.1 s and DC-link capacitor voltage reaches to steady-state value after SAPF is turned on.

THD of source current without SAPF is measured 29.86%, which is reduced to 2.98% with the application of conventional PI controller method and reduced to 2.88% with the application of proposed MRAC as shown in Fig. 9.

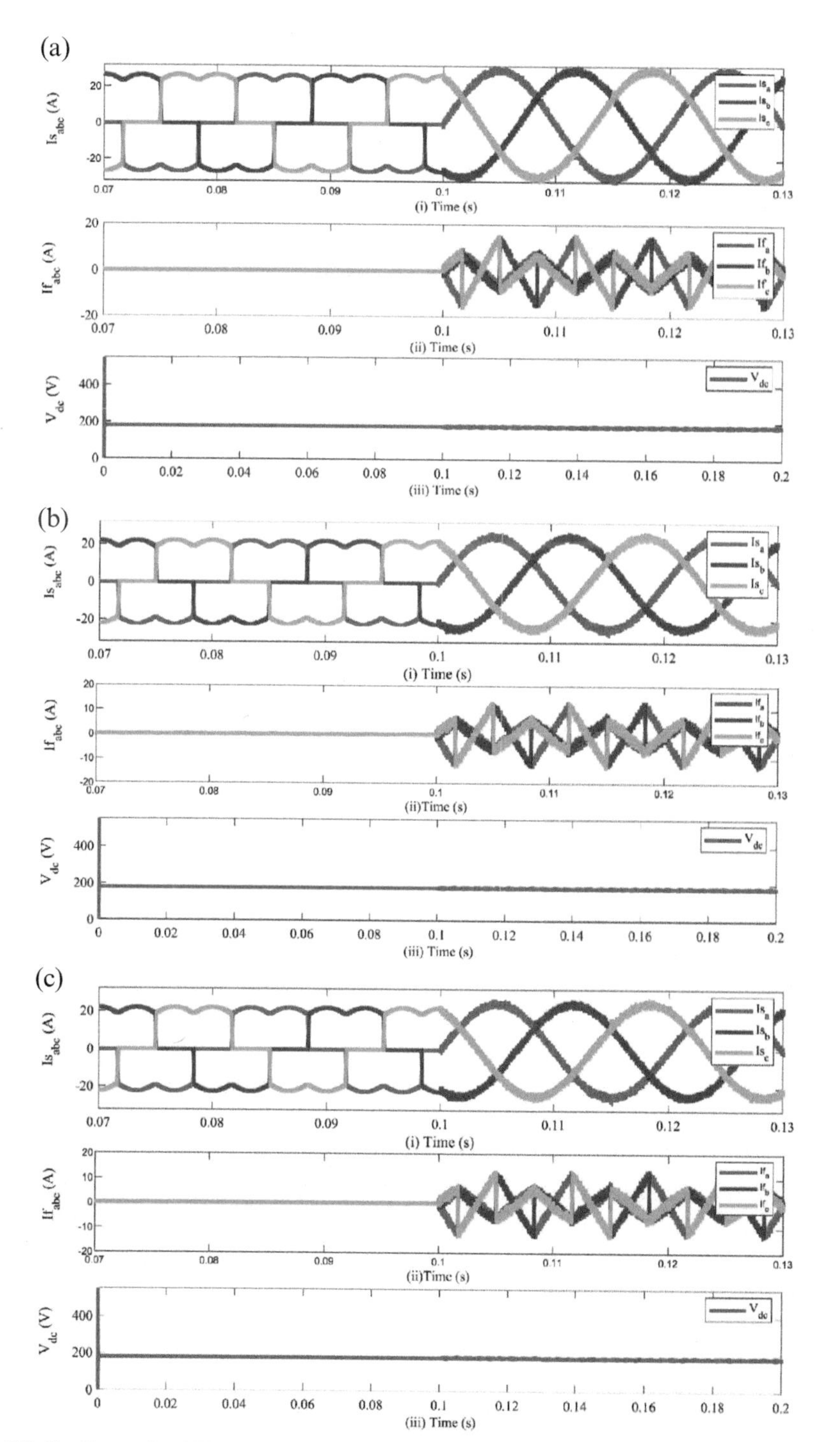

FIG. 7 Conventional PI controller for (A) Case I, (B) Case II, and (C) Case III. (i) Source current, (ii) filter current, and (iii) DC-link capacitor voltage.

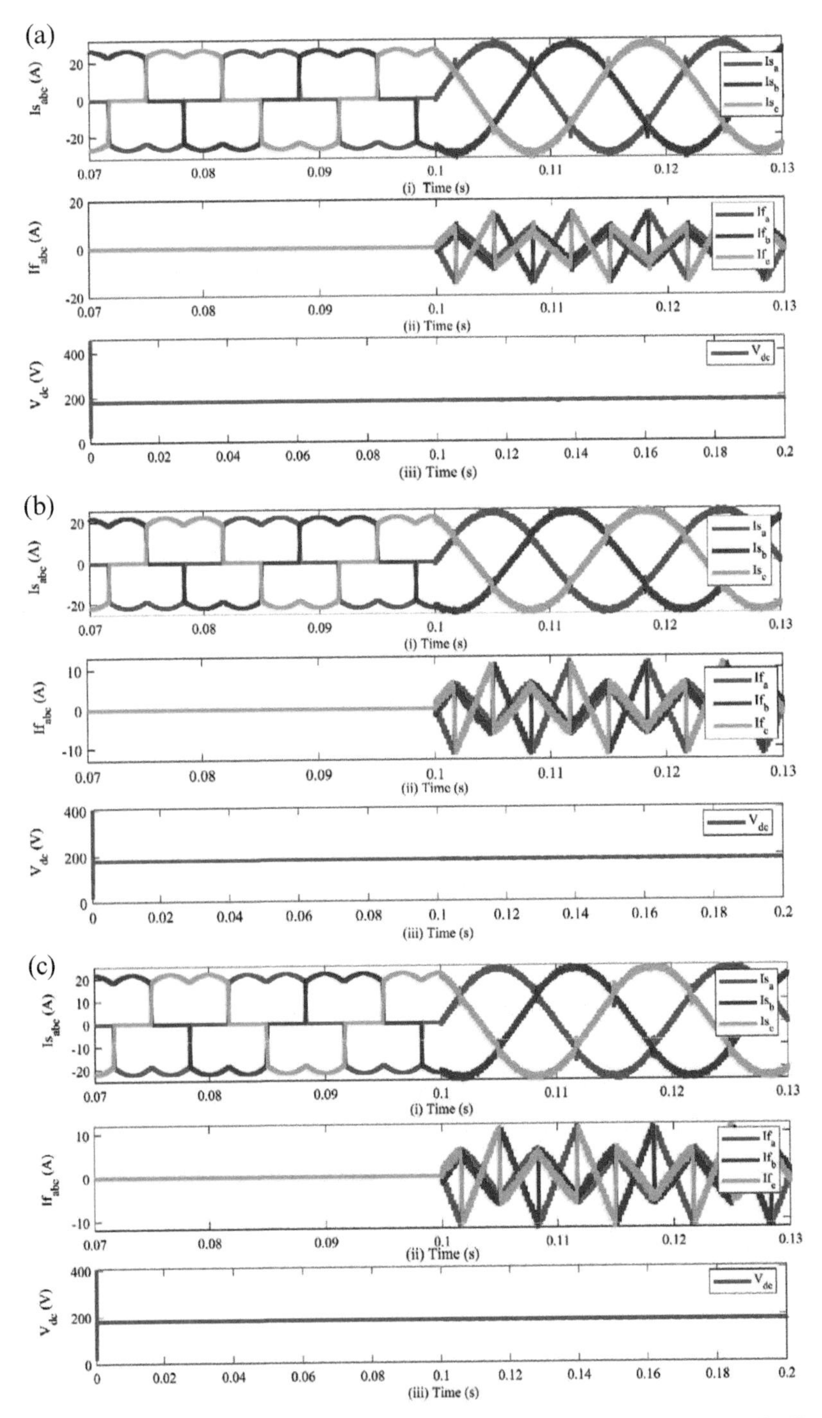

FIG. 8 Proposed MRAC for (A) Case I, (B) Case II, and (C) Case III. (i) Source current (ii), filter current, and (iii) DC-link capacitor voltage.

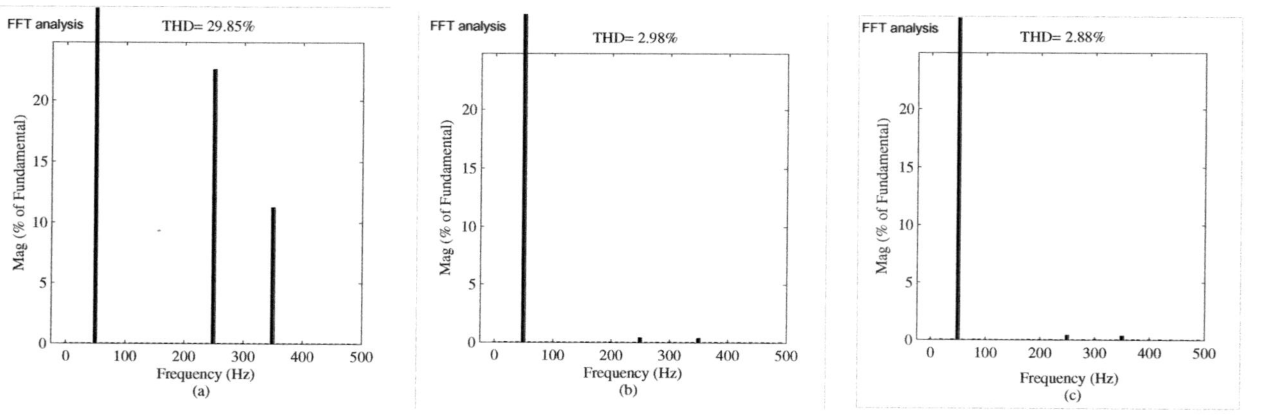

FIG. 9 Case I. (A) THD without SAPF, (B) THD with SAPF PI controller, and (C) THD with MRAC.

6.2 Case II: Balanced supply with varying load condition

In the second case, both the controllers are applied under varying load conditions. Three-phase source current, filter current, and DC-link capacitor voltage for conventional PI controller and proposed MRAC are shown in Fig. 7B and Fig. 8B, respectively. We can observe mitigation of harmonics in source current after 0.1 s, when SAPF is turned on. Also, SAPF compensating filter current can be observed after 0.1 s.

For this case, with the application of conventional PI controller, THD is reduced to 4.24% and it reaches 4.10% when proposed MRAC is used as shown in Fig. 10. In this case, also MRAC provides better compensation with lesser THD.

6.3 Case III: Balanced supply with parameter variation of SAPF

In this case, designing parameters of SAPF coupling resistor (R_c) and coupling inductor (L_c) are changed from $R_f = 3.14\,\Omega$, $L_f = 10\,\text{mH}$ to $L_f = 12\,\text{mH}$, $R_f = 6.27\Omega$. In this case, proposed MRAC provides better compensation than conventional PI controller as shown in Fig. 7C and Fig. 8C. THD is reduced to 4.62% for proposed MRAC and it reduced to 4.87% for PI controller. THD without SAPF and with the SAPF along with PI controller and MRAC is shown in Fig. 11.

Simulation results for the above three cases evident that the proposed MRAC provides better compensation and lesser THD as compared to conventional PI controller.

7. Experimental results

The real validation of this proposed MRAC has been implemented by a laboratory developed prototype. The proposed MRAC and conventional PI controller prototype have been developed in dSPACE/MATLAB interactive common platform and implemented to DS1104 controller board, which is compatible with any version of MATLAB. The switching signal is generated from the controller board, passed to SIMIKRON inverter. The hardware prototype developed in the laboratory includes SIMIKRON inverter, DS1104, voltage and current sensors, and three-phase bridge rectifier as nonlinear load. The block diagram of the laboratory prototype is shown in Fig. 12. Three-phase source voltage and load current are shown in Fig. 13A and B, respectively. Fig. 13C represents the three-phase filter current provided by SAPF, which compensates the harmonics in source current. Table 1 represents the system parameters used for experimentation. The experimental analysis of SAPF with the proposed algorithm is given as follows.

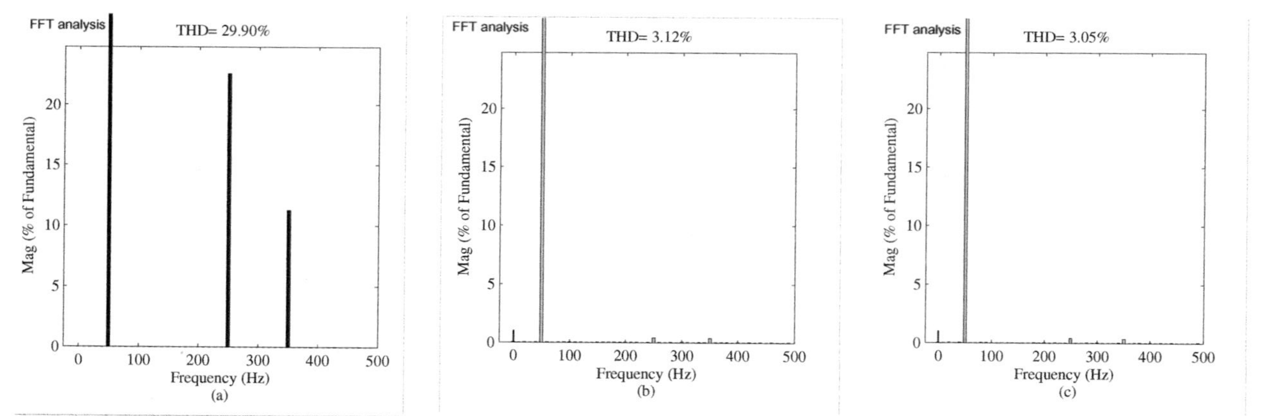

FIG. 10 Case II. (A) THD without SAPF, (B) THD with SAPF existing method, and (C) THD with SAPF proposed method.

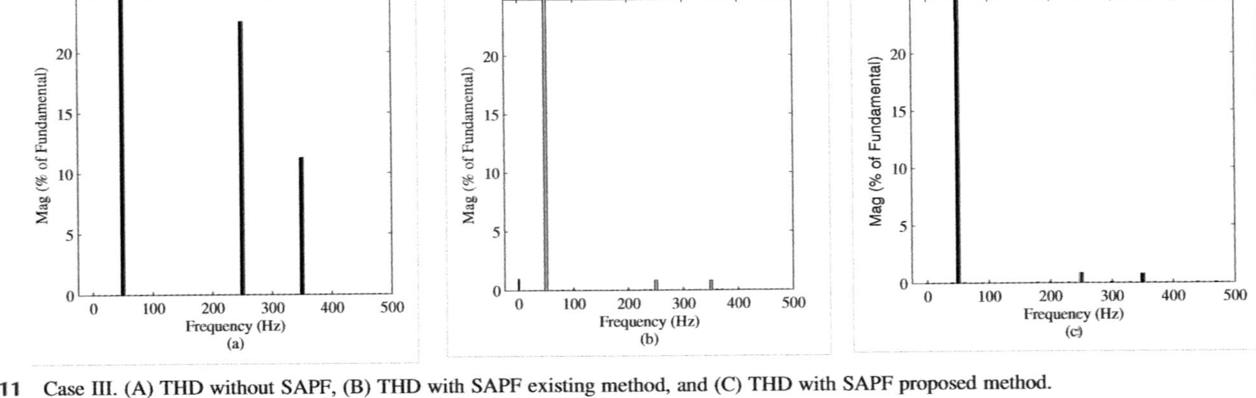

FIG. 11 Case III. (A) THD without SAPF, (B) THD with SAPF existing method, and (C) THD with SAPF proposed method.

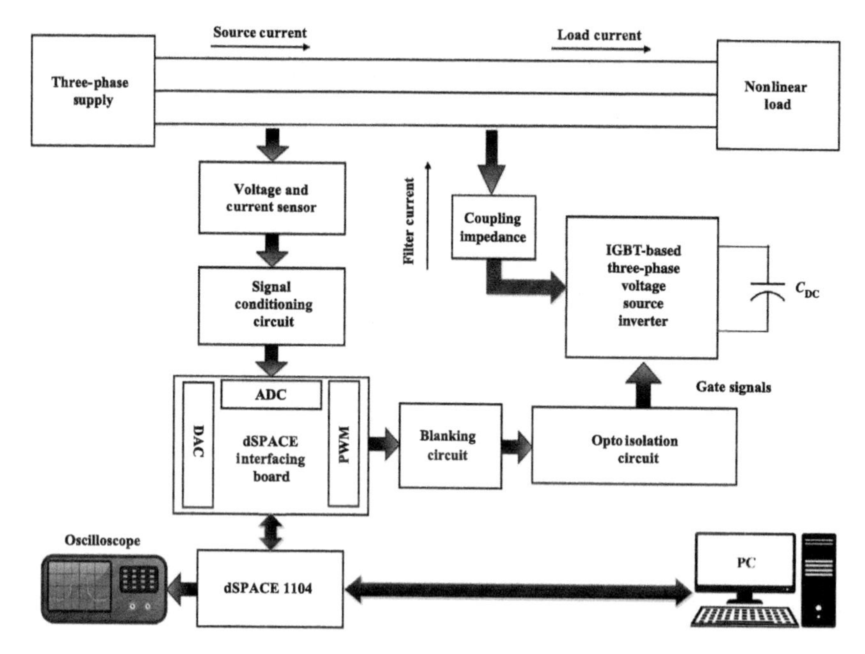

FIG. 12 Experimental setup.

7.1 Case I: Balanced supply with steady-state load condition

Resistive-inductive (*R-L*) load bridge rectifier injects harmonics to the distribution network. Three-phase source current after the application of SAPF with conventional PI controller and proposed MRAC is shown in Fig. 14A and B, respectively. It is observed that the source current is harmonic free and sinusoidal. DC-link capacitor voltage for conventional PI and MRAC is shown in Fig. 14C and D, respectively. THD before and after compensation for both conventional PI controller and proposed MRAC is shown in Fig. 15. Without compensation, THD is as high as 21.81% as shown in Fig. 15A. It is observed in Fig. 15B and C that THD of proposed MRAC (THD = 1.57%) is lower than existing PI controller (THD = 3.32%). This is due to the better compensation of the proposed controller over existing controller.

7.2 Case II: Balanced supply with varying load condition

In this case, three-phase bridge rectifier with varying load conditions is considered for SAPF performance observation under both controllers. Three-phase source current after the application of the conventional PI controller and proposed MRAC is shown in Fig. 16A and B, respectively. Fig. 16C and D shows DC-link capacitor voltage for conventional PI controller and proposed MRAC, respectively. THD of load current is shown in Fig. 17A. THD of source current

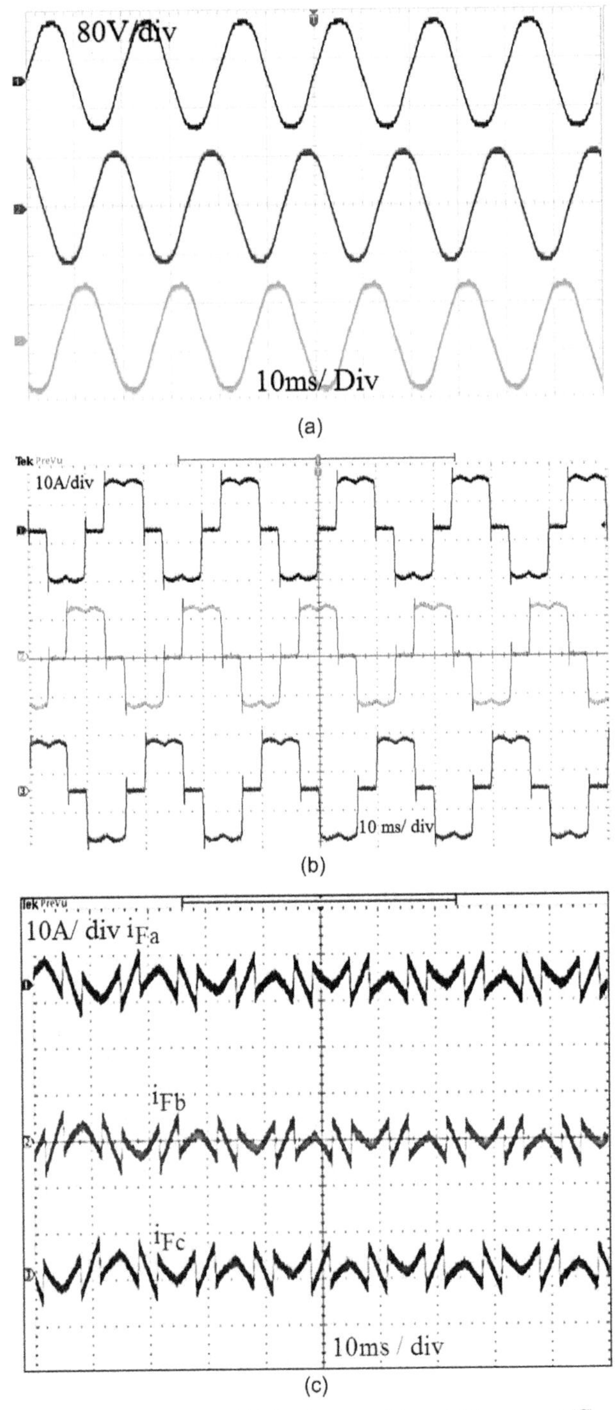

FIG. 13 Three-phase source voltage (A), load current (B), and filter current (C).

TABLE 1 System parameters.

System parameters	Values
3ϕ supply voltage	$V_s = 100\,V$ (peak)
Frequency	$f_s = 50\,Hz$
Line resistance	$R_s = 3\,\Omega$
Load resistance	$R_l = 100\,\Omega$
Load inductance	$L_l = 50\,mH$
Filter resistance	$R_f = 3.14\,\Omega$
Filter inductance	$L_f = 10\,mH$
DC-link capacitance	$C_{dc} = 2200\,\mu F$
Switching frequency	$f_{sw} = 10\,kHz$
Proportional constant	$K_p = 0.0058$
Integral constant	$K_i = 0.0036$

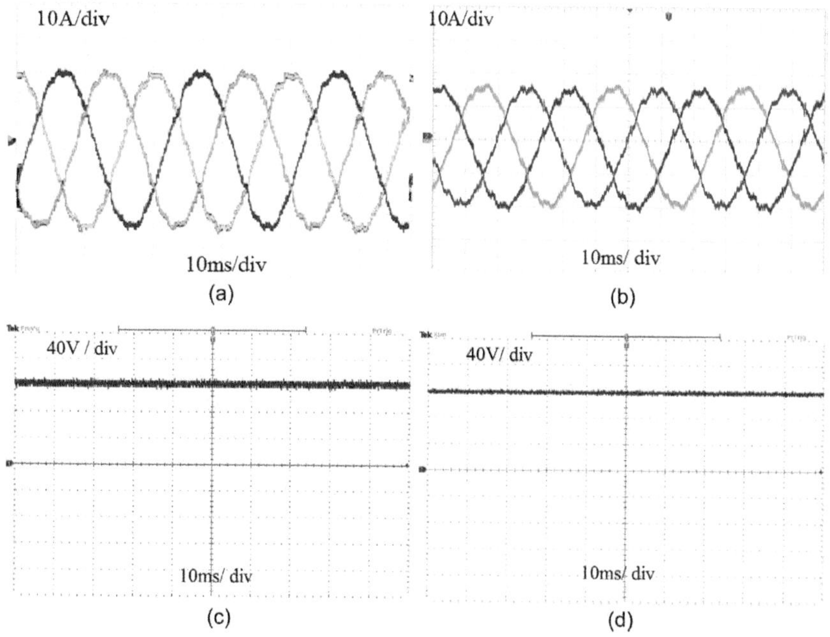

FIG. 14 Case I: Balanced supply with steady-state load condition. (A) Source current in conventional PI controller; (B) source current in proposed MRAC; (C) DC-link capacitor voltage (V_{dc}) for conventional PI controller; and (D) DC-link capacitor voltage (V_{dc}) for proposed MRAC.

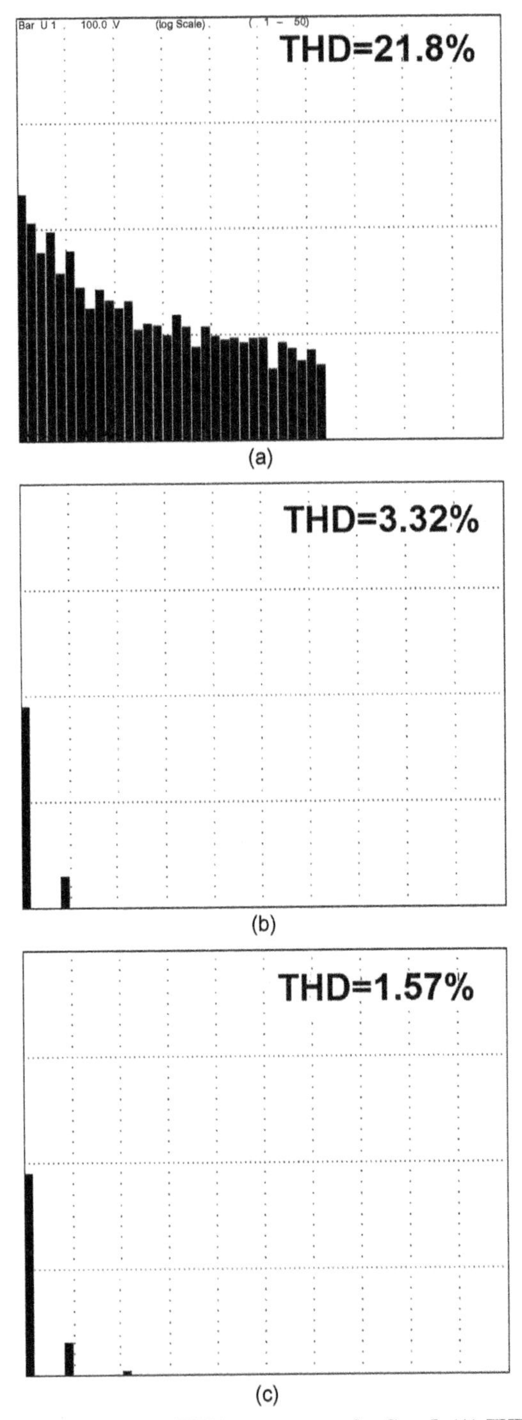

FIG. 15 Total harmonic distortion (THD) measurement for Case I. (A) THD without applying SAPF; (B) THD with SAPF in conventional PI controller; and (C) THD with SAPF in proposed MRAC.

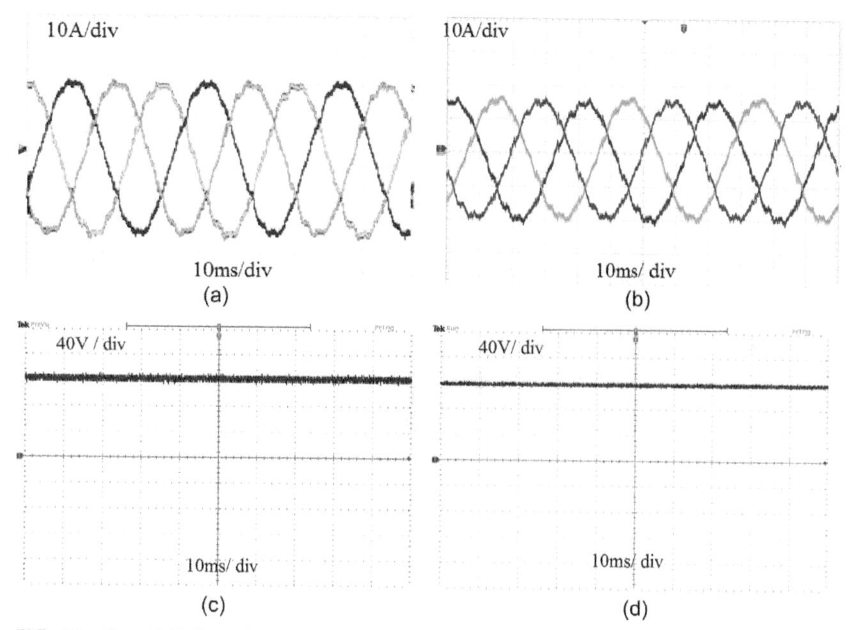

FIG. 16 Case II: Balanced supply with steady-state load condition. (A) Source current in conventional PI controller; (B) source current in proposed MRAC; (C) DC-link capacitor voltage (V_{dc}) for conventional PI controller; and (D) DC-link capacitor voltage (V_{dc}) for proposed MRAC.

with conventional and proposed controller is shown in Fig. 17B and C. In this case, load current harmonics reaches 22.32% without SAPF. With the application of the proposed MRAC, THD reduces to 1.62% over in conventional PI controller THD = 3.45%. Hence, proposed controller enhanced the performance of SAPF.

7.3 Case III: Balanced supply with parameter variation of SAPF

In this case, designing parameters of SAPF coupling resistor (R_c) and coupling inductor (L_c) is changed from $R_f = 3.14\,\Omega$, $L_f = 10\,\text{mH}$ to $L_f = 12\,\text{mH}$, $R_f = 6.27\,\Omega$. Under this case, validation of the proposed MRAC is verified over conventional PI controller. It is noted from Fig. 18A and B that source current is better and harmonics is less in proposed controller as compared to conventional controller. Fig. 18C and D represents DC-link capacitor voltage for conventional and proposed controller, respectively. THD of load current is 27.91% as shown in Fig. 19A. Source current THD for proposed MRAC is 1.96%, which is better than the conventional PI controller THD of 3.57% as shown in Fig. 19B and C, respectively.

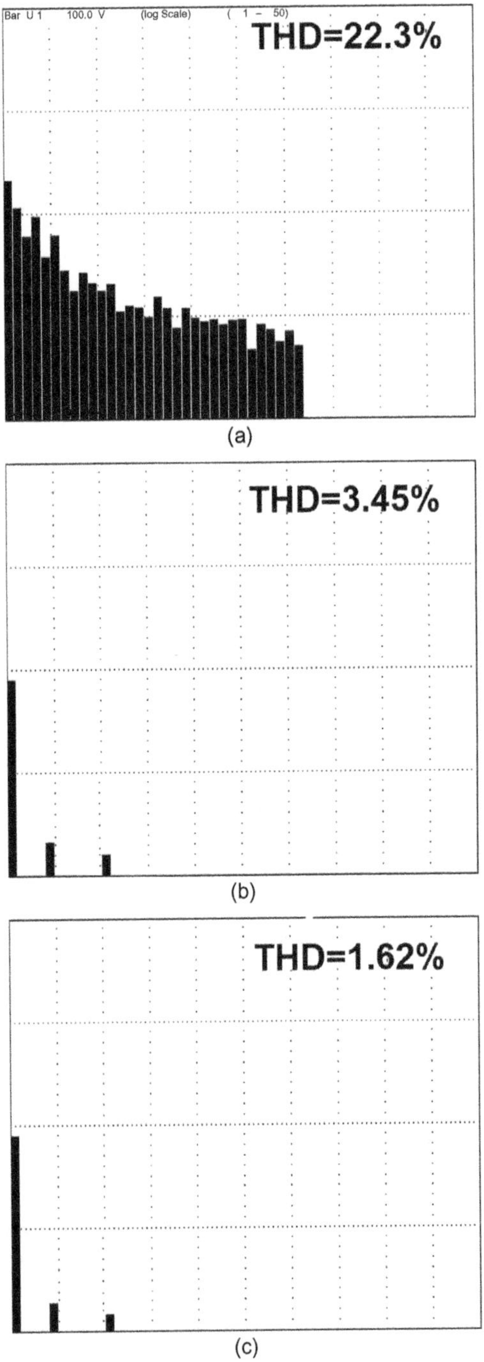

Bar U 1 100.0 V (log Scale) (1 - 50)

THD=22.3%

(a)

THD=3.45%

(b)

THD=1.62%

(c)

FIG. 17 Total harmonic distortion (THD) measurement for Case II. (A) THD without applying SAPF; (B) THD with SAPF in conventional PI controller; and (C) THD with SAPF in proposed MRAC.

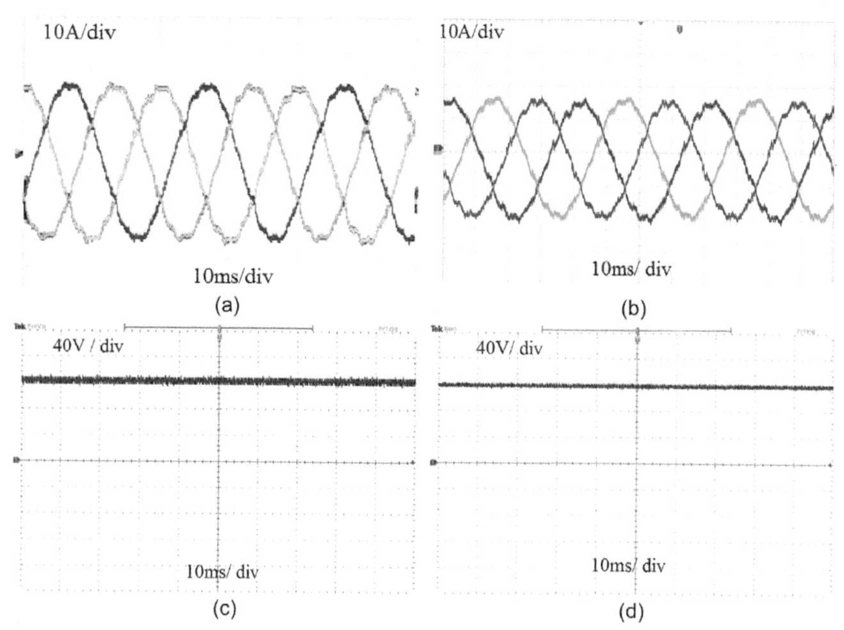

FIG. 18 Case III: Balanced supply with steady-state load condition. (A) Source current in conventional PI controller; (B) source current in proposed MRAC; (C) DC-link capacitor voltage (V_{dc}) for conventional PI controller; and (D) DC-link capacitor voltage (V_{dc}) for proposed MRAC.

To measure the robustness of SAPF, harmonics compensation ratio (HCR) is calculated in Table 2. HCR is the ratio of total harmonics distortion after and before compensation as shown in Eq. (66):

$$\%HCR = \frac{\%THD_{after\_compensation}}{\%THD_{before\_compensation}} \times 100 \tag{66}$$

Here, %THD signifies the mean of all three-phase currents %THD as shown in Eq. (67):

$$\%THD = \sqrt{\frac{\%THD_a^2 + \%THD_b^2 + \%THD_c^2}{3}} \tag{67}$$

where

$\%THD_a$ = %THD of Phase a source current;
$\%THD_b$ = %THD of Phase b source current; and
$\%THD_c$ = %THD of Phase c source current.

The difference in %HCR is calculated for both proposed MRAC and existing conventional PI controller. It is clear from Table 2 that there is very less

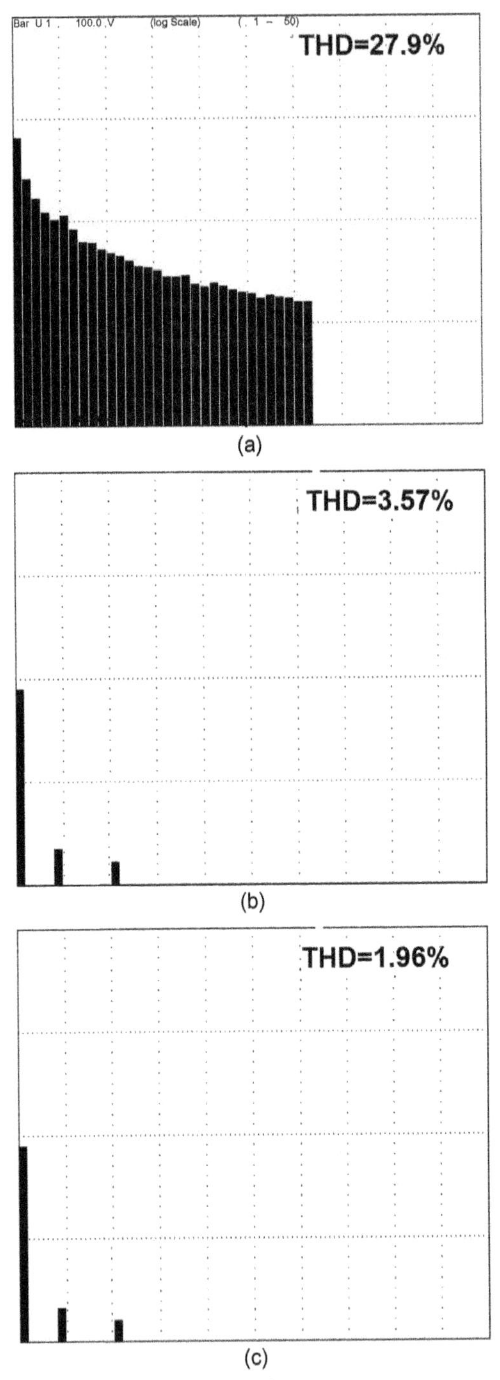

FIG. 19 Total harmonic distortion (THD) measurement for Case III. (A) THD without applying SAPF; (B) THD with SAPF in conventional PI controller; and (C) THD with SAPF in proposed MRAC.

TABLE 2 THD for both the controller-based SAPF (experimental).

Controller	Cases	%THD without SAPF	%THD with SAPF	%HCR
PI	I	21.81	3.32	15.22
	II	22.32	3.45	15.47
	III	27.91	3.57	12.79
MRAC	I	21.81	1.57	7.20
	II	22.32	1.62	7.26
	III	27.91	1.96	7.02

deviation in %HCR in proposed MRAC as compared to existing conventional PI controller for different cases. It proves the superiority of proposed MRAC over conventional PI controller.

8. Conclusions

In this chapter, harmonics compensation of proposed MRAC and conventional PI controller has been studied under various conditions, and permanence of both the controller has been compared by HCR as shown in Table 2. Both the controllers improve power quality by providing harmonics compensation, but as we can observe from Table 2, the proposed MRAC has less %THD and lesser variation in HCR as compared to conventional PI controller. Hence for all the different varying environmental conditions, the proposed MRAC is better than the conventional PI controller in terms of %THD and %HCR.

References

[1] P.S. Harmonics, Power system harmonics: an overview, IEEE Trans. Power Appar. Syst. PAS-102 (8) (1983) 2455–2460.

[2] D. Sharma, Y.K. Nagar, S. Agrawal, M. Kumar, Performance analysis of hybrid filter to mitigate harmonics, Int. J. Electron. Electr. Comput. Syst 6 (3) (2017) 235–238.

[3] J.C. Das, Passive filters—potentialities and limitations, IEEE Trans. Indus. Appl. 40 (2003) 232–241.

[4] B. Singh, K. Al-Haddad, A. Chandra, A review of active filters for power quality improvement, IEEE Trans. Ind. Electron. 46 (5) (1999) 960–971.

[5] Q. Liu, Y. Li, L. Luo, Y. Peng, Y. Cao, Power quality management of PV power plant with transformer integrated filtering method, IEEE Trans. Power Deliv. 34 (3) (2018) 941–949.

[6] E. Koutroulis, K. Kalaitzakis, N.C. Voulgaris, Development of a microcontroller-based, photovoltaic maximum power point tracking control system, IEEE Trans. Power Electron. 16 (1) (2001) 46–54.

[7] B. Subudhi, R. Pradhan, A comparative study on maximum power point tracking techniques for photovoltaic power systems, IEEE Trans. Sustain. Energy 4 (1) (2013) 89–98.

[8] R. Kumar, H.O. Bansal, Real-time implementation of adaptive PV-integrated SAPF to enhance power quality, Int. Trans. Electr. Energy Syst. 29 (5) (2019) e12004.

[9] P. Karuppanan, K.K. Mahapatra, PI and fuzzy logic controllers for shunt active power filter—a report, ISA Trans. 51 (1) (2012) 163–169.

[10] H. Akagi, E.H. Watanabe, M. Aredes, Instantaneous Power Theory and Applications to Power Conditioning, John Wiley & Sons, 2017.

[11] M. Ashraf, A. Rehman, Performance analysis of current injection techniques for shunt active power filter, in: IOP Conference Series: Earth and Environmental Science, vol. 168, IOP Publishing, 2018, p. 012014.

[12] N. Gotherwal, J.K. Nama, S. Ray, N. Gupta, Performance comparison of reference current extraction techniques for indirect current control based shunt active filter, in: 2016 IEEE 7th Power India International Conference (PIICON), IEEE, 2016, pp. 1–6.

[13] P. Dey, S. Mekhilef, Current controllers of active power filter for power quality improvement: a technical analysis, automatika 56 (1) (2015) 42–54.

[14] K.-K. Shyu, M.-J. Yang, Y.-M. Chen, Y.-F. Lin, Model reference adaptive control design for a shunt active-power-filter system, IEEE Trans. Indus. Electron. 55 (1) (2008) 97–106.

[15] P.K. Ray, S.D. Swain, Performance enhancement of shunt active power filter with the application of an adaptive controller, IET Gener. Transm. Distrib. 14 (20) (2020) 4444–4451.

[16] A. Bag, B. Subudhi, P.K. Ray, Grid integration of PV system with active power filtering, in: 2016—2nd International Conference on Control, Instrumentation, Energy Communication (CIEC), 2016, pp. 372–376.

[17] R. Faranda, S. Leva, V. Maugeri, MPPT techniques for PV Systems: energetic and cost comparison, in: 2008 IEEE Power and Energy Society General Meeting—Conversion and Delivery of Electrical Energy in the 21st Century, 2008, pp. 1–6.

[18] H. Akagi, New trends in active filters for power conditioning, Indus. Appl. 32 (6) (1996) 1312–1322.

Chapter 9

Issues and challenges in microgrid protection

Nikhil Kumar Sharma and Subhransu Ranjan Samantaray
School of Electrical Sciences, Indian Institute of Technology Bhubaneswar, Bhubaneswar, India

1. Introduction

The thermal power stations are one of the most important sources for fulfilling the electricity demand of the industry and consumers for many decades. However, the thermal power stations produce electricity by combustion of coal, which is a nonrenewable source depleting at faster rates and also a major source of environmental pollution [1]. Therefore, the utilization of renewable energy resources has been encouraged globally in recent years. The generation by the renewable energy sources are scattered and not centralized in a single location, and thus they have also been named distributed generations (DGs) [2]. This type of distributed energy system possesses numerous benefits. One of the major advantages is the capability to disconnect and function independently from the main grid in the event of a catastrophic failure at the upstream side. Additionally, the proximity of the energy production at the end-user allows for the excess energy produced during the power generation process to be leveraged into a parallel heating/cooling cycle, which is generally dissipated. This increases the energy efficiency of the entire process.

The integration of distributed energy resources with the existing power network gave birth to the concept of microgrid. The microgrid is defined as a coordinated group of DGs that supply a set of loads through a distribution system and has the ability to operate in either grid-connected mode (GCM) or islanding mode (IM) and has the capability of smooth transition between the GCM and the IM [3]. The microgrid provides a lucrative clean and green alternative to the conventional fossil fuel-based power generating sources. Further, microgrids possess numerous benefits such as improving the grid resiliency, supplying the growing power demand, and reducing the financial losses suffered by industries due to power cuts. Furthermore, the microgrids can also be utilized to supply power to remote locations, hilly areas, and isolated islands where supply by the existing grid is economically not viable. A microgrid can be used to meet the

Microgrid Cyberphysical Systems. https://doi.org/10.1016/B978-0-323-99910-6.00004-9

electricity demand of consumers locally and hence, it is beneficial from the customer's point of view. Thus, there has been a significant increase in research activities for the implementation and development of microgrids in recent years.

Microgrids are also prone to faults due to man-made or natural disasters. It has been reported that the frequency of fault occurrence in overhead lines is the highest compared to the other components. Therefore, the faults in overhead lines need to be addressed carefully. Power system protection is the art and science of detecting problems such as short circuits, equipment failure, and any other abnormal condition with power system components and their isolation. The main objective of the protection system is to isolate a faulty section from the rest of the power system so that the nonfaulty segment can continue functioning satisfactorily without any severe damage due to possible propagation of the fault transient. It helps in maintaining personal and public safety in the event of a fault and prevent damages to critical equipments such as generators, transformers, and transmission lines. The protection system also prevents power system stability problems by stopping the cascaded tripping. It is essential to protect the microgrid in both GCM and IM against all types of symmetrical and unsymmetrical faults. The microgrid protection has the following performance requirements [4]:

(i) Detection of the internal and external faults of the microgrid even if the fault current injection is low.
(ii) When the fault is within the microgrid, the protection system should clear the fault without affecting the utility and other parts of the microgrid.
(iii) There should be a main and back-up protection strategy for ensuring reliability in isolating the fault.
(iv) The protective scheme should not mal-operate during no-fault situations, including grid and islanding resynchronization.
(v) There should be an acceptable compromise between the speed of operation and the accuracy of the protection scheme.

2. Microgrid protection challenges

Due to operational changes in the microgrid topologies, the existing schemes used in conventional power distribution systems find limitations and may not be reliable. The adoption of the microgrid concept can disrupt the conventional hierarchical approach to the protection of typical radial power distribution networks with large unidirectional fault currents. The complexity of developing an efficient protection scheme for microgrid increases due to interaction between utility and different types of DGs. The following are the challenges associated with microgrid protection [4–7]:

1. **Bidirectional power flow:** One of the major challenges in the protection of the microgrid arises due to the bidirectional flow of power. As discussed

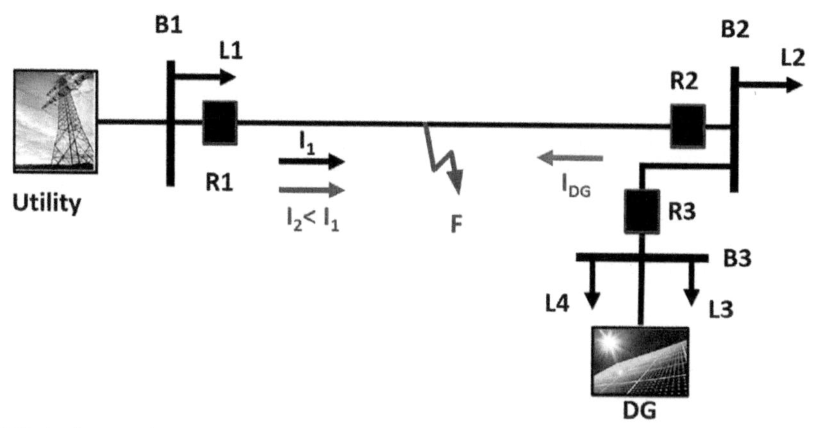

FIG. 1 Impact of bidirectional power flow in microgrid.

earlier, the microgrid is an active network and there may exist several generating sources in different parts of the network. These generating sources may contribute to feeding power to the load, which leads to the bidirectional flow of power. The existing protective schemes designed for the unidirectional power flow may not be able to provide protection measures under these circumstances. The impact of bidirectional fault on the microgrid protection scheme is shown in Fig. 1.

For a fault in the distribution line joining buses B1 and B2, the fault current is contributed by the utility and thus is unidirectional. Let us assume that a DG is connected to the bus B2 and for the same fault scenario, the DG will also contribute along with the utility. Therefore, the flow of the current becomes bidirectional and the fault current seen by the relay R1 is lesser than without DG case. There is a possibility that the current seen by relay R1 may go below the set threshold resulting in delayed or no operation of the relay and corresponding circuit breaker.

II. **Blinding of the protection system:** When the DG is connected somewhere between the feeding substation and fault location, it also contributes to the fault current. Due to the DG contribution to the fault, the relay placed at the feeding substation finds lower current compared to the situation when no DG is present. The protection system blinding phenomenon is illustrated in Figs. 2–4. The magnitude of fault current affects the reach of the overcurrent relay. The original reach of the relay when only utility is feeding the fault covers the entire line length and the relay detects the faults occurring at the far end of the line. However, if a DG is connected anywhere between B1 and B2, then for the same fault condition, the DG will also contribute to the fault current. The fault current with and without DG in the grid and the corresponding expression of fault currents are presented below.

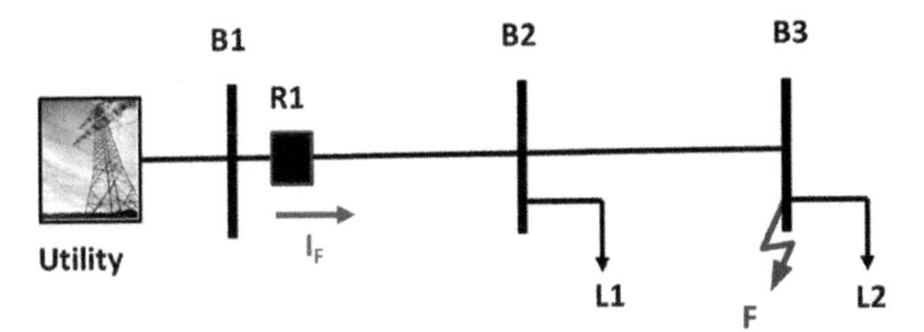

FIG. 2 Power flow in the absence of DG in the grid.

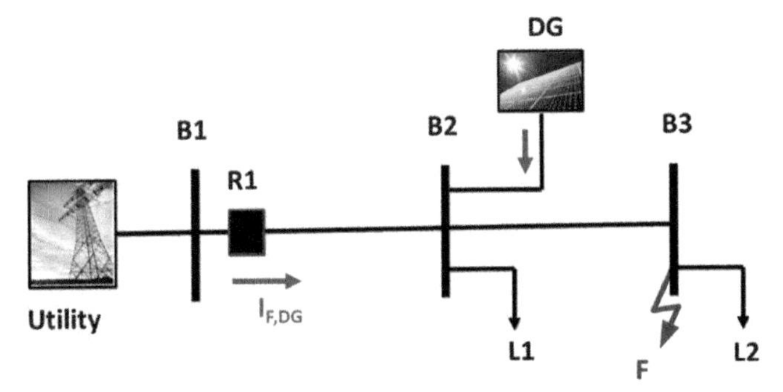

FIG. 3 Power flow with the presence of DG in the grid.

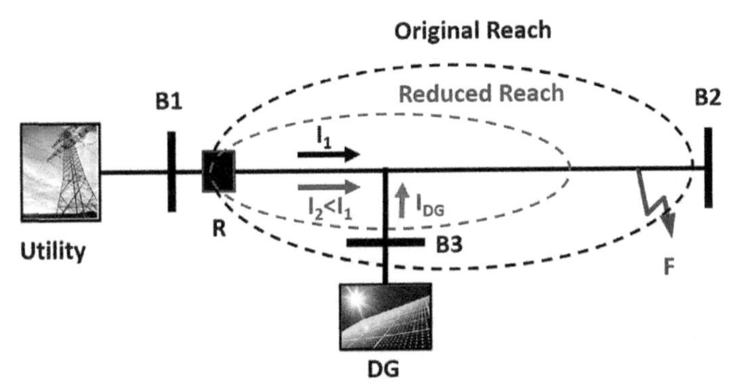

FIG. 4 Blinding of protection system in microgrid.

The fault current without and with DG can be found out as,

$$I_F^{AB} = \frac{V_{B\_pre}}{Z_{\text{eq\_B}} + Z_{BC}} \tag{1}$$

$$I_F^{AB} = \frac{V_{B\_pre}}{Z_{\text{eq\_B}}\left(1 + \frac{Z_{BC}}{Z_{DG}}\right) + Z_{BC}} \tag{2}$$

Impact of blinding: It is observed from Eqs. (1) and (2) that the fault current in the presence of DG decreases as DG also start contributing to the fault. Therefore, the current contributed by the utility decreases and relay R1 records lesser fault current than without DG scenario. This lower magnitude of current leads to reduction in the reach of the relay and thus, the relay fails to detect the remote end fault. This undesirable phenomenon is known as relay blinding and may result in the delayed operation of the relay, or the relay may not operate. Hence, the introduction of DG in the grid may disturb the operation of the relay.

III. **Variation in fault level:** The fault level in the microgrid is also affected by the network topology, type of distributed generators, mode of operation (GCM or IM), and load ratings. The fault current contribution varies from 8 to 10 per unit (pu) from the synchronous generator to 7–8 pu in induction generator and to 1.5 pu in the case of inverter-based DGs [8]. The DGs are integrated into the grid with the help of converters and these converters have the inherent property of restricting the fault current just above the rated value, which results in much less fault current. The radial and mesh topology brings a significant difference in fault current seen by the relay. The level of DG penetration is also a contributing factor to the fault current. Further, the most important issue is the mode of operation and the fault current magnitudes differ between grid-connected and islanded modes of operation. Due to the aforementioned reasons, the fault level in the microgrid is quite different from a conventional grid, which creates many problems for existing protection schemes. The variation on fault current with different operating modes, types of DGs, and microgrid topologies are presented in [8].

IV. **False tripping of the relays:** When the fault takes place in the feeder close to the feeder where DG is connected or in the case when a fault is occurring in the neighboring feeder to the DG, the DG contributes to the fault through its feeder. Under this condition, the nondirectional relay of the healthy feeder may falsely detect the fault and may trip the nonfaulty feeder, which is undesirable. As shown in Fig. 5, for the fault incepting at the line B1-B2, the DG also contribute. When the renewable energy source is a strong source (such as synchronous generator-based DGs), the current contribution is significant. This may cause the relay R2 to trip, resulting in the outage of the healthy line B1-B3. This is known as false/sympathetic tripping of the protection scheme in which relay fails to discriminate the external faults from the internal faults due to the presence of DGs in the grid.

V. **Auto reclosing failure:** The function of the auto recloser is to close the circuit breaker (CB) after a momentary interruption. Since the breakers open many times during small and large disruptions, which are not faults, a recloser improves the service continuity by automatically

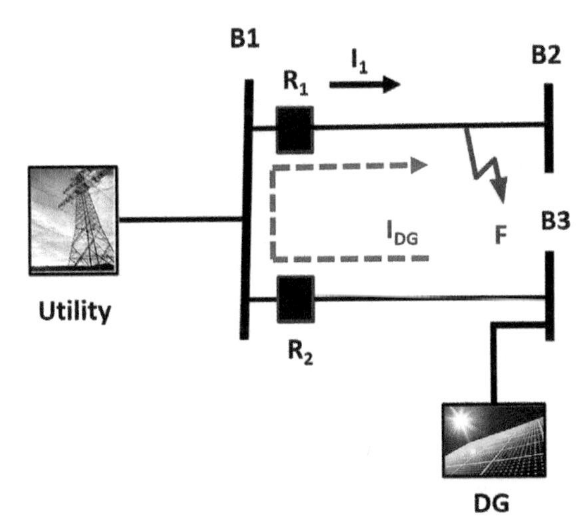

FIG. 5 False tripping of protection system.

restoring the power in such cases. This fault clearing method may fail in the presence of DG in the system. If the protection system of DG does not trigger during the operating time of recloser, the temporary fault will not be cleared and will become permanent. Hence, the presence of DG may also affect the performance of the auto reclosing mechanism.

VI. **Unintentional Islanding:** Due to large switching incidents or severe fault conditions, the protection scheme may shut down the DG and disconnect the microgrid from the main power grid. For the fault in line B1-B2, the current contribution of the DG may cause the relay R3 to operate, as shown in Fig. 6. This results in the islanding of the DG, which may further cause loss of critical loads. This situation is known as unintentional islanding. It compromises the reliability of the system because any kind of temporary fault may lead to service interruption to the end-users.

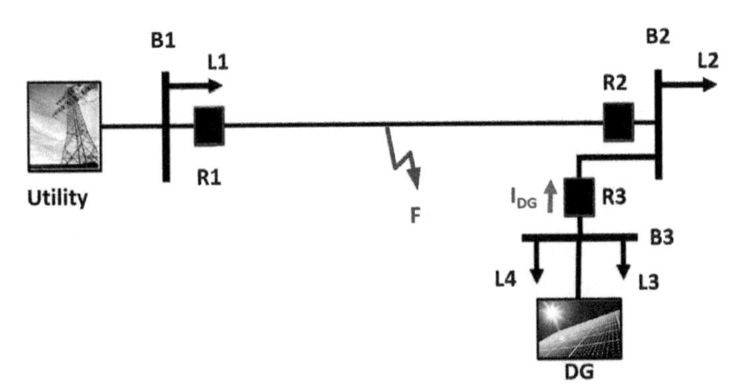

FIG. 6 Unintentional Islanding due to presence of DG in microgrid.

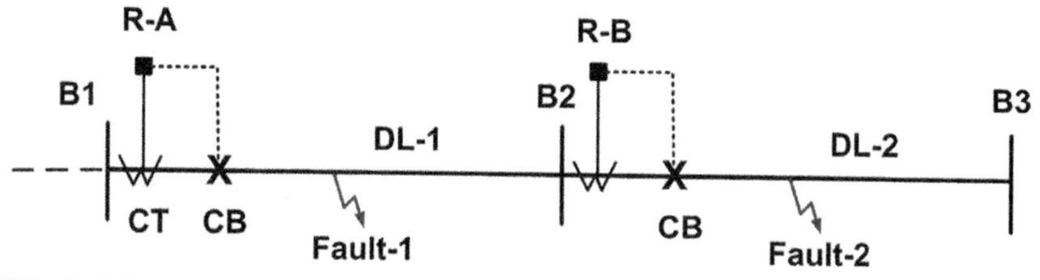

FIG. 7 Primary and back-up overcurrent protection in the distribution grid.

VII. **Protective Relay Coordination:** In power distribution networks with higher operating voltage levels, overcurrent relays are used as a back-up to more sophisticated and faster acting main protection systems. The back-up protection should be graded to achieve selective tripping, if the main protection fails to operate. However, it must be noted that this is not always possible on highly interconnected networks involving widespread generation sources, especially in microgrids. The operating conditions and network topology of microgrid may frequently change to achieve the economic and operational targets and as a result, controllable islands of different sizes and content are formed. In such circumstances, the fault current magnitude may change and the existing overcurrent relay with a single fixed setting may fail. There may be a loss of relay coordination as well and it will not guarantee a selective operation for all possible faults (shunt faults). Fig. 7 shows a simple relay scheme for a distribution line where relay 1 (back-up) and relay 2 (primary) should be coordinated based on their operation curves shown in Fig. 8, where the coordination time interval is represented between the two curves.

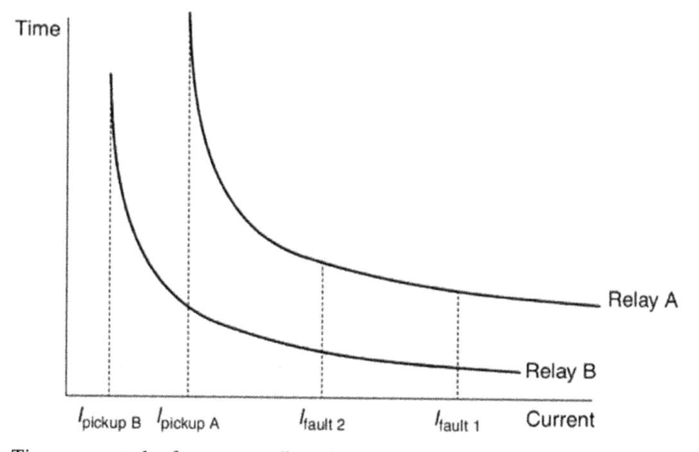

FIG. 8 Time–current plot for two coordinated overcurrent relay.

VIII. **Absence of zero crossing point:** Microgrid protection challenges in DC microgrids: This problem mainly exists in DC microgrids and the DC side of the hybrid microgrids. The arc extinction in the electromechanical circuit breaker is being carried out at natural zero-crossing points, leaving the faulty section isolated, which is absent in the DC microgrid. Due to the absence of a zero-crossing point, the interruption of current produces severe arcing. Extinguishing this arc is a challenging task and mandates fast clearing of the faults.

3. Review of microgrid protection schemes

The protection requirements of the microgrids are different from the conventional protection schemes that are being used for decades. The challenges involved in microgrid protection are discussed in the previous section. The varying operating condition in the microgrid environment makes the existing relaying techniques unreliable. The research work for the development of a protection scheme specific to the microgrid is being carried out extensively in the last few decades. These schemes are broadly clustered into overcurrent protection, differential protection, and distance protection.

3.1 Overcurrent protection schemes

The overcurrent relaying (OCR) scheme has been primarily utilized for the protection of distribution lines for many decades. The analysis of the existing overcurrent protection on the microgrid scenario was carried out in [5]. The overcurrent relay is the simplest form of protection that triggers the trip command when the current crosses the set threshold. The placement and mathematical equation of the OCR operation and its time of operation is shown below (Fig. 9)

$$I_{relay} \geq I_{threshold} \tag{3}$$

$$T = \tau_s \left[\left(\frac{I}{I_P} \right)^2 - 1 \right] - K_d \left(\frac{\partial \theta}{\partial t} \right) \tag{4}$$

where τ_s = restraining spring torque; I = current seen by the relay; I_P = pick up current; K_d = damping factor of the rotating disk; θ = angle of the disk rotation.

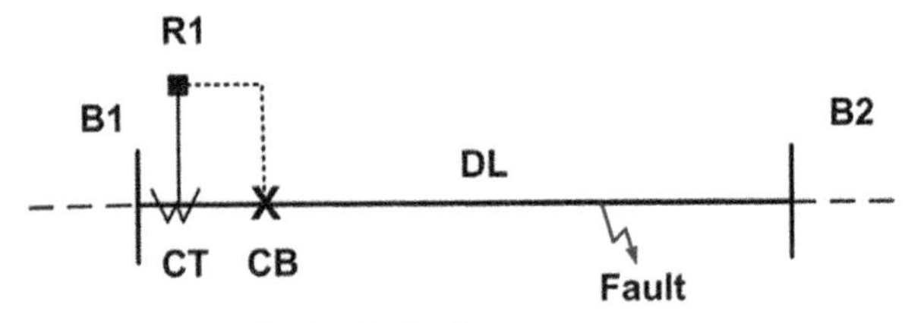

FIG. 9 Overcurrent protection for distribution line.

Eq. (4) is the mathematical expression for the disk type OCR. The first term of Eq. (4) represents the operating torque and the second term denotes the restraining torque. The disk starts rotating when the operating torque is more than the restraining torque and closes the trip contact after a time delay. Further, the expression of the trip time of the relay is presented below;

$$\text{Trip time} = \text{TDS}\frac{\frac{K_d}{\tau_s}}{\left(\frac{I}{I_P}\right)^2 - 1} \tag{5}$$

where TDS = time dial setting.

It is observed that the nondirectional OCR used in the distribution system may fail to detect the fault in the microgrid as the power flow is bidirectional compared to unidirectional in the case of the conventional power distribution grid. Further, directional OCR has limited performance due to significant differences in fault level in GCM and IM of microgrid operation. Furthermore, the current magnitude for high resistance fault cases decreases significantly and thus, the OCR either fails to detect the high resistance faults or detects with delayed response time.

Many research works presented modified OCR-based relaying schemes for the microgrid to address the aforementioned issues. Mahat et al. [9] proposed an adaptive OCR that addresses the issue of significant variation in fault current in GCM and IM of microgrid operation. The proposed relaying scheme updates the tripping characteristics with microgrid operating conditions and the faulted section. However, this technique has a high response time. A voltage-controlled OCR technique that curtails the DG current magnitude during faults to restore the microgrid to the steady-state was proposed in [10]. The OCR scheme restricts the current contributed by the DGs in the event of a fault and also restores the microgrid following the fault clearance. The overload feature limits the power by the voltage-controlled DG units. This scheme is suitable for microgrids operating in stand-alone mode. Jones et al. [11] demonstrated the limitations of nondirectional OCR and improperly configured directional OCR. It may not be able to provide adequate security and sensitivity toward the fault occurring at the remote end of the line. To overcome these issues, the authors have proposed a load encroachment supervised directional OCR scheme. This scheme is suitable for microgrids with high penetration of doubly-fed induction generators (DFIGs).

3.2 Differential protection schemes

Differential protection has attracted many interests for application in distribution networks and microgrids. It is one of the simplest forms of the protection technique, which operates when the differential current across both ends of the protected device becomes nonzero or exceeds the set pickup value, also known as bias. The differential protection principle is shown in Fig. 10.

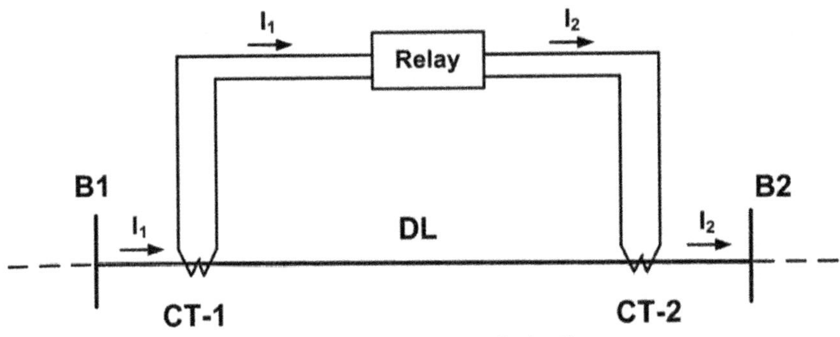

FIG. 10 Differential current protection scheme for distribution line.

$$I_1 - I_2 \geq K\frac{I_1 + I_2}{2} \tag{6}$$

where I_1 and I_2 are the current measured by the CTs placed on both ends of the line and K is the bias of the relay.

Many researchers have proposed modifications in the differential scheme to make it deployable to the microgrids. Gururani et al. [12] have suggested the utilization of signal processing techniques to improve the performance of differential relays. They proposed a nonstationary signal processing algorithm—Hilbert-Huang transform based differential relaying scheme. This research work advocates different relay settings for GCM, IM, and high impedance faults that make the relay nonadaptive in nature and also, changing the settings for different scenarios increases the response time of the relaying operation. Kar et al. [13] proposed a data mining model based differential protection scheme that extracts the most sensitive parameter from fault voltage and current signal, and uses the differential feature as a fault detector. This scheme uses both end information of the line to produce a decision tree-based index. However, the proposed scheme has not been tested for the intermittent nature of the DGs, where the current magnitude changes significantly with the DG penetration.

Gil et al. [14] suggested the improvement in the existing differential protection scheme by the addition of a new feature extracted from differential current using S-transform analysis. The new extracted feature, as well as the magnitude of the differential and restraint currents, are utilized to increase the security of the existing differential protection scheme. Besides, the Support Vector Machine (SVM) technique is used for the discrimination of internal faults from external faults. A virtual multiterminal current differential Scheme [15] was proposed in which communication is required only between the relays and done by the pilot wires. However, a differential scheme that can provide overall protection for different operating conditions and all types of DGs with fast response time is not established yet.

3.3 Distance protection schemes

The distance relaying scheme is one of the most robust protection techniques used for transmission lines. The feasibility of employing distance relays in the distribution system is also examined by many researchers. The impact of integrating renewable energy sources on the distance relays is thoroughly inspected in [16]. The fault level in IM is much lower than the GCM and therefore, the distance relay records higher impedance in such cases and fails to detect the fault. The authors in this research work proposed solutions to some of the issues such as variable generation and problems due to current infeed by the DGs.

Nikolaidis et al. [17] examined the impact of intermediate infeed, which affects the accuracy of impedance calculation and fault resistance on the distance relays employed in a medium voltage radial microgrid testbed. Arroudi et al. [18] suggested a modified distance algorithm that performs satisfactorily in the case of the microgrid integrated with wind farms, despite the intermittent nature of wind turbines. However, the study did not include the performance of the microgrid, including synchronous-based DGs. Tsimtsios et al. [19] examined the effect of the zero-sequence compensation factor on the operation accuracy of distance relays protecting distribution feeders. They proposed a scheme to set the zero-sequence compensation factor to improve the accuracy of fault detection of the distance relay.

Considering the aforementioned challenges and limitations of the existing protection schemes, it is necessary to develop a protective relaying scheme, which should alleviate the limitations of the existing techniques and be able to protect the microgrid with significant penetration of DGs. The relaying scheme should be fast, accurate, and selective in detecting the faults with wide variation in operating and fault conditions in the microgrid and should remain stable during temporary transient and critical no-fault conditions.

3.4 New microgrid protection techniques

The challenges described in the previous section restrict the operation of the existing protection schemes and make them unreliable to be utilized in the microgrid. This section discusses some new protection schemes developed to provide robust and reliable solutions to the microgrid protection challenges.

3.5 Rate of angle difference-based protection scheme

This scheme presents a wide-area rate of angle (RAD) difference-based [20] fault detection scheme for AC microgrids. The absolute rate of change of phase voltage angle difference between the point of common coupling (PCC) and the bus closest to the faulted line is used as a fault detection index in the proposed research work. The phasor measurement units (PMUs) [21] are utilized for the accurate and time-stamped measurement of the voltage signals. The performance of the RAD-based relaying scheme is carried out for wide variation in fault and operating conditions of the microgrid. The proposed scheme is validated on a medium voltage AC microgrid modeled on MATLAB/SIMULINK platform. The operation of the RAD-based relaying scheme is presented in Fig. 11. The RAD for the fault occurring on any point of the distribution line joining two buses can be mathematically defined as;

$$\mathrm{RAD} = \frac{d\{\mathrm{abs}(\phi_{V1} - \phi_{V2})\}}{dt} \tag{7}$$

where ϕ_{V1} = phase angle of the voltage recorded at bus-1 and ϕ_{V2} = phase angle of the voltage recorded at bus-2.

The medium voltage microgrid modeled and studied for the proposed research work is shown in Fig. 12. The voltage level of this system is 12.47 kV at an operating frequency of 60 Hz. In this microgrid, the utility is connected to PCC through a CB to create the GCM and IM. This microgrid consists of three DGs—including wind farm, solar farm, and hydro generator—five transformers, and several loads connected to the buses. The PMUs are placed on buses B14, B11, B9, B6, B3, and PCC. The capacity of the utility grid is fixed to 100 MW. As the power flow in the circuit is bidirectional, the CBs are

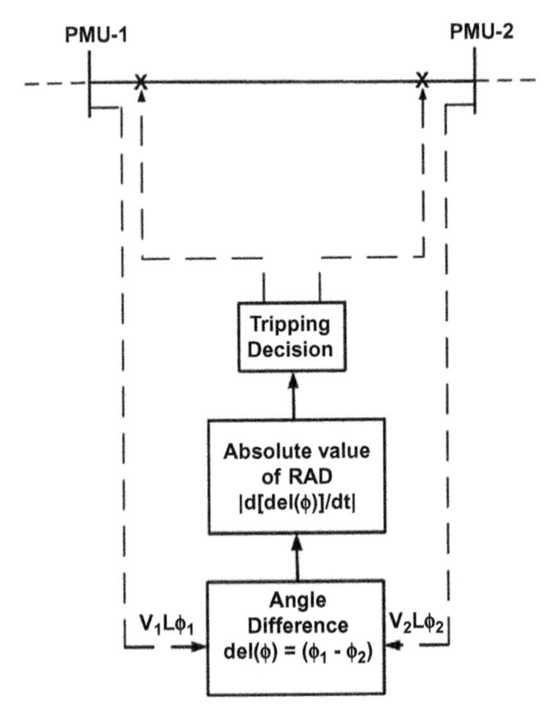

FIG. 11 Operating principle of RAD-based protection scheme.

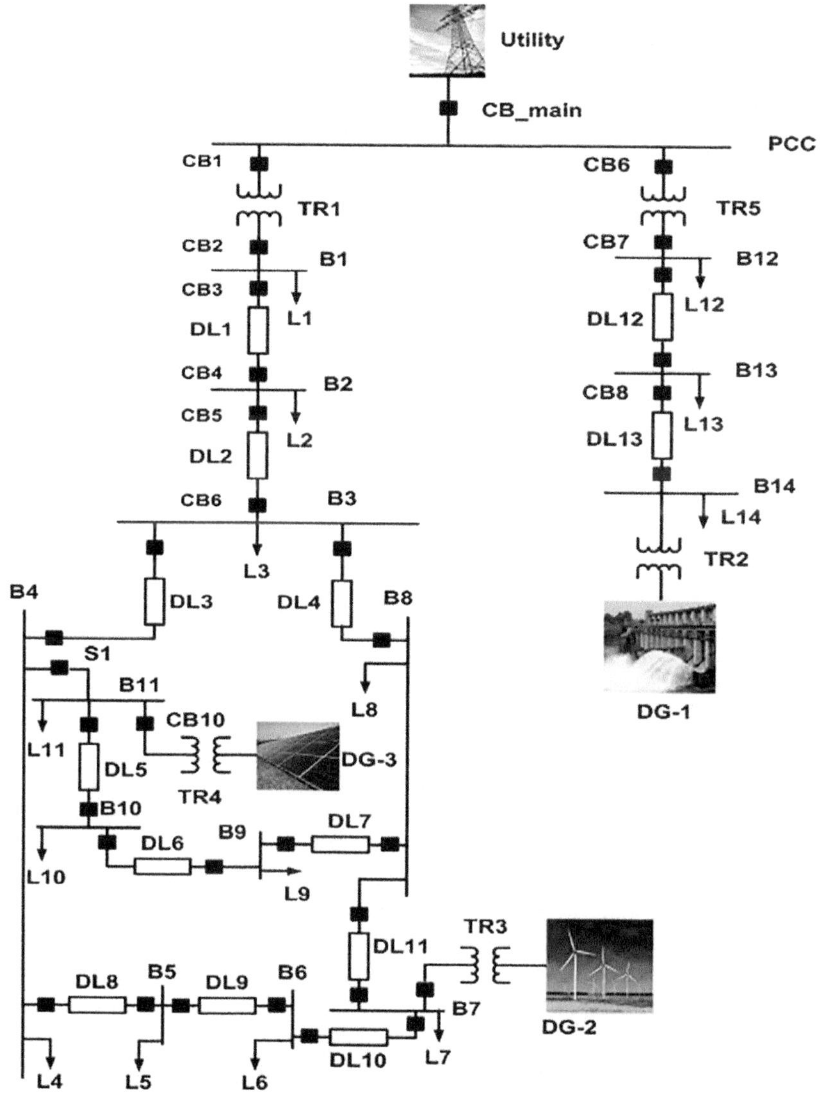

FIG. 12 Studied microgrid testbed.

installed on both ends of the lines to isolate the faulty section. The details of utility, transformer ratings, DGs voltage level, and power ratings are kept the same as [22]. The SIMULINK implementation of the DG are detailed models and the loads considered are constant impedance loads. This complete setup is modeled and simulated using SimPower System blocks with the power component in the SIMULINK platform.

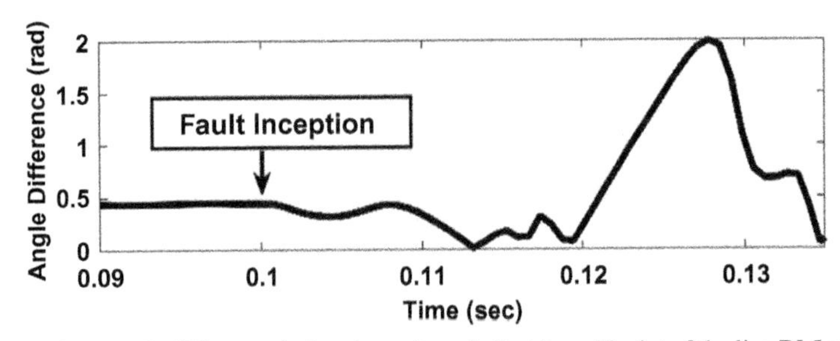

FIG. 13 Phase angle difference during three-phase fault at the midpoint of the line DL5 with the measurement at PCC and B11 in islanding mode.

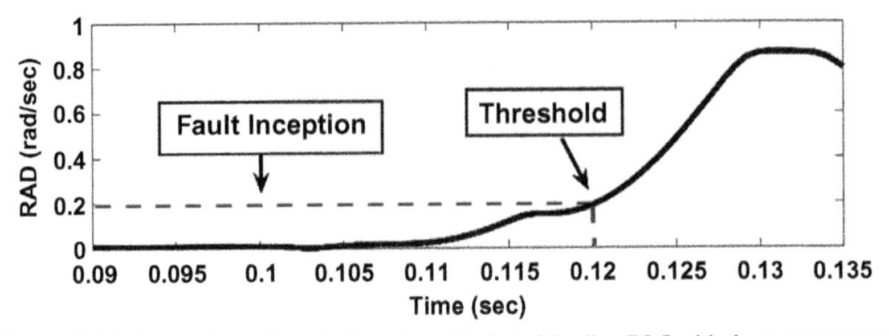

FIG. 14 RAD during three-phase fault at the midpoint of the line DL5 with the measurement at PCC and B11 in islanding mode.

The performance analysis is carried out by creating a three-phase to ground (LLLG) fault at the midpoint of the line DL5 at 0.1 s (sec) in the IM and fault resistance is fixed to 0.001 Ω. The phase angle difference and the absolute value of RAD between PCC and bus B11 are shown in Figs. 13 and 14. It is observed from the results that the absolute phase angle jump reaches 1.95 rad (maximum) and RAD reaches 0.89 rad/sec (maximum) following the fault inception. The threshold of this relay is set to 0.2 rad/sec after analyzing extensive test cases. Thus, the relay will trigger the trip command when the RAD crosses the threshold value. The change in the RAD following the fault inception is significant, which is utilized for the fault detection in overhead lines of the microgrid.

3.6 Integrated impedance angle-based protection scheme

This subsection describes an integrated impedance angle (IIA)-based [23] microgrid protection scheme. The integrated impedance is defined as the ratio of summation of positive sequence voltages to positive sequence currents of both ends of the faulted line. The IEEE C37.118.1 complied PMUs [21] are deployed at both ends of the line to retrieve the positive sequence voltages and current signals. This scheme utilizes the positive sequence voltage and current signals to compute the IIA, which is used as a key indicator for identifying the faults in the microgrid. The pure fault component network shown in Fig. 15 is used for the development of the mathematical model of the IIA. This network includes a source and a DG. The positive sequence component of the currents and voltages at the buses and the fault points are denoted by V_1^+, I_1^+, V_2^+, I_2^+, V_F^+, and I_F^+, respectively. The symbols Z_{S1}^+, Z_{S2}^+, Z_{d1}, and Z_{d2} represent equivalent positive sequence impedances of sources and distribution line segments. The fault impedance is represented by Z_F^+.

The IIA is mathematically defined as:

$$\mathrm{IIA} = \arg\left(\frac{V_1^+ + V_2^+}{I_1^+ + I_2^+}\right) \tag{8}$$

The expression for IIA is derived for fault occurring near bus 1, at the middle of the line, and near bus 2 using Fig. 15. The results for these cases are shown in Eqs. (9)–(11).

$$\mathrm{IIA} = \arg\left[-\left(Z_{S1}^+ + \frac{Z_1^+ Z_{S2}^+}{Z_2^+}\right)\right] \tag{9}$$

$$\mathrm{IIA} = \arg\left[-\frac{1}{2}\left(Z_{S1}^+ + \frac{Z_1^+ Z_{S2}^+}{Z_2^+}\right)\right] \tag{10}$$

$$\mathrm{IIA} = \arg\left[-\frac{I_1^P}{I_2^P}\left(Z_{S1}^+ + \frac{Z_1^+ Z_{S2}^+}{Z_2^+}\right)\right] \tag{11}$$

where $Z_1^+ = Z_{S1}^+ + Z_{d1}$ and $Z_2^+ = Z_{S2}^+ + Z_{d2}$. It can be analyzed from the above equations that following the fault inception, the IIA will cross 0° and go into the negative zone. The flowchart of this protection scheme is shown in Fig. 16. The IIA computed by the PMUs located at both line ends is continuously monitored. The dead zone of ±5° is fixed as the dead zone to avoid nuisance tripping and thus, the threshold of the proposed scheme is fixed to −5°. Therefore, as the IIA crosses the set threshold, the line will be isolated from the healthy part of the grid.

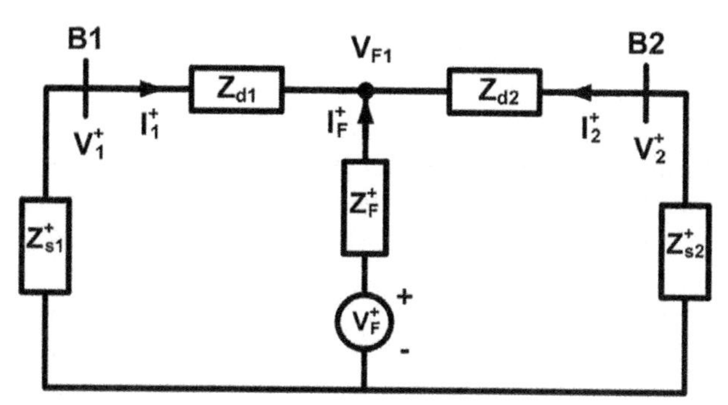

FIG. 15 The fault model for an internal fault condition.

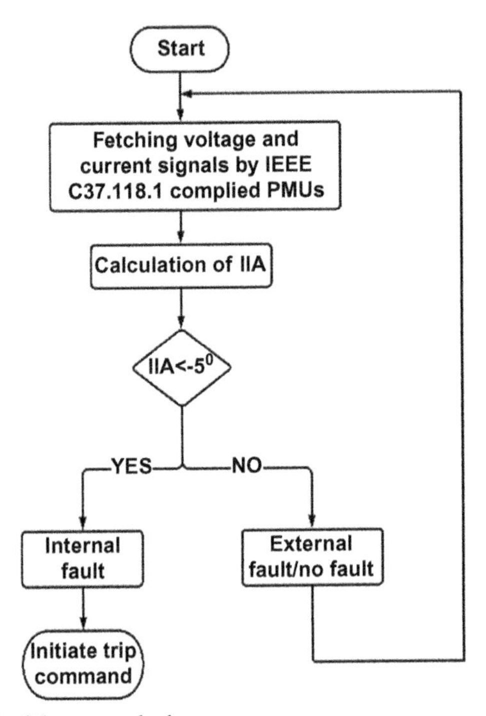

FIG. 16 Flowchart of the proposed scheme.

The proposed scheme is validated on the initial test system shown in Fig. 12, including DGs and loads (cumulative load rating is 14.15 MW and 2.34 MVAr). The PMUs are assumed to be present at both ends of the buses and PMU information of both ends of the line are fetched at one end to compute IIA, which identifies the faulty line segment in the microgrid. For this proposed scheme, the breakers are installed at both ends of all the lines considering bidirectional power flow in the microgrid and to ensure the reliability of isolating the line in case of faults. The operating zone of IIA for the internal fault is $-5°$ to $-90°$ and the no-operating zone is $5°$ to $90°$.

The performance of the proposed protection scheme is tested for the different types of faults for GCM and the results are shown in Fig. 17. The faults are created at the midpoint of the line DL9 at 0.2 s and fault resistance is fixed to 0.1 Ω. It is found that the IIA crosses the threshold in 22 milliseconds (ms) for LG fault. The communication delay of 8.33 ms (½ cycle) is considered for the medium voltage testbed simulated for the present work. Thus, the response time for the aforementioned cases will be 30.33 ms for GCM. The response time for other cases is close to this range. It is observed that the IIA crosses the threshold for all cases indicating the reliability of the proposed scheme in identifying different types of faults.

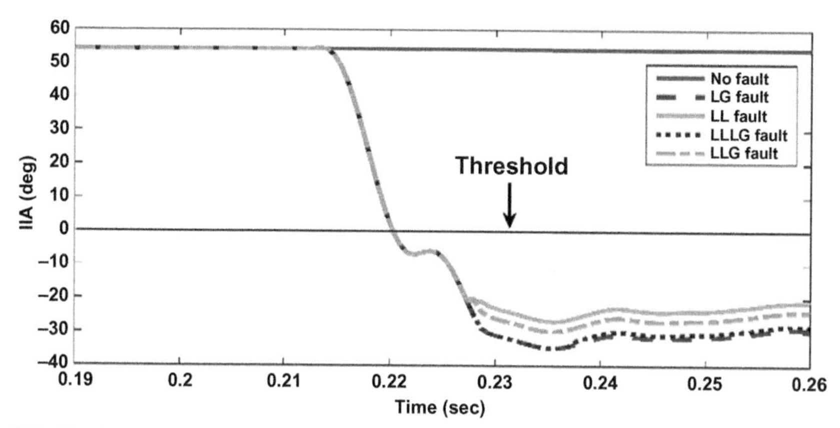

FIG. 17 Variation in IIA with fault types at the midpoint of the line DL9 in GCM with $R_F = 0.1\,\Omega$.

3.7 Impedance Difference-Based microgrid protection scheme

This research work introduces an impedance difference (ID)-based protection scheme [24] for the microgrid. To obtain accurate, faithful, and time-stamped measurement signals, the PMUs [21] are deployed at both ends of the line for monitoring purposes. The positive sequence components of the voltages and currents extracted by the PMUs are used to compute ID. A composite plane combining both the magnitude and angle of the ID is utilized for fault detection in the lines of the microgrid. The difference of the positive sequence impedances of both line ends is termed as impedance difference. The ID can be defined mathematically as:

$$\text{ID} = \frac{V_1^+}{I_1^+} - \frac{V_2^+}{I_2^+} \tag{12}$$

The mathematical expression of the ID, its magnitude and angle, presented from Eqs. (13)–(15) is derived from the fault component model shown in Fig. 15.

$$\text{ID} = \frac{Z_d^+\left(Z_{S2}^+ - Z_{S1}^+\right)}{Z_{S1}^+ + Z_{S2}^+ + Z_d^+} \tag{13}$$

$$\text{MID} = \text{abs}\left(\frac{Z_d^+\left(Z_{S2}^+ - Z_{S1}^+\right)}{Z_{S1}^+ + Z_{S2}^+ + Z_d^+}\right) \tag{14}$$

$$\text{AID} = \arg\left(\frac{Z_d^+\left(Z_{S2}^+ - Z_{S1}^+\right)}{Z_{S1}^+ + Z_{S2}^+ + Z_d^+}\right) \tag{15}$$

It is observed from Eqs. (13)–(15) that ID is dominated by the denominator term as the numerator is relatively smaller. Thus, the MID will be very small and AID will be in the negative quadrant for internal fault cases. The combined

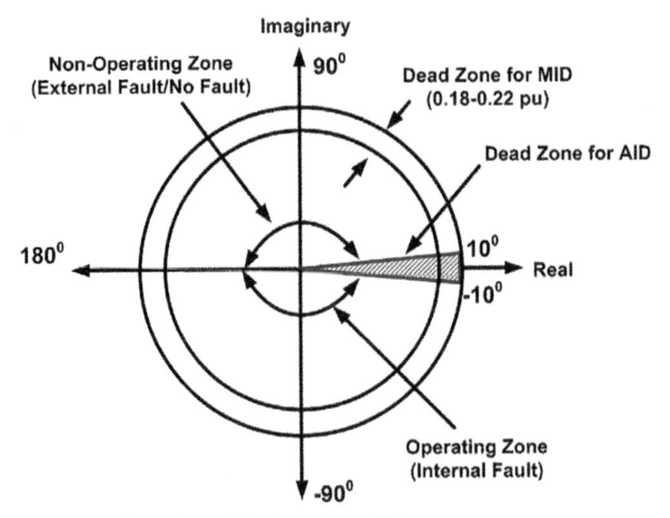

FIG. 18 Magnitude-phase plane of the impedance difference.

magnitude phase plot with the operating and restraining zones of the proposed relaying scheme is presented in Fig. 18. Further, the flowchart describing the operation of the proposed relaying technique is presented in Fig. 19.

The testing is carried out for different types of symmetrical and asymmetrical faults at the midpoint of the line DL9 of the microgrid shown in Fig. 12.

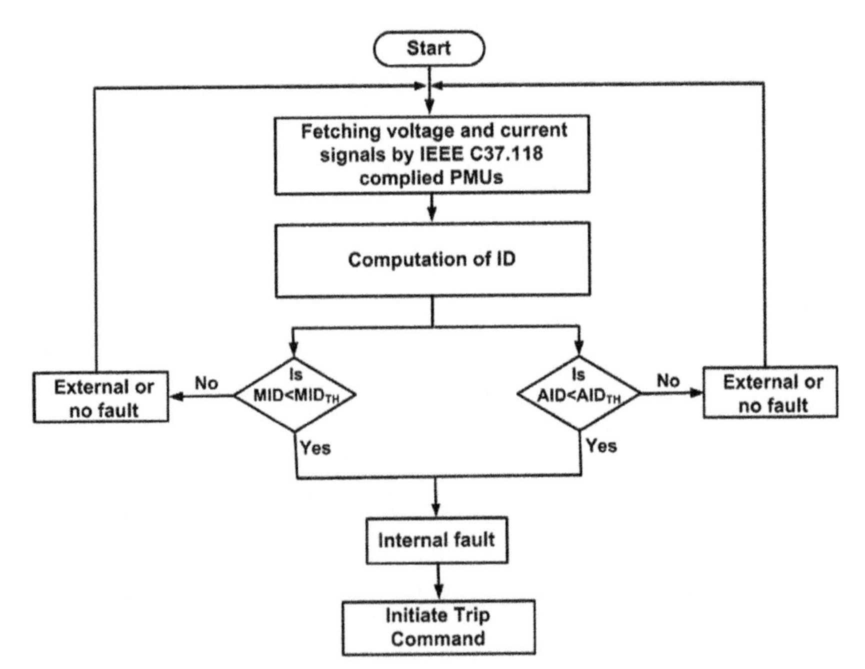

FIG. 19 Flowchart of the ID-based scheme.

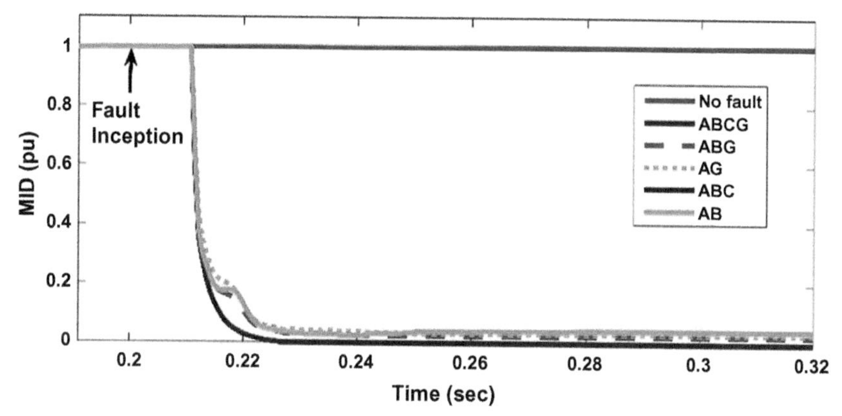

FIG. 20 MID for different fault types in GCM.

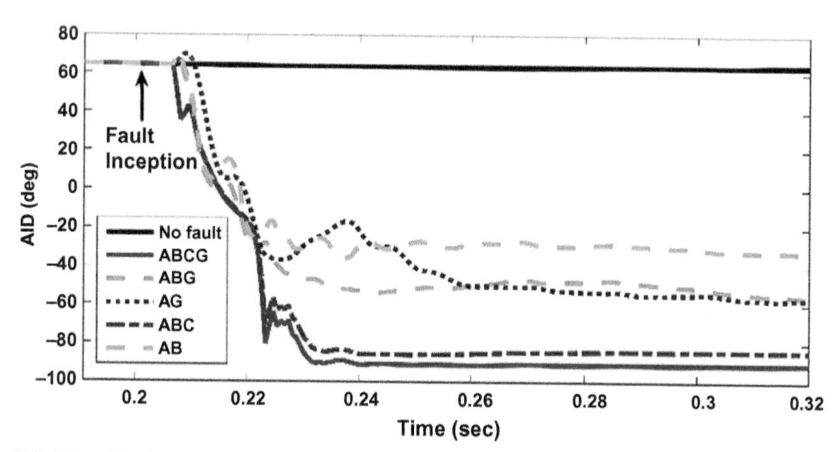

FIG. 21 AID for different fault types in GCM.

The fault is initialized at 0.2 s and the fault resistance is fixed to 0.1 Ω. The IEEE C37.118.1 complied PMUs are placed at the buses B5 and B6. The relay placed at the starting of the line monitors the ID values continuously. The MID, AID, and combined composite magnitude phase plot for the IM is shown in Figs. 20–22. It is observed that the MID falls from a pre-fault value of 1 pu to 0.02 pu for a three-phase to ground (LLLG) fault in the GCM. Hence, the change in the MADI in case of LLLG fault is found out to be 0.98 pu. Similarly, the AID falls from 65° to −89° during the fault. The jump in the proposed indices is a very significant and strong indicator of the fault condition in the line. The composite plane clearly shows that the operating points of AID and MID during the occurrence of the fault lies in the third quadrant, which is the operating zone of the relay. Therefore, the ID relay detects all types of line and ground faults in the lines of the microgrid.

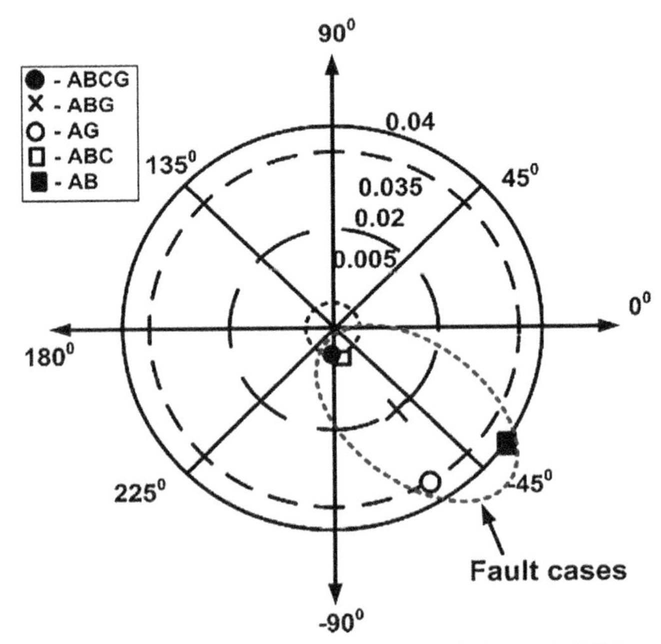

FIG. 22 The composite magnitude-phase plane for different fault types in GCM.

4. Conclusions

A robust protection scheme is one of the most important requirements for the proper functioning of the grid. The major challenges faced by the protection schemes in microgrids are variation in fault level with fault location and microgrid operating condition, current limiting properties of inverter-based DGs, and bidirectional power flow. This chapter presents new protection indices devised for microgrid protection. The proposed relaying techniques are examined for all types of line–line and line–ground with wide variation in fault resistance in GCM and IM. Further, the presented protection methodologies are examined for critical no-fault, transient, and external fault cases. These relaying techniques provide the latest developments in the area of microgrid protection. The proposed relaying schemes can be potential solutions for the overall protection of future smart microgrids.

References

[1] M. Stadler, A. Siddiqui, C. Marnay, H. Aki, J. Lai, Control of greenhouse gas emissions by optimal DER technology investment and energy management in zero-net-energy buildings, Eur. T. Electr. Power 21 (2) (2011) 1291–1309.

[2] N. Jenkins, R. Allan, P. Crossley, D. Kitschen, G. Strbac, Embedded generation, in: IEE Power and Energy Series, The Institution of Electrical Engineers, London, United Kingdom, 2004.

[3] Microgrid-1, engineering, economics and experience, CIGRE, Oct. 2015. Tech. Rep. 635.
[4] M.A. Zamani, T.S. Sidhu, A. Yazdani, Investigations into the control and protection of an existing distribution network to operate as a microgrid: a case study, IEEE Trans. Ind. Electron. 61 (4) (April 2014) 1904–1915.
[5] C.R. Bayliss, B.J. Hardy, Transmission and Distribution Electrical Engineering (Third Edition), Newnes, 2007, pp. 269–340.
[6] V. Telukunta, J. Pradhan, A. Agrawal, M. Singh, S.G. Srivani, Protection challenges under bulk penetration of renewable energy resources in power systems: a review, CSEE Int. J. Power Energy Syst 3 (4) (Dec. 2017) 365–379.
[7] N.K. Sharma, S.R. Samantaray, C.N. Bhende, VMD enabled current based fast fault detection scheme for DC Microgrid, IEEE Syst. J. 16 (1) (2022) 933–944.
[8] K. Susmita, S.R. Samantaray, M.R.D. Zadeh, Data-mining model based intelligent differential microgrid protection scheme, IEEE Syst. J. 11 (2) (2015) 1161–1169.
[9] P. Mahat, Z. Chen, B. Bak-Jensen, C.L. Bak, A simple adaptive overcurrent protection of distribution systems with distributed generation, IEEE Trans. Smart Grid 2 (3) (Sept. 2011) 428–437.
[10] A.H. Etemadi, R. Iravani, Overcurrent and overload protection of directly voltage-controlled distributed resources in a microgrid, IEEE Trans. Ind. Electron. 60 (12) (Dec. 2013) 5629–5638.
[11] D. Jones, J.J. Kumm, Future distribution feeder protection using directional overcurrent elements, IEEE Trans. Ind. Appl. 50 (2) (2014) 1385–1390. March-April.
[12] A. Gururani, S.R. Mohanty, J.C. Mohanta, Microgrid protection using Hilbert–Huang transform based-differential scheme, IET Gener. Transm. Distrib. 10 (15) (2016) 3707–3716.
[13] S. Kar, S.R. Samantaray, M.D. Zadeh, Data-mining model based intelligent differential microgrid protection scheme, IEEE Syst. J. 11 (2) (2017) 1161–1169.
[14] M. Gil, A.A. Abdoos, Intelligent busbar protection scheme based on combination of support vector machine and S-transform, IET Gener. Transm. Distrib. 11 (8) (2017) 2056–2064.
[15] B. Han, H. Li, G. Wang, et al., A virtual multi-terminal current differential protection scheme for distribution networks with inverter-interfaced distributed generators, IEEE Trans. Smart Grid 9 (5) (2018) 5418–5431.
[16] V. Telukunta, J. Pradhan, A. Agrawal, M. Singh, S.G. Srivani, Protection challenges under bulk penetration of renewable energy resources in power systems: a review, CSEE J. Power Energy Syst 3 (4) (Dec. 2017) 365–379.
[17] V.C. Nikolaidis, A.M. Tsimtsios, A.S. Safigianni, Investigating particularities of infeed and fault resistance effect on distance relays protecting radial distribution feeders with DG, IEEE Access 6 (2018) 11301–11312.
[18] K. El-Arroudi, G. Joós, Performance of interconnection protection based on distance relaying for wind power distributed generation, IEEE Trans. Power Delivery 33 (2) (2018) 620–629.
[19] A.M. Tsimtsios, V.C. Nikolaidis, Setting zero-sequence compensation factor in distance relays protecting distribution systems, IEEE Trans. Power Delivery 33 (3) (2018) 1236–1246.
[20] N.K. Sharma, S.R. Samantaray, Assessment of PMU based wide-area angle criterion for microgrid protection, IET Gener. Transm. Distrib. 13 (19) (2019) 4301–4310.
[21] I. Kamwa, S.R. Samantaray, G. Joos, Compliance analysis of PMU algorithms and devices for wide-area stabilizing control of large power systems, IEEE Trans. Power Syst. 28 (2) (2013) 1766–1778.
[22] S.R. Samantaray, I. Kamwa, G. Joos, Phasor measurement unit based wide-area monitoring and information sharing between micro-grids, IET Gener. Transm. Distrib. 11 (5) (2017) 1293–1302.

[23] N.K. Sharma, S.R. Samantaray, PMU assisted integrated impedance angle-based microgrid protection scheme, IEEE Trans. Power Delivery 35 (1) (2020) 183–193.
[24] N.K. Sharma, S.R. Samantaray, A composite magnitude-phase plane of impedance difference for microgrid protection using synchzrophasor measurements, IEEE Syst. J. 15 (3) (2021) 4199–4209.

Chapter 10

Protection schemes in microgrid

Kartika Dubey, Sanat, and Premalata Jena
Department of Electrical Engineering, Indian Institute of Technology Roorkee, Roorkee, India

1. Introduction

The energy crisis and various environmental issues associated with the electrical power systems motivated us to research in the area of distributed generations (DGs). These issues are prominently increasing leading us to develop low carbon economy and to construct environmentally friendly civilization to achieve sustainable development. The production of green and clean energy is now the primary concern for the development of country, which is supported by producing the power using distributed energy resources (DERs) and hence increasing system reliability. DER includes solar PV panels, wind plants, combined heat and power plants, various electricity storage devices, small natural gas-fueled generators, electric vehicles (EV), and loads. These resources support small-scale power generations to serve the demand. Thus, DGs allow to produce small power near to the distribution load end resulting in the formation of microgrids which are capable of operating in grid-connected and islanded modes. Thus, the microgrid, as defined by the U.S. Department of Energy, is "A group of interconnected loads and distributed energy resources (DERs), with clearly defined electrical boundaries, that acts as a single controllable entity with respect to the grid and can connect and disconnect from the grid to enable it to operate in both grid-connected or island modes" [1,2]. The microgrid has become more interesting since last decade because of its potential benefits to provide reliable, secure, efficient, environmentally friendly, and sustainable electricity from renewable energy sources (RES). The microgrids are the future of electric distribution networks. They are of decentralized and self-sustainable nature, also capable of fulfilling the local demand by utilizing local DGs. Microgrids have diverse structure which includes DC, AC, and hybrid configurations [3]. These schemes consist of power electronic circuits, controllers, storage devices and are designed to operate in grid connected mode and autonomous or islanded mode of operation. Both the sides of microgrid, i.e., supply and demand, are of controllable nature despite the intermittent characteristics of DERs. Higher

Microgrid Cyberphysical Systems. https://doi.org/10.1016/B978-0-323-99910-6.00003-7

penetration of intermittent DERs into the distribution networks will force the existing utilities to evolve in a smart and intelligent way with good connectivity for reliable system operation. Microgrids provide better management of peak loads, because of local power generation, virtual inertial support, demand management, load shifting, etc. This however improves the quality and security of power supply [4]. Thus, the microgrid structure is very much needed in the present power system scenarios, as they are the promising solution to many challenges due to conventional energy resources:

1. It helps in reducing the gaseous emissions due to coal-based generations
2. It is energy efficient technology
3. Deregulation or competition policy
4. Diversification of energy sources
5. Provide relief from the climate change problems
6. Meeting the national power requirement
7. Helps in the peak load reduction

Thus, the secure and reliable power system operation is essential requirement of any country and India is third largest consumer of electricity in world. Due to the limited infrastructure, limited space, large population, environmental concern and fluctuation in demand and supply, it is not possible to increase installed capacity day by day. Microgrid helps us to fulfill our electricity demand, which fluctuates day by day in optimized way and better utilizing our existing infrastructure. Despite all the advantages microgrid provide, and it seems to be very prominent solution for the future grid, they have few limitations and challenges as well. One of the biggest challenges in management of microgrid is the maintaining the power supply for reliable system. This can be achieved by choosing appropriate storage devices integrated at the optimal locations and optimal size meaning that the storage units should meet the local energy demands perfectly. Further, microgrid protection is also one of the most prominent issues in the distribution network. As the integration of DGs in existing distribution system changes its radial nature, the suggested conventional protection techniques in literature may fail to operate for these microgrid structure. However, few of the major challenges are listed below addressing the protection issue of microgrid [5]:

1. The change in fault current levels of microgrids is different for the two operating modes, i.e., grid-connected and islanded modes.
2. The presence of types of DGs (i.e., wind generation, solar generation, etc.) also affects the protection units because of their intermittent nature. Also, the placement of DGs plays a huge role in the protection area as they will inject the amount of power demanded and changes the radial flow at the buses they are integrated. However, the failure of a microgrid during a fault does not impact a wide area, thereby limiting the investment in microgrid protection.

3. The integration of DGs is carried out by interfacing the power electronic units (like voltage source converters (VSCs)), following the IEEE power quality standards [6], which results in the unconventional fault behavior like introduction of small faults currents or limiting the fault currents.

Thus, the protection of microgrids is very important rea of research for providing better reliability and security of power supplies. Further, the chapter discusses about types of microgrid configuration and its hierarchical control architecture.

2. Classification of microgrids

This section reviews the types and of microgrid on the basis of their location, capacity, and power supply. These types are classified for better understanding of the microgrids, their characteristics and other control and monitoring aspects. Thus, the microgrids can be classified on the different basis stated below:

(i) On the basis of geographical locations.
(ii) On the basis of power capacity.
(iii) On the basis of electrical power supply.

2.1 Classification of microgrids based on location

2.1.1 Urban and rural microgrid

In urban microgrids, the feeders are loaded densely and they are located in the populated area. The feeders, laterals, or main trunks are of short lengths with low degree of imbalance. Such types of microgrid offers the short circuit ration normally greater than 25, due to which less fluctuations of voltages or frequencies are noticed providing flat voltage profile. However, the feeders are located in sparsely populated area in case of rural microgrids. Thus, the loads associated are scattered in the area resulting in the longer feeder or lateral lengths. Thus, the short circuit ration is comparatively lesser than the urban microgrids and they do not have flat voltage profile. These types of microgrids have significant fluctuations of voltage and frequency.

2.1.2 Utility microgrid

These microgrids are capable of exchanging the power with the main grid as they are connected to the main grid at point of common coupling. They consist of distribution feeders and substations within its spread, and they have different types of power generating resources and loads connected to the nearby buses. Grid management system is also provided in these types of microgrids for monitoring and controlling the energy exchange at different segments. As these are operated in both the mode of operation of microgrid (grid connected and islanded), as required, so islanding detection unit is compulsory.

2.1.3 Community or campus microgrid

The campus or community type of microgrids are usually defined for the standalone buildings or campus like universities, schools, commercial complexes, military camps, corporate offices, etc. These microgrids are connected with the main grid as a backup power supply. They usually have smart metering and control systems for better monitoring of entire infrastructure.

2.1.4 Industrial microgrid

These microgrids are formed for supplying the power to industrial customers like automated manufacturing and processing units. Since the loss of electricity in such units results in huge economic loss; thus, the reliable and secure power supply is provided using such industrial microgrids.

2.1.5 Stand-alone or off-grid microgrid

The sites or area where it is difficult to receive the power from the main grid due to lack of infrastructure resources or transmission capabilities are benefitted from off-grid type of microgrids. The loads located at the sites which are not easily accessible such as islands, grid-deprived remote locations like tribal hamlets and isolated colonies are benefitted from these types of microgrids. Such microgrids operate in standalone mode and are not connected to any local electric utility. Therefore, off-grid microgrids will potentially be the fastest growing microgrid type and remain as a member of microgrid family in terms of requirements and challenges.

2.2 Classification based on capacity

On the basis of the capacity of microgrids they are further classified as: Minigrid, Microgrid, and Nanogrid. These are classified on the basis of their capacity and size.

- According to World Bank *minigrid* is defined as an "Isolated, small-scale distribution network typically operating below 11 kV that provides power to a localized group of customers and produce electricity from small generators, potentially coupled with energy storage,"
- As per Lawrence Berkeley National Laboratory, *nanogrids* are defined as, "A small electrical domain connected to the grid of no greater than 100 kW and limited to a single building structure or primary load or a network of off-grid loads not exceeding 5 kW, both categories representing devices capable of islanding and/or energy self-sufficiency through some level of intelligent distributed energy resources management or controls." Nanogrid is the smallest discrete network unit with the capability to operate independently like a building-level circuit with building-integrated power generation source(s).

These microgrids are classified as per their capacity, size, and configuration. They operate in both off-grid or grid-connected mode of operation as per the requirement.

2.3 Classification of microgrids based on power supply

2.3.1 AC microgrids

The microgrids who consists of AC bus as their backbone structure are AC microgrids. These are the most used topology and investigation on this concept have been started since early 2000s. These microgrids are the most straightforward solution to integrate DGs in the existing power system infrastructure. Fig. 1A illustrates the configuration of a typical AC microgrid consists of different types of generation, energy storage systems, and loads.

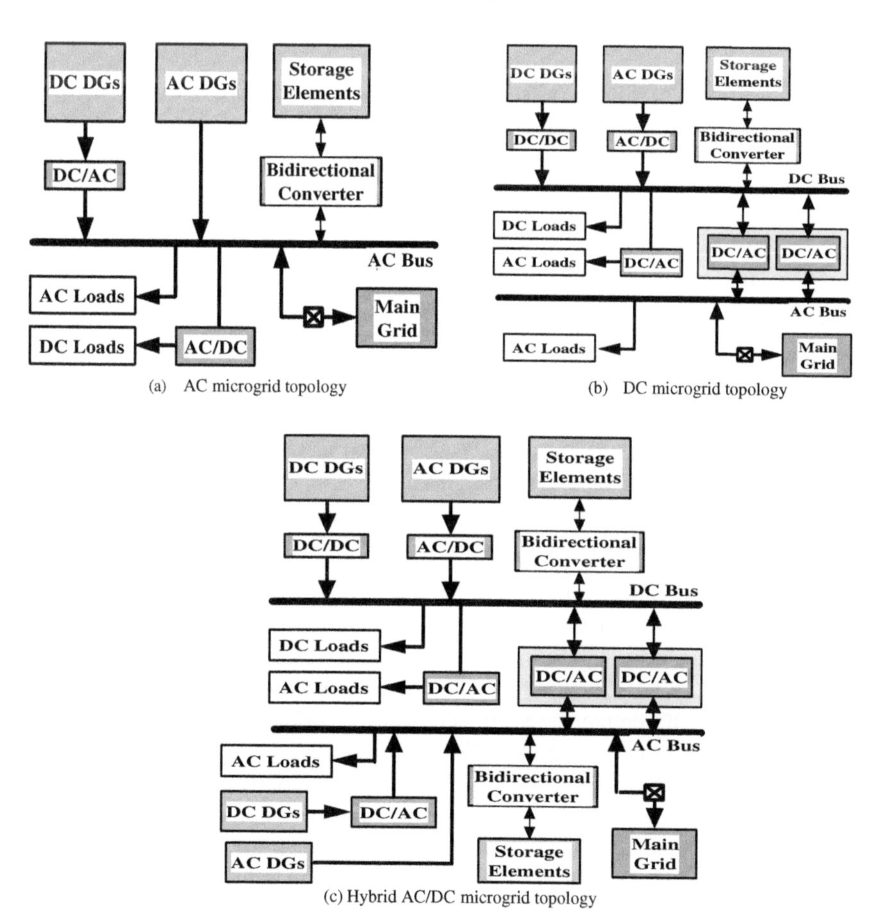

FIG. 1 Structure of various types of microgrids on the basis of electrical supply.

One of the reasons for employing AC grids is that since their early stages, several devices have been based on AC and therefore can be directly connected to the grid. A clear example is the classical synchronous generators employed for energy generation in nuclear, thermal, or hydroelectric power plants. The principal advantages of AC grids can be summarized as follows:

- *Voltage level transformation*: The conversion of voltage levels can be easily achieved using transformers. The high maturity of this field has led to very efficient and reliable devices at very low costs, enabling their use in very diverse applications apart from electric grids.
- *Protection*: The knowledge gained over the years over AC systems has facilitated the development of different types of protection devices oriented to all kind of applications. Moreover, the zero-crossings of the AC current suppose an advantage in comparison to DC grids, as they allow the opening of faulty lines when the current is nearly zero.
- *Maintenance*: According to Hossain et al. the maintenance of AC grids is easier and less expensive than DC ones [7].
- *Inertial device*: Synchronous machines generating electrical power, are directly connected to AC grid, and the loads which provide inertial response under power perturbations as a result dynamic response, help in improving the stability of AC grid.

2.3.2 DC microgrid

The need of DC microgrid is because of availability of RES and modern DC load which requires DC interface. Even if most distribution grids operate in AC, the evolution of semiconductor technologies and power electronics during the last decades has enabled the use of DC grids as a competitive alternative for the integration of DG systems. Fig. 1B shows the same type of devices as in the previous section connected to a DC microgrid.

The main reasons for considering the use of DC grids instead of classical AC ones are summarized below:

- *Integration of devices*: in the last decades, the number of devices based on DC or that require a DC stage—e.g., photovoltaic-based DG, li-ion batteries or supercapacitors, light emitting-diode (LED)-based lighting, or variable speed drives—has increased significantly. By employing DC grids the quantity of AC-DC conversions are reduced, providing a more efficient and robust way of integrating DG, ESS, or loads.
- *Synchronization*: unlike at AC-based distribution networks, in DC there is no need to synchronize the output current of devices before being connected to the grid. The lack of synchronizing algorithms in the controllers such as phase-locked loops (PLLs) simplify the control and reduces the risk of a system becoming unstable under certain grid conditions [8,9].

- *Reactive power*: as the current circulating in this kind of grids is DC, there is no reactive current, which means no reactive power circulation. This is directly reflected in the overall efficiency of the system, as electrical losses are reduced. Moreover, there is no need to include additional regulators in the control strategies of devices connected to DC grids, simplifying their design and dynamic behavior.
- *Power transmission capacity*: usually AC cables must be designed for the peak voltage value, but the transferable power is determined with the root mean square value of this voltage. This is 1.414 times less power that if we employed DC current, so at a DC microgrid we could either transfer more power in a conductor of the same diameter or reduce the size of the conductor for the same power transmission. DC microgrids provide several advantages over AC-based.

2.3.3 Hybrid AC/DC microgrid

Hybrid AC/DC microgrids hybrid AC/DC (from now on hybrid) microgrids are an interesting alternative because they enable the integration of DC-based systems through a DC subgrid while maintaining the current AC infrastructure. By doing this, the advantages of AC as well as DC grids can be combined toward an efficient and reliable integration of DG systems [10]. In Fig. 1C, we have integrated the same devices as in the previous sections in a hybrid microgrid to illustrate a typical configuration.

In addition to the advantages of AC or DC microgrids, hybrid AC/DC systems improve even more the integration of AC- or DC-based devices, because they are directly connected to the network with the minimum number of interface elements, reducing the conversion stages and therefore the energy losses. This structure is suitable for integrating distributed storages such as the emerging electric-vehicle energy-storage systems (EV-ESS) for vehicle-to-grid (V2G) applications. EV storages can effectively improve the overall performance of a hybrid microgrid in terms of voltage and frequency regulation, system stability, active and reactive power support and fault robustness. However, the optimized coordination of EV storages within microgrids is an intricate issue due to their different control and configuration structures along with inadequate standards regarding V2G applications. The centralized and distributed control structures are viable options to coordinate spatially dispersed EV storages within microgrids of different geographical sizes Moreover, the integration of interlinking AC-DC converters through the grid increases the degrees of freedom in the management and regulation of the overall grid, enabling the active control of power flows—which cannot be done with passive low-frequency transformers—and providing auxiliary services such as dynamic response support, active fault management, and power quality or stability improvement. Configurations of Interlinking Converters Hybrid microgrids can be further classified depending on the type of connection of interlinking converters with the main AC grid and between AC and DC subgrids. These connections can be further divided into two main subgroups, namely, coupled and decoupled configurations. The former corresponds to hybrid microgrids whose AC subgrids are connected through a transformer to the main AC grid, whereas the latter refers to topologies whose AC subgrids are decoupled via power converters—which would enable for instance the interconnection of asynchronous AC grids. Moreover, these connections can be subdivided depending on whether they are completely or partially isolated from the main grid through a transformer and the stages of the power converter.

3. Hierarchical control architecture of microgrids

In microgrids, the different power generation systems are interconnected together. For ensuring stable and economic operation of microgrid, the active and reactive powers of DGs must be shared simultaneously. The operation of a microgrid is established under hierarchical architecture in a similar way as the main utility grid. The hierarchical control structure is implemented in order to have minimum operation cost, maximum efficiency, reliability, and controllability.

This structure is divided into three level of controls: primary, secondary, and tertiary [11,12]. These control levels basically deal with performance (stationary and dynamic) of the microgrids managing the considering the economic aspects as well. The proper control structure of microgrid will help in stable and economical operation of the microgrid system.

In this scheme, decision layers at higher level in hierarchy define the goals and coordinate lower-level layers, but do not override their decisions. This increase's reliability of the system so that it can survive when one of the control units is disconnected or decomposed because the system as a whole is less sensitive to disturbance if local units can respond more quickly than a remote central decision unit. Microgrids are expected to fulfill the following demands in a stable manner:

(a) Active and reactive power support to loads
(b) Voltage and frequency regulation at their nominal values
(c) Power flow control with the grid
(d) Optimal and economic operation of the microgrid

The above-mentioned roles are managed by the microgrid with a hierarchical control structure in which different objectives are achieved in different time frames. The lower level takes account of the most important functioning, i.e., proper power sharing with voltage and frequency control and is implemented with fast control loops. Similarly process such as optimal operation requires analysis on a large time frame so is implemented in slower loops. In between these two different time frames of operation, operation such as removal of

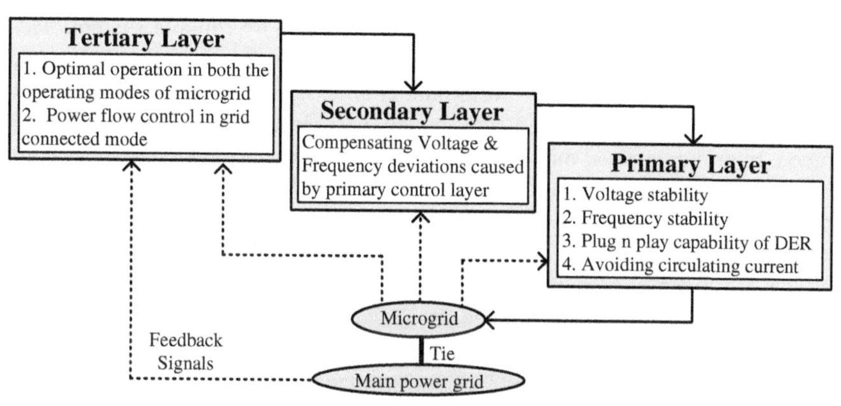

FIG. 2 Hierarchical control architecture of microgrid.

steady state errors from voltage and frequency are implemented on secondary level. In this way, a hierarchical control strategy is able to fulfill all the requirements of microgrid operation.

Hierarchical architecture for the control operation of a microgrid depicting three control levels is given in Fig. 2. Primary control response is fastest and use droop control techniques to share the load between converters by mimicking the synchronous generator inertia concepts [13]. Secondary control is in authority for removal of any steady-state error, which is introduced by the primary level control on the lines of automatic generation control or load frequency control. The secondary control is designed with slower dynamic response as compared to primary, so that the dynamics of primary and secondary controls can be decoupled for stable operation. Tertiary control is the slowest control level and it provides the power references to the lower control levels but do not override their responses as per the optimal operation of microgrid. Primary control depends on local signal measurements which removes the requirement of complex communication networks between the DG controllers. The main design parameters for the primary controller are:

(a) Voltage and frequency stabilization.
(b) Proper sharing of power preferably without the communication between units.
(c) Removal of circulating currents that else can damage the DC link capacitor [14].

4. Protection issues in microgrids

The advantages, need, and types of the microgrids with the hierarchical control architecture are already been discussed in the sections above. Even though the advantages offered by the microgrid are very huge but it does have few limitations which are needed to be addressed for more reliable system. The major protection issues faced by the microgrid are discussed in Che et al. [15], Hosseini et al. [16], and Sortomme et al. [17]. As it is already discussed that the microgrid operates in two modes, i.e., Grid connected and Islanded mode. In grid connected mode microgrid is directly connected to grid with help of Static Connector Switch (SCS) while in islanded mode of operation fault is mainly associated with inverter side of generating sources. The conventional protecting devices have ability to clear fault having limited fault current magnitude. However, in case of microgrid the fault current levels changes resulting in the requirement of the protection strategies offering better reliability for AC microgrids [18–25], DC microgrids [26–30] and Unamuno and Barrena [10].

Based on modern era of research the major classification of microgrid is on the basis of electrical supply, i.e., they are classified as AC, DC, and Hybrid AC/DC microgrid, as discussed in the above section. Each of these types of microgrids have their own protection issues and solutions in literature. Thus, it is required to understand the need of different protection strategies for different types of microgrids. The AC and DC types of microgrid have major protection concerns due to different issues arises which are listed in the Table 1.

TABLE 1 **Protection issues in types of microgrids classified on the basis of electrical supply.**

S.no.	Type of microgrid	Issues
1.	AC microgrids	• Short circuit level or fault current level [25] • False tripping of feeder [22] • Blindness of protection [20,21,24] • Automatic reclosing [18,23] • Bidirectional power flow • Unsynchronized reclosing [18,19] • Sympathetic tripping
2.	DC microgrids	• Change in fault current magnitude [26] • Absence of natural zero crossing [27] • Direction of fault current • Dynamic fault current value due to common power load [27,28] • Prohibition of Automatic Reclosing • Grounding Issues [30] • High Sensitivity of fault current in Inverter • Nonsuitability of AC Circuit Breaker
3.	Hybrid AC/DC microgrids [10]	Combination of issues arising in both AC and DC type of microgrids

5. Protection schemes for AC microgrids

The protection schemes suggested for the AC microgrids can be majorly classified in adaptive and nonadaptive methods. The conventional protection schemes which are not capable of modifying themselves according to the modifications made in the power system network like change in network configuration or sudden change of network parameters, are usually nonadaptive protection methods. However, the adaptive schemes have in-built feature of modifying themselves as per the network changes for giving the better performance.

A good protection method should essentially have the following features: (i) Selectivity: The method should only isolate the faulty part of the utility, (ii) Redundancy: For the backup protection, the relays must have redundant functionalities, (iii) Security: The unwanted tripping of breakers must be avoided for the conditions, which does not qualify as a fault for secure power supply. Thus, a protection method must have these features for providing reliable protection of microgrid.

5.1 Nonadaptive protection for microgrids

The nonadaptive methods suggested for the protection of AC microgrids are given in the Table 2 [17,31–35]. Different techniques used for the protection of microgrid are discussed with their advantages and disadvantages. The protection schemes discussed utilizes various parameters for the technique like voltage, current, harmonics, etc. The methods based on distance, voltage, symmetrical components, inverse time admittance, pattern recognition, total harmonic distortion (THD)-based methods, current traveling wave, and differential current are given for the AC microgrids in the tabular form below. The word nonadaptive is used because these techniques are proposed specifically for the predefined network parameters or characteristics and they will not operate for changing system conditions. Thus, most of the conventional schemes are nonadaptive in nature.

5.2 Adaptive protection for microgrids

The most common protection scheme employed in the distribution grid is time overcurrent protection method because of its simplicity and economic reasons. However, when DERs are integrated in the system, the short-circuit current magnitude and direction changes which depends on the location, size, and type of DG connected. Thus, in such modified networks, the value of fault current continuously varies as the microgrid operating conditions varies due to the intermittent features. Also, during faults the decision for controllable islands should be made depending upon of the size of microgrid islanded clusters. Such circumstances require the proper relay coordination, which is not possible with

TABLE 2 Nonadaptive protection schemes for AC microgrids.

S.no.	Type of protection scheme	Principal of fault detection	Advantage	Disadvantage
1.	Protection based on distance [31]	Calculation of impedance with help of fault current and voltage along line at various location of relay.	Applicable for all type of network	Adaptivity problem for various DG unit
2.	Protection based on voltage [32]	Disturbance in the d-q transformation of DER terminal voltages	Applicable with low fault currents	High impedance fault and single-pole tripping is major concern
3.	Protection based on symmetrical component [33]	By calculating residual current and zero sequence current	Free from communication requirement	Overload and overvoltage, 3-phase and LLG fault is main concern
4.	Protection based on inverse time admittance [34]	Measuring line admittance and compare it to set value of admittance	Free from communication requirement	Tripping Process is slow in High impedance fault. Harmonics and transient are main issue
5.	Protection based on pattern recognition	Based on time frequency transform	Sensitivity is too less	Economic concern, high cost
6.	Protection based on THD [35]	Increasing THD level of voltage available at terminal of DG	Resolve issue of low fault current	Dynamic loading is one of concern Finding of perfect THD threshold is major concern
7.	Protection based on differential current [17]	In this compare current available at opposite end of protected part of system	Resolve high impedance fault	Economic concern, high cost, transient cause issue
8.	Protection based on current traveling wave	Bus bar residual voltage for fault detection and comparison of wavelet transform of zero sequence current traveling waves for fault location	No effect of power flow, fault current, and unbalance	Practical implementation is one of concern

the fixed relay settings (as suggested in the nonadaptive methods). Thus, it is essential for a guaranteed selective operation that these settings are modified or adaptive in nature as per the dynamics of the microgrid operation, characteristics and configuration. The word "adaptive" means ability to make available for changing condition. Thus, in microgrids, adaptive protection scheme provides online solution for change in operating condition of microgrid such as transition between islanded and grid connected with help of externally generated signal or control action [36]. For the implementation of adaptive protection schemes, a proper communication system is primarily required. The strong communication between microgrid control center and relay, which are located at different positions, is mandatory for implementation of this protection scheme.

5.2.1 Adaptive overcurrent protection scheme

In adaptive overcurrent protection scheme, the Central Protection Unit (CPU) is responsible for data storing and updating of even based data, action-based data, and fault current based data in tabular form by monitoring continuously. The variation of data table is due to different position of DG sources placed in the network. An adaptive overcurrent protection scheme with the help of directional over current relay is proposed by Oudalov and Fidigatti as a most suitable adaptive protection scheme for microgrid [24]. In this, offline analysis and online operation are performed. During offline analysis, proposer generate event record for all possible configuration of microgrid and DER unit also consider location of circuit breaker. This scheme is suitable for fault analysis when fault occurs at different locations in microgrid network by updating status of CB or DER unit. Data obtained from this fault analysis for different fault, calculating direction and magnitude of current for each relay and design suitable setting for operation. During online operation, a central controller is proposed to monitor the microgrid state by polling individual relays (1) periodically and (2) after triggered by an event like the tripping of a CB or a protection alarm. This state information is then compared to the event table, and the precalculated relay settings corresponding to the current system state are retrieved and uploaded to the relays. This method is complicated and complex for determining setting of controller.

Other methods are discussed in Alam [37], Conti et al. [38], Etemadi and Iravani [20], and Muda and Jena [39] where the fault current is calculated using event-based data table, from different sites with help of suitable relay and the data are stored in tabular form. With reference to this calculated data, researcher developed action table with time delay. Based on these data table, CPU generate tripping signal to relay at those locations. Now when relay fails to trip researchers has deployed a secondary protection system with help of upward and downward relay operation. For overcoming the false tripping issue due to changing operating mode of microgrid, when fault occurs near to the grid side, a method is proposed in Muda and Jena [39] based on positive and negative sequence currents. Another method is proposed in Alam [37], for real-time protection of the microgrid with communication channel interconnecting the relay and central unit. There are several issues in this method; some of them are complexity, higher cost, and complicated structure.

5.2.2 *Adaptive differential scheme*

The basic differential protection scheme is based on the principle of Kirchhoff's current law. In this method, whenever the threshold value of protection device is less than that of difference between incoming and outgoing current across feeder, the relay sends trip signal. In this type of protection strategy, the current is measured using measuring current transformer located on both the sides of line, transformer, and bus bar, which is sent to the relay. If the threshold value of current passing through the relay is less than that of measured value, the trip signal is generated by relay and the healthy part is disconnected from the faulty part.

In Conti et al. [38], propose a communication-based adaptive differential protection scheme for the microgrid, which is capable to protect low voltage microgrid associated with synchronous-based or inverter-based DGs in balanced condition. For unbalanced condition, Sortomme et al. proposed a scheme in Sortomme et al. [17], using Phasor Measurement Unit (PMUs), Digital Relay and Communication Channel, which has three layers of protection (i.e., primary protection, secondary protection, and backup protection). However, this method is very costly. In Dewadasa et al. [40], a method is proposed for the protection of line and bus bar with differential current and DG sources under overvoltage and under voltage conditions. Using differential relay as secondary protection, a method is proposed in Ustun et al. [41] based on the upward and downward relay. Another method based on the feature extraction technique is proposed in Lei et al. [42], which provides accurate protection scheme fulfilling the selectivity and accuracy features. However, the differential techniques always require a strong communication system which makes it expensive. Also, for unbalanced loading conditions and switching transients this protection scheme is little complicated in structure.

5.2.3 *Symmetrical component based adaptive protection scheme*

This scheme utilizes the symmetrical or sequence components for the protection of microgrids. A method is suggested in Elneweihi et al. [43], which differentiation between Line-to-Line fault and line to ground fault with utilizing zero-sequence current and negative-sequence current for the islanded mode of operation. This scheme requires communication link and is unable to detect high impedance fault which is the main concern. A method is proposed in Zamani et al. [44], which overcomes the issue stated above, using a microprocessor-based relay which utilize zero-sequence current and negative-sequence current for low voltage microgrid application. This scheme does not require any communication link which make its cost effective and simple. However, it is not useful in case of feeder having mesh network. For overcoming this issue, a method is proposed in Mirsaeidi et al. [45], utilizing microprocessor-based relay with PMU and digital communication-based system which uses the positive sequence current and give protection support to both radial and mesh network feeder. Communication system and PMU make this method costly. Another method is given in Ravyts et al. [46], which overcomes the economic concern and uses a sequence component to detect and classify the types of faults. If there is positive sequence current, relay get reset with help of communication channel, if there is negative sequence current, communication system generates trip signal with help of tripping time of overcurrent relay. However, the major drawback of this method is that it is unable to protect the looped microgrids. In Dubey and Jena [47], a method using sequence impedance is suggested for the unbalanced type of microgrid, which detects the high impedance faults accurately.

6. Protection schemes for DC microgrids

The DC grids have different behavior under faults as compared to the AC grids. Thus, the protection challenges faced by the DC grids due to the DC fault behavior of VSCs are needed to be resolved. The conventional grid is of AC types and it is associated with the AC type circuit breakers in practice, and so the need of DC breakers is justified in the real-time applications for the protection of DC microgrids. Thus, the DC breakers are mandatorily required in the DC grids. So, the protection challenges in DC grids primarily concern with the lack of presence of cost-effective DC breakers which is fast and poses the selective and sensitive qualities. Since the DC microgrids are now evolving day by day over the past few years, thus, its protection methods must be well-defined for secure and reliable system operation, keeping the feature of the DC breakers in mind.

Similar to the AC microgrid, the DC microgrid technology is also facing the unavoidable issues arises during its operation and protection. During the faults in DC grids, the stringent rise in initial value of DC fault current over a shorter duration results in the complex protection strategy. This further increases the size of the DC breaker and its capacity. Moreover, lack of standards and regulations proposed for the DC microgrids make the designing of protection strategy more complicated. The various organizations like IEEE, IEC, etc. are working for developing the set of standards for the DC microgrid. Some major protection challenges for the DC type of microgrids are stated below for the better understanding of protection issues associated with the DC microgrids:

1. Due to the presence of dynamic and bidirectional fault current behavior, development of protection strategy becomes little complex. The adaptive protection scheme must be designed for tackling the dynamic behavior of

the DC fault current and to enhance the selectivity due to the bidirectional fault current, a directional feature should be added to the protection method.
2. The grounding in DC microgrid plays a very important role while designing the protection scheme. The fault current magnitude is affected by the grounding configurations, grounding resistance and fault type. The transient over voltages are also affected by the grounding configurations. Thus, all these grounding factors should be considered while developing the protection strategy of DC microgrid.
3. Unlike AC systems, DC microgrids do not have a natural zero-crossing fault current which creates fault interruption problem in the DC microgrids.
4. The DC microgrid system has less virtual inertia and low system stability as compared to AC systems. The oscillation problems on AC side, generally propagates on the DC side in the interlinked systems, resulting in the momentary faults.

Thus, the challenges and issues faced by DC microgrids are already discussed and the need of protection schemes is justified. The protection scheme suggested for the DC microgrids can be majorly classified in two methods: (i) Unit protection schemes and (ii) Nonunit protection schemes:

(a) *Unit protection methods:* These methods are specifically suggested for the protection of fixed zones in DC microgrids. They are mostly used for the protection of DC bus, energy storage units, converters, DC loads, etc. The current differential protection is the most common type of unit protection method. However, these are not much suitable for the backup protection but they are very accurate and fast in nature [48,49].

(b) *Nonunit protection methods:* The nonunit protection methods have large coverage area and it relies on the predefined threshold settings of the electrical parameters for their operation. These include the current and voltage derivatives methods, over/under-voltage methods, etc. These protection schemes assure the protection reliability [48,49].

6.1 Current-based protection scheme

The most common type of protection scheme used for the DC microgrids are the overcurrent-based relays [50]. Another method suggested in Saleh et al. [51], is proposed for the protection of DC microgrid consisting of a hybrid passive over current relay having an inductor and a capacitor for detecting high current magnitude for low impedance faults. Also, for detecting high resistance faults, the voltage transient is generated by the inductor and capacitor having damped frequency which is known using real-time discrete wavelet transform technique. Another method using derivative of current derivative is proposed in Meghwani et al. [52], which utilizes the rate of change of fault current for detecting and locating the faults in DC microgrid.

6.2 Voltage-based protection scheme

These schemes are based on the local measurements and do not require any communication link [53]. Thus, these are the single-ended methods. Since it only depends on the voltage magnitude, so it is one of the fast methods. However, it is unable to discriminate the permanent and temporary faults.

6.3 Traveling wave-based protection scheme

The traveling waves of the initial current and voltages occurs as soon as the fault are initiated, and they travel through the line until the point circuit has been interrupted. Further, the fault location is identified by analyzing the feature of the high frequency transient wave like polarity, time interval, and magnitude for both AC and DC systems [53–55].

6.4 Interruption in DC current-based protection scheme

The fuse is simple, cheap, standardized device having low steady-state losses. DC microgrid network has high time constant and time delay, due to this fuse is not suitable. For overcoming this, a method is suggested in Tang and Ooi [56], in which the entire network is divided in the small zones and for interruption in flow of fault current, an isolator is used. However, the drawback of this method is that it is not economic due to high installation cost of circuit breaker.

7. Protection scheme for hybrid AC/DC microgrid

Hybrid AC/DC microgrid configurations are the combination of AC and DC subgrids and they combine the advantages of the AC and DC architectures. Their main advantages are that the AC and DC networks are combined in the same distribution network facilitating the direct connection of AC and DC-based DGs, loads, and energy storage devices. For developing the protection scheme of this type of microgrid, on the AC side we deploy protection scheme, which is used in case of AC microgrid and DC side, we deploy DC microgrid protection strategies. However, a separate protection scheme is required for the bidirectional power flow in case of interlinking converters.

7.1 Fuse-based short-circuit protection of converter

Fuse is simple, cheap, standardized device having low steady state losses, and is considered as an alternative of DC breaker. It is suitable for protection of power system having less time constant. A fuse-based protection strategy requires additional capacitance to clear faults compared to the necessary capacitance for system stability. In Ravyts et al. [46], the use of fuses as a protection device is suggested for low voltage DC grid.

7.2 Protection scheme based on islanding detection

Hybrid AC/DC microgrid consist of various DG sources, some of them are inverter-based DG which has high droop gain and low inertia due to this disturbance in terms of variation of electrical parameter that makes system unstable. This problem is associated with stand-alone Hybrid AC/DC microgrid. To overcome such issue, researcher proposed an autoencoder-based detection method of abnormal condition in Arif et al. [57]. Autoencoder is a combination of encoder and decoder, where encoder act as detection network and decoder act as generation network, is capable to recreate the data using neural network. This autoencoder has ability to reduce complex nonlinear function with help of neural network. For reliable detection, there is offline and online validation of proposed technique. This method is reliable, robust, dependable, and fast.

7.3 Protection scheme for DC lines in hybrid AC/DC microgrid with multilevel converter

In DC line protection, identification of fault zone at same time coordination between upstream delay and downstream delay is major concern. For overcoming this concern, there is classification of protection zone named as main protection zone and backup protection zone proposed in Dai et al. [58]. At the time of main protection speed of operation plays vital role. Main protection consists of current differential protection, the overcurrent protection, and the unbalanced current protection. When threshold value of resistance is less than that of fault resistance, protection system unable to find fault, to identify this there is need of backup protection. Back up protection consists of directional current protection and the low-voltage protection devices, in which the minor effect of fault resistance on fault voltage and direction of fault current during faults are studied and observed for identification of faults. Due to this, independent nature of voltage and current protection of DC line is in efficient manner and line will not trip under undesirable conditions.

8. Conclusion

The article discusses the need of the microgrid protection strategies and provides the solutions for the same. As the development of the microgrid is important part of the power distribution network, its protection strategies are also needed to be updated eventually for better system reliability and performance. The article highlighted the protection problems faced by different types of microgrids like AC, DC, and Hybrid AC/DC microgrid and also suggested the techniques to rectify the problems.

References

[1] R.H. Lasseter, Smart distribution: coupled microgrids, Proc. IEEE 99 (6) (2011) 1074–1082, https://doi.org/10.1109/JPROC.2011.2114630.

[2] Office of Electricity Delivery and Energy Reliability Smart Grid R&D Program, Summary Report: 2012 DOE Microgrid Workshop, U.S. Department of Energy, 2012, pp. 1–33. http://energy.gov/sites/prod/files/2012MicrogridWorkshopReport 09102012.pdf.

[3] A. Hirsch, Y. Parag, J. Guerrero, Microgrids: a review of technologies, key drivers, and outstanding issues, Renew. Sust. Energ. Rev. 90 (2018) 402–411, https://doi.org/10.1016/j.rser.2018.03.040.

[4] S. Chowdhury, S.P. Chowdhury, P. Crossley, Microgrids and active distribution networks, in: Microgrids and Active Distribution Networks, 2009, https://doi.org/10.1049/pbrn006e.

[5] A.A. Memon, K. Kauhaniemi, A critical review of AC microgrid protection issues and available solutions, Electr. Power Syst. Res. 129 (2015) 23–31, https://doi.org/10.1016/j.epsr.2015.07.006.

[6] Institute of Electrical and Electronics Engineers, IEEE Std 1159—IEEE recommended practice for monitoring electric power quality, in: IEEE Std 1159-2009 (Revision of IEEE Std 1159-1995), vol. 2009, 2009, https://doi.org/10.1109/IEEESTD.2009.5154067. Issue June.

[7] M. Khederzadeh, M. Sadeghi, Virtual active power filter: a notable feature for hybrid ac/dc microgrids, IET Gener. Transm. Distrib. 10 (14) (2016) 3539–3546, https://doi.org/10.1049/iet-gtd.2016.0217.

[8] P.C. Loh, D. Li, Y.K. Chai, F. Blaabjerg, Autonomous operation of ac-dc microgrids with minimised interlinking energy flow, IET Power Electron. 6 (8) (2013) 1650–1657, https://doi.org/10.1049/iet-pel.2012.0567.

[9] P.C. Loh, D. Li, Y.K. Chai, F. Blaabjerg, Autonomous operation of hybrid microgrid with ac and dc subgrids, IEEE Trans. Power Electron. 28 (5) (2013) 2214–2223, https://doi.org/10.1109/TPEL.2012.2214792.

[10] E. Unamuno, J.A. Barrena, Hybrid ac/dc microgrids—part I: review and classification of topologies, Renew. Sust. Energ. Rev. 52 (2015) 1251–1259, https://doi.org/10.1016/j.rser.2015.07.194.

[11] R. Jadeja, A. Ved, T. Trivedi, G. Khanduja, Control of power electronic converters in AC microgrid, Power Syst. 27 (11) (2020) 329–355, https://doi.org/10.1007/978-3-030-23723-3_13.

[12] O. Palizban, K. Kauhaniemi, J.M. Guerrero, Microgrids in active network management—part I: hierarchical control, energy storage, virtual power plants, and market participation, Renew. Sust. Energ. Rev. 36 (2014) 428–439, https://doi.org/10.1016/j.rser.2014.01.016.

[13] K. De Brabandere, B. Bolsens, J. Van Den Keybus, A. Woyte, J. Driesen, R. Belmans, A voltage and frequency droop control method for parallel inverters, in: PESC Record—IEEE Annual Power Electronics Specialists Conference, vol. 4, 2004, pp. 2501–2507, https://doi.org/10.1109/PESC.2004.1355222. 4.

[14] A. Bidram, A. Davoudi, Hierarchical structure of microgrids control system, IEEE Trans. Smart Grid 3 (4) (2012) 1963–1976, https://doi.org/10.1109/TSG.2012.2197425.

[15] L. Che, M.E. Khodayar, M. Shahidehpour, Adaptive protection system for microgrids: protection practices of a functional microgrid system, IEEE Electrif. Mag. 2 (1) (2014) 66–80, https://doi.org/10.1109/MELE.2013.2297031.

[16] S.A. Hosseini, H.A. Abyaneh, S.H.H. Sadeghi, F. Razavi, A. Nasiri, An overview of microgrid protection methods and the factors involved, Renew. Sust. Energ. Rev. 64 (2016) 174–186, https://doi.org/10.1016/j.rser.2016.05.089.

[17] E. Sortomme, S.S. Venkata, J. Mitra, Microgrid protection using communication-assisted digital relays, IEEE Trans. Power Delivery 25 (4) (2010) 2789–2796, https://doi.org/10.1109/TPWRD.2009.2035810.

[18] P.P. Barker, R.W. De Mello, Determining the Impact of Distributed Generation on Power Systems: Part 1—Radial Distribution Systems, 3, Power Engineering Society Summer Meeting (Cat. No. 00CH37134), 2000, pp. 1645–1656.

[19] A. Darlington, K. Birt, S. Boutilier, W. Hartmann, D. Sevcik, S.S. Venkata, Impact of distributed resources on distribution relay protection, in: A Report to the Line Protection Subcommittee of the Power System Relay Committee of The IEEE Power Engineering Society, August, 2004, pp. 1–24.

[20] A.H. Etemadi, R. Iravani, Overcurrent and overload protection of directly voltage-controlled distributed resources in a microgrid, IEEE Trans. Ind. Electron. 60 (12) (2013) 5629–5638, https://doi.org/10.1109/TIE.2012.2229680.

[21] H. Laaksonen, D. Ishchenko, A. Oudalov, Adaptive protection and microgrid control design for Hailuoto Island, IEEE Trans. Smart Grid 5 (3) (2014) 1486–1493, https://doi.org/10.1109/TSG.2013.2287672.

[22] E. Lakervi, E.J. Holmes, Electricity distribution network design, in: Electricity Distribution Network Design, 2003, https://doi.org/10.1049/pbpo021e.

[23] K. Mäki, S. Repo, P. Järventausta, Effect of wind power based distributed generation on protection of distribution network, IEE Conf. Publ. 1 (2004) 327–330, https://doi.org/10.1049/cp:20040129.

[24] A. Oudalov, A. Fidigatti, Adaptive network protection in microgrids, ABB Int. J. Distrib. Energy Resour. 5 (2009) 201–227.

[25] H.M. Sharaf, H.H. Zeineldin, E. El-Saadany, Protection coordination for microgrids with grid-connected and islanded capabilities using communication assisted dual setting directional overcurrent relays, IEEE Trans. Smart Grid 9 (1) (2018) 143–151, https://doi.org/10.1109/TSG.2016.2546961.

[26] P. Cairoli, R. Rodrigues, H. Zheng, Fault current limiting power converters for protection of DC microgrids, in: Conf. Proc.—IEEE SoutheastCon, 2017, https://doi.org/10.1109/SECON.2017.7925392.

[27] A. Chandra, G.K. Singh, V. Pant, Protection techniques for DC microgrid—a review, Electr. Power Syst. Res. 187 (2020), https://doi.org/10.1016/j.epsr.2020.106439, 106439.

[28] R.M. Kamel, K. Nagasaka, Effect of load type on standalone micro grid fault performance, Appl. Energy 160 (2015) 532–540, https://doi.org/10.1016/j.apenergy.2015.09.044.

[29] P. Magne, B. Nahid-Mobarakeh, S. Pierfederici, Dynamic consideration of dc microgrids with constant power loads and active damping system—a design method for fault-tolerant stabilizing system, IEEE J. Emerg. Sel. Top. Power Electron. 2 (3) (2014) 562–570, https://doi.org/10.1109/jestpe.2014.2305979.

[30] L. Meng, Q. Shafiee, G.F. Trecate, H. Karimi, D. Fulwani, X. Lu, J.M. Guerrero, Review on control of DC microgrids and multiple microgrid clusters, IEEE J. Emerg. Sel. Top. Power Electron. 5 (3) (2017) 928–948, https://doi.org/10.1109/JESTPE.2017.2690219.

[31] S. Voima, K. Kauhaniemi, H. Laaksonen, Novel protection approach for MV microgrid, in: 21st International Conference on Electricity Distribution, 0430, 2011, pp. 1–4.

[32] H. Al-Nasseri, M.A. Redfern, F. Li, A voltage based protection for micro-grids containing power electronic converters, in: 2006 IEEE Power Engineering Society General Meeting, PES, 2006, pp. 1–7, https://doi.org/10.1109/pes.2006.1709423.

[33] A. Hooshyar, R. Iravani, Microgrid protection, Proc. IEEE 105 (7) (2017) 1332–1353, https://doi.org/10.1109/JPROC.2017.2669342.

[34] M. Dewadasa, R. Majumder, A. Ghosh, G. Ledwich, Control and protection of a microgrid with converter interfaced micro sources, in: 2009 International Conference on Power Systems, ICPS '09, 1, 2009, pp. 25–30, https://doi.org/10.1109/ICPWS.2009.5442654.

[35] H. Al-Nasseri, M.A. Redfern, Harmonics content based protection scheme for micro-grids dominated by solid state converters, in: 2008 12th International Middle East Power System Conference, MEPCON 2008, 2008, pp. 50–56, https://doi.org/10.1109/MEPCON.2008.4562361.

[36] T.S. Ustun, C. Ozansoy, A. Zayegh, Fault current coefficient and time delay assignment for microgrid protection system with central protection unit, IEEE Trans. Power Syst. 28 (2) (2013) 598–606, https://doi.org/10.1109/TPWRS.2012.2214489.

[37] M.N. Alam, Adaptive protection coordination scheme using numerical directional overcurrent relays, IEEE Trans. Ind. Inf. 15 (1) (2019) 64–73, https://doi.org/10.1109/TII.2018.2834474.

[38] S. Conti, L. Raffa, U. Vagliasindi, Innovative solutions for protection schemes in autonomous MV micro-grids, in: 2009 International Conference on Clean Electrical Power, ICCEP 2009, 2009, pp. 647–654, https://doi.org/10.1109/ICCEP.2009.5211985.

[39] H. Muda, P. Jena, Real time simulation of new adaptive overcurrent technique for microgrid protection, in: 2016 National Power Systems Conference, NPSC 2016, 2017, pp. 3–8, https://doi.org/10.1109/NPSC.2016.7858897.

[40] M. Dewadasa, A. Ghosh, G. Ledwich, Protection of microgrids using differential relays, in: 2011 21st Australasian Universities Power Engineering Conference, AUPEC 2011, 2011, pp. 1–6.

[41] T.S. Ustun, C. Ozansoy, A. Zayegh, Modeling of a centralized microgrid protection system and distributed energy resources according to IEC 61850-7-420, IEEE Trans. Power Syst. 27 (3) (2012) 1560–1567, https://doi.org/10.1109/TPWRS.2012.2185072.

[42] L. Lei, C. Wang, J. Gao, J. Zhao, X. Wang, A protection method based on feature cosine and differential scheme for microgrid, Math. Probl. Eng. 2019 (2019), https://doi.org/10.1155/2019/7248072.

[43] A.F. Elneweihi, E.O. Schweitzer, M.W. Feltis, Negative-sequence overcurrent element application and coordination in distribution protection, IEEE Trans. Power Delivery 8 (3) (1993) 915–924, https://doi.org/10.1109/61.252618.

[44] M.A. Zamani, T.S. Sidhu, A. Yazdani, A protection strategy and microprocessor-based relay for low-voltage microgrids, IEEE Trans. Power Delivery 26 (3) (2011) 1873–1883, https://doi.org/10.1109/TPWRD.2011.2120628.

[45] S. Mirsaeidi, D.M. Said, M.W. Mustafa, M.H. Habibuddin, A protection strategy for micro-grids based on positive-sequence component, IET Renew. Power Gener. 9 (6) (2015) 600–609, https://doi.org/10.1049/iet-rpg.2014.0255.

[46] S. Ravyts, G. Van Den Broeck, L. Hallemans, M.D. Vecchia, J. Driesen, Fuse-based short-circuit protection of converter controlled low-voltage DC grids, IEEE Trans. Power Electron. 35 (11) (2020) 11694–11706, https://doi.org/10.1109/TPEL.2020.2988087.

[47] K. Dubey, P. Jena, Impedance angle-based differential protection scheme for microgrid feeders, IEEE Syst. J. (2020) 1–10, https://doi.org/10.1109/jsyst.2020.3005645.

[48] S.D.A. Fletcher, P.J. Norman, S.J. Galloway, P. Crolla, G.M. Burt, Optimizing the roles of unit and non-unit protection methods within DC microgrids, IEEE Trans. Smart Grid 3 (4) (2012) 2079–2087, https://doi.org/10.1109/TSG.2012.2198499.

[49] A. Meghwani, S.C. Srivastava, S. Chakrabarti, A non-unit protection scheme for DC microgrid based on local measurements, IEEE Trans. Power Delivery 32 (1) (2017) 172–181, https://doi.org/10.1109/TPWRD.2016.2555844.

[50] J.P. Brozek, DC overcurrent protection—where we stand, IEEE Trans. Ind. Appl. 29 (5) (1993) 1029–1032, https://doi.org/10.1109/28.245730.

[51] K.A. Saleh, A. Hooshyar, E.F. El-Saadany, Hybrid passive-overcurrent relay for detection of faults in low-voltage DC grids, IEEE Trans. Smart Grid 8 (3) (2017) 1129–1138, https://doi.org/10.1109/TSG.2015.2477482.

[52] A. Meghwani, S. Chakrabarti, S.C. Srivastava, A fast scheme for fault detection in DC microgrid based on voltage prediction, in: 2016 National Power Systems Conference, NPSC 2016, 2017, https://doi.org/10.1109/NPSC.2016.7858867.

[53] Y. Chen, D. Liu, B. Xu, Wide-area traveling wave fault location system based on IEC61850, IEEE Trans. Smart Grid 4 (2) (2013) 1207–1215, https://doi.org/10.1109/TSG.2012.2233767.

[54] Z. Galijasevic, A. Abur, Fault area estimation via intelligent processing of fault-induced transients, IEEE Trans. Power Syst. 18 (4) (2003) 1241–1247, https://doi.org/10.1109/TPWRS.2003.814854.

[55] Y. Zhang, N. Tai, B. Xu, Fault analysis and traveling-wave protection scheme for bipolar HVDC lines, IEEE Trans. Power Delivery 27 (3) (2012) 1583–1591, https://doi.org/10.1109/TPWRD.2012.2190528.

[56] L. Tang, B.T. Ooi, Locating and isolating DC faults in multi-terminal DC systems, IEEE Trans. Power Delivery 22 (3) (2007) 1877–1884, https://doi.org/10.1109/TPWRD.2007.899276.

[57] A. Arif, K. Imran, Q. Cui, Y. Weng, Islanding detection for inverter-based distributed generation using unsupervised anomaly detection, IEEE Access 9 (2021) 1, https://doi.org/10.1109/access.2021.3091293.

[58] Z. Dai, X. Liu, C. Zhang, H. Zhu, Protection scheme for DC lines in AC/DC hybrid distribution grids with MMCs, in: 2018 International Conference on Power System Technology, POWERCON 2018 - Proceedings, 201804270000263, 2019, pp. 2518–2523, https://doi.org/10.1109/POWERCON.2018.8602350.

Index

Note: Page numbers followed by *f* indicate figures and *t* indicate tables.

A

AC microgrid, 1–2
  configurations, 125–126, 127*f*
  electrical supply, 259, 259*f*
  nonadaptive protection for, 265, 266*t*
  protection schemes for, 265–269
Active power filter (APF), 136, 203
Adaptive control scheme
  active power control, 169–171
  adaptive frequency, 169–171
  adaptive voltage, 166–169
  reactive power control, 166–169
  simulation analysis, 172–176
Adaptive protection
  differential scheme, 268
  overcurrent protection scheme, 267–268
  symmetrical component based scheme, 268–269
Ancillary services, 67, 179, 181, 184
Artificial intelligence, 67–68
Asymptotic convergence, 162
Automatic voltage control (AVC), 184–185
AVC. *See* Automatic voltage control (AVC)

B

Boost-type DC-DC power electronic converter
  circuit diagram, 94, 94*f*
  experimental validation, 108–112
  feedback control structure, 95–96, 96*f*
  frequency response, 95, 95*f*
  frequency selection
    design validation, 101–104
    high frequency (HF) specifications, 97
    loop-shaping technique, 99–101
    stability margin bounds, 97–99
    template generation, 97
  linear, 105–106
  MATLAB/Simulink, 105
  nominal boost converter transfer function, 95
  nominal resistance value, 94–95
  nonlinear, 106–107
  specifications
    robust external disturbance rejection problem, 96
    robust stability margin, 97
Buck-boost-type converters, 90

C

CHR method, 91–92
Classification, microgrid, 257
  based on capacity, 258–259
  based on location
    community/campus, 258
    industrial, 258
    stand-alone/off-grid, 258
    urban and rural, 257
    utility, 257
  based on power supply
    AC microgrids, 259–260
    DC microgrids, 260–261
    hybrid AC/DC microgrids, 261–262
Clear sky library (CSL), 69–70, 71*f*
Closed-loop system, 89
Cloud determination algorithm, 69, 70*f*
Cloud distribution, 68
Cloud pixel classification algorithm, 73, 74*f*
Cohen-Coon methods, 91–92
Commercial/industrial microgrids (CI-μ grids), 125
Consensus-based technique, 161–162
Controllers, grid-side inverter
  current controllers, 136–139
    modified *dq*-current controller, 137–139
    multi-functional current controller (MFCC), 139
  PQ controller, 135–136
    advanced open-loop, 136
    conventional closed-loop, 135–136
    conventional open-loop, 135
  synchronization techniques, 139–144
    decoupling network in αβ-domain PLL (DNabPLL), 141–143
    dq-PLL, 141

Controllers, grid-side inverter *(Continued)*
harmonic-interharmonic DC offset PLL (HIHDOPLL), 143–144
moving average filter-based PLLs (MAFPLL), 144
Conventional fossil fuels, 67
Converter control, in EV-based system
distributed control, 60–62
case study, 61–62
EV chargers model, 60
framework, 60
modified droop control, 57–59
CPNs. *See* Cyber-physical networks (CPNs)
CSL. *See* Clear sky library (CSL)
Current control block, 164*f*
Current controller, 54
Current control loop (CCL), 160
Current-mode control (CMC) approach, 90
Cyber-physical microgrid structure, 1
Cyber-physical networks (CPNs), 2–4
Cyber safety, 1

D

DC-DC power electronic converter, 89
DC microgrid
component, 89
internal model control, 90
linear time invariant (LTI) model, 91–92
quantitative feedback theory (QFT), 91
sliding mode control (SMC), 90
voltage mode control (VMC), 90
Ziegler and Nichols method, 91–92
DC voltage stabilization techniques, 204–205
Decoupling network in αβ-domain PLL (DNabPLL), 141–143
Denial-of-service (DOS)
algorithms, 4
event-based control (*see* Event-based control, DOS attack)
impact, 4
Lyapunov-based stability analysis, 4–5
microgrid communication network, 4
microgrid control design
consensus-based frequency error, 8–9
consensus-based voltage error, 8
control objective, 6–7
distributed generators, 4–5
event-based SMC scheme, 10–11
graph theory, 7–8
secondary control, 7
sliding mode-based control strategy, 9–10
microgrid stability, 4–5
DER. *See* Distributed energy resource (DER)
DHI. *See* Diffuse horizontal irradiance (DHI)
Differential protection schemes, 241–242
Diffuse horizontal irradiance (DHI), 68
Digital real-time simulators
dSPACE, 147–151
OPAL-RT, 151–152
RTDS, 152–153
speedgoat, 152
Direct normal irradiance (DNI), 68
Distance protection schemes, 243
Distributed energy resource (DER), 31, 161–162, 255–256
Distributed generating system
converter control in EV-based system (*see* Converter control in EV-based system)
energy storage system, 29–30
grid following mode
grid-imposed frequency VSC system, 47
phase locked loop (PLL), 50–51
real and reactive power controller, 48–49
VSC current control-SRF, 49–50
grid forming mode (*see* Grid forming mode (GFM))
grid supporting mode
frequency support, 36–38
inertial support, 43–47
standards, 32–36
voltage support, 38–43
IEEE standards, 31–32
power/current control mode, 30–31
power systems, 29–30
synchronization process, 31
transitions, 30
voltage source inverter, 30
Distributed generations (DGs), 2–4, 29, 233, 255–256
Doubly-fed induction generators (DFIGs), 241
Droop control loop, 162–163

E

Electrical power systems, 255–256
Electric vehicles (EV), 115, 255–256
charging algorithms, 145–147
modeling techniques, 145
Event-based control, DOS attack
close loop control, 11
parameter design
convergence rate, 14–16
divergence rate, 16–20
preliminaries time limitations, 12–14
EV market growth, 118–120

F

Fast and flexible resources (FFR), 182
Feedback linearization, 161–162
Forecast horizon (FH), 67–68, 87
Forecasting errors, 179
Frequency controller, 56
Frequency support, 36–38

G

Global green energy, 179
Global horizontal irradiance (GHI), 68
Global warming, 67
Grid-connected mode (GCM), 233
Grid connected system
AC side control, 128–135
αβ reference frame, 132–133
abc reference frame, 132
dq reference frame, 133–135
DC side control, 128
Grid forming mode (GFM), 30
conventional droop control approach, 52–54
current controller, 54
power controller, 52–54
voltage controller, 54
virtual synchronous machine (VSM) based control approach, 55–56
frequency controller, 56
voltage controller, 56

H

Harmonic-interharmonic DC offset PLL (HIHDOPLL), 143–144
Hybrid microgrid, 126

I

ICC-based model reference adaptive control (ICC-MRAC), 204–205
IEEE C37.118.1, 246–247
IEEE standards, 31–32
Impedance difference (ID), 249
Impedance difference-based microgrid protection scheme, 249–251
Indirect current controller (ICC), 204
Ineichen and Perez clear sky model, 77–78
Inertial support, 43–47
Instantaneous reactive power theory or *p-q* theory (IRPT), 204
Integrated impedance angle (IIA), 246–247
Integrated impedance angle based protection scheme, 246–248
Internal model control (IMC), 90
Islanded/remote microgrid (IR-μ girds), 124–125
Islanding mode (IM), 233

L

Linear time invariant (LTI) model, 91–92
Linke turbidity co-efficient, 78
Loop-shaping technique, 99–101
Lyapunov functions, 169–170

M

Machine learning techniques, 67–68
Macro-scale measurements, 68
Market-oriented operation, 180–181
Master controller, 161
Maximum power point tracking (MPPT) technique, 204
Microgrids, 1
AC microgrid, 125–126
characteristics, 29
conventional energy resource, 255–256
classification, 159
based on capacity, 258–259
based on location, 257–258
based on power supply, 259–262
commercial/industrial microgrids (CI-μ grids), 125
islanded/remote microgrids (IR-μ girds), 124–125
university campus/institutional microgrids (UCI-μ grids), 124
communication link, 4
control architecture
controllers (*see* Controllers, grid-side inverter)
grid connected system (*see* Grid connected system)
power grid challenges, 126–128
control strategies, 2–4
cyber components, 2–4
cyber threats, 3*f*
DC microgrid, 126
distributed generation (DG) devices, 159
droop-based control system, 162–164
electricity demand, 233–234
first level control system, 160
hierarchical control architecture, 262–263
hybrid microgrid, 126
layout, 162–163, 163*f*
model predictive control (MPC), 160–161
second level control system, 160
three-layer hierarchical structure, 160

Microgrids *(Continued)*
man-made/natural disasters, 234
operation, 29
power system operation, 256–257
protection, 234 (*see also* Microgrid protection)
protection issues, 256–257, 263–264
Microgrid cyber-physical structure, 1
Microgrid protection
challenges
bidirectional power flow, 234, 235*f*
blinding, 235, 236*f*, 237–240
schemes, 240
differential protection, 241–242
distance protection, 243
impedance difference-based microgrid protection, 249–251
integrated impedance angle based protection, 246–248
new microgrid protection, 243
overcurrent protection, 240–241
rate of angle difference-based protection, 243–246
Micro-scale measurements, 68
Model predictive control (MPC), 160–162
Model reference adaptive controller (MRAC), 204–205
Modified *dq*-current controller, 137–139
Moving average filter-based PLLs (MAFPLL), 144
MRAC. *See* Model reference adaptive controller (MRAC)
Multiagent system, 161–162
Multi-functional current controller (MFCC), 139

N

New microgrid protection techniques, 243
Nominal loop-shaping method, 92
Nonnoise vectors, 76
Numerical weather prediction (NWP) method, 67–68

O

Optical airmass, 78
Optical flow algorithm, 75–76
Overcurrent protection schemes, 240–241
Overcurrent relaying (OCR) scheme, 240

P

Phase locked loop (PLL), 50–51
Phasor measurement units (PMUs), 243–244
Photovoltaic (PV) systems, 29, 115
PMU-based wide area measurement system (WAMS), 187–188
Point of common coupling (PCC), 203, 243–244
Power controller, 52–54, 163*f*
Power grid resiliency, 187
Proactive defense methodology, 187–189
case studies, 193–199
correlation coefficients, 192–193
e-Business based market, 191
implementation, 189–190, 190*f*
voltage variation, 191
WACS-based execution, 193
Proportional–integral (PI) controllers, 162–163
Protection schemes, microgrid
AC
adaptive (*see* Adaptive protection)
nonadaptive, 265
DC, 269–271
current-based protection, 270
interruption in DC current-based protection, 271
traveling wave-based protection, 271
voltage-based protection, 271
hybrid AC/DC microgrid, 271–272
Protective relay coordination, 239
Pulse width modulation (PWM), 204
PV and EV impacts, on electricity grid
electricity market, 123
power quality, 122–123
power system stability, 120–122
PV market penetration, 116–117
Pyranometer, 82

Q

Quantitative feedback theory (QFT), 91
boost-type DC-DC power electronic converter (*see* Boost-type DC-DC power electronic converter)
control specification, 93
linear time invariant (LTI) model, 93
loop transmission function, 93
Nichols chart (NC), 92–93
NMP system, 93–94
two degree of freedom (2-DOF) method, 92

R

Rate of angle difference-based protection scheme, 243–246
R/B ratio values, 70

Reactive power, 181–183
Recursive least square (RLS) algorithm, 83–84
Renewable energy sources (RES), 29, 179, 204
Renewable energy (RE) systems, 115
Renewable intermittency instigated disturbance propagation, 185–187
RES. *See* Renewable energy sources (RES)
Resiliency implementation phases, 181*f*
Right-half-plane (RHP), 90
RLS. *See* Recursive least square (RLS) algorithm
Robust control methods, 90
Robust external disturbance rejection problem, 96
Robust stability margin, 97

S

Shunt active power filter (SAPF), 203
  balanced supply
    parameter variation, 220
    with steady-state load condition, 216–219
    with varying load condition, 220
  experimental results, balanced supply
    with parameter variation, 227–231
    with steady-state load condition, 223
    with varying load condition, 223–227
  filter configuration, 205–206
  model reference adaptive controller (MRAC), 213–216
  MPPT algorithm, 205–206
  PI controller, 211–213
    closed-loop, 209, 209*f*
    DC-link capacitor, 210
    VSI, 210–211
  reference current estimation, 206
  three-phase three-wire modeling, 206–209
Slave controller, 161
Sliding mode control (SMC), 90
Solar irradiance forecasting, 67–68, 82–83
  cloud movement determination
    optical flow algorithm, 75
    velocity vector computation, 75–76
  cloud pixel determination, 69–74
    clear sky library, 69–70
    cloud pixels classification, 73–74
    threshold value computation, 70–73
  forecast solar irradiance
    clear sky index, 78–80
    clear sky model, 77–78
    horizon of 2min, 81
  preprocess sky images, 69
  recursive least square (RLS) algorithm, 83–84
  variable leaky least mean square algorithm, 85–86
Solar PV generation, 67
SPC. *See* System protection centers (SPC)
Stability margin bounds, 97–99
State-feedback signal, 1–2
Step-up converters, 112
Support vector machine (SVM), 242
Synchronization process, 31
System protection centers (SPC), 187–188

T

Thermal power stations, 233
Three-phase grid-connected power converter, 44–45
Total Sky Imagers (TSIs), 68
Two degree of freedom (2-DOF) method, 92

U

UCI-μ grids. *See* University campus/institutional microgrids (UCI-μ grids)
Unintentional islanding, 238
University campus/institutional microgrids (UCI-μ grids), 124

V

Variable leaky least mean square (VLLMS) algorithm, 85–86
Vehicle to grid (V2G) concept, 115
Virtual inertia control mechanism, 46–47
Virtual synchronous machine (VSM) based control approach, 55–56
VLLMS. *See* Variable leaky least mean square (VLLMS) algorithm
VMC. *See* Voltage mode control (VMC) approach
Voltage control area formation, 183–184
Voltage control block, 164*f*
Voltage controller, 54, 56
Voltage control loop (VCL), 160
Voltage mode control (VMC) approach, 90
Voltage source converters (VSCs), 29–30
Voltage source inverter (VSI), 30, 162–163, 203
Voltage support, 38–43

W

Wind energy conversion systems, 115

Z

Zero crossing point, 240
Ziegler and Nichols method, 91–92

Printed and bound by CPI Group (UK) Ltd, Croydon, CR0 4YY
08/06/2025
01896869-0004